Nonlinear Systems: Analysis and Design

Nonlinear Systems: Analysis and Design

Edited by
Hazel Johnson

MURPHY & MOORE
www.murphy-moorepublishing.com

Published by Murphy & Moore Publishing,
1 Rockefeller Plaza,
New York City, NY 10020, USA

ISBN: 978-1-63987-399-9

Cataloging-in-Publication Data

Nonlinear systems : analysis and design / edited by Hazel Johnson.
p. cm.
Includes bibliographical references and index.
ISBN 978-1-63987-399-9
1. Nonlinear systems. 2. System theory. 3. Nonlinear theories. 4. Mathematics. I. Johnson, Hazel.
QA427 .N66 2022
515--dc23

For information on all Murphy & Moore Publications
visit our website at www.murphy-moorepublishing.com

MURPHY & MOORE

Contents

Preface

It is often said that books are a boon to mankind. They document every progress and pass on the knowledge from one generation to the other. They play a crucial role in our lives. Thus I was both excited and nervous while editing this book. I was pleased by the thought of being able to make a mark but I was also nervous to do it right because the future of students depends upon it. Hence, I took a few months to research further into the discipline, revise my knowledge and also explore some more aspects. Post this process, I begun with the editing of this book.

A system where the changes in the output and input are not proportional is called a nonlinear system. Most systems are inherently nonlinear in nature, and hence nonlinear problems are studied in various fields such as engineering, biology, physics, mathematics, etc. Unlike simple linear systems, nonlinear dynamical systems that describe changes in variables over time may appear chaotic, unpredictable or counterintuitive. Nonlinear systems are commonly approximated by linear equations since nonlinear dynamical equations are difficult to solve. This process of approximation is called linearization. This book provides significant information of this discipline to help develop a good understanding of nonlinear systems and related fields. From theories to research to practical applications, case studies related to all contemporary topics of relevance to this field have been included herein. This book is a vital tool for all researching or studying nonlinear systems as it gives incredible insights into emerging trends and concepts.

I thank my publisher with all my heart for considering me worthy of this unparalleled opportunity and for showing unwavering faith in my skills. I would also like to thank the editorial team who worked closely with me at every step and contributed immensely towards the successful completion of this book. Last but not the least, I wish to thank my friends and colleagues for their support.

Editor

1

A Reduced-Order Gauss-Newton Method for Nonlinear Problems Based on Compressed Sensing for PDE Applications

Horacio Florez and Miguel Argáez

Abstract

A global regularized Gauss-Newton (GN) method is proposed to obtain a zero residual for square nonlinear problems on an affine subspace built by wavelets, which allows reducing systems that arise from the discretization of nonlinear elliptic partial differential equations (PDEs) without performing a priori simulations. This chapter introduces a Petrov-Galerkin (PG) GN approach together with its standard assumptions that ensure retaining the q-quadratic rate of convergence. It also proposes a regularization strategy, which maintains the fast pace of convergence, to avoid singularities and high nonlinearities. It also includes a line-search method for achieving global convergence. The numerical results manifest the capability of the algorithm for reproducing the full-order model (FOM) essential features while decreasing the runtime by a significant magnitude. This chapter refers to a wavelet-based reduced-order model (ROM) as WROM, while PROM is the proper orthogonal decomposition (POD)-based counterpart. The authors also implemented the combination of WROM and PROM as a hybrid method referred herein as (HROM). Preliminary results with Bratu's problem show that if the WROM could correctly reproduce the FOM behavior, then HROM can also reproduce that FOM accurately.

Keywords: Gauss-Newton method, line search, Petrov-Galerkin direction, data compression, wavelets

1. Introduction

The Newton method for solving square nonlinear problems is one of the most popular techniques used in engineering applications due to its simplicity and fast convergence rate [1–3]. However, the quality of the final numerical results is affected by the possible Jacobian's singularity and high nonlinearity. Another drawback is that the method depends on the initial point. Therefore, it is necessary to implement a globalization strategy to get the solution independently of the initial guess. One such approach is the line-search method that relies on a suitable merit function that yields that the iterations progress toward a solution of the problem.

From the numerical point of view, the technique requires solving square linear systems several times, and it is necessary to carry out function evaluations of the order of the problem, as well as first-order information of the square law of the function to compute the Jacobians. In the case of high-dimensional nonlinear problems, the method can overcome the capacity of the computer memory or decrease speed for solving these linear systems, even in the case of a few iterations. One of the current research activities focuses on solving large-scale square nonlinear problems in real time. The purpose of this chapter is to provide an algorithm for solving large-scale square nonlinear problems, in real time, while retaining the fast convergence rate. One strategy for addressing such challenges is to characterize an affine subspace, of much lower dimension than the original one, that contains the initial solution and thus reproduces the problem's principal features.

One procedure to characterize the affine subspace consists of solving the full-order model (FOM) in several input points whose solutions are called snapshots, then using a principal component analysis method such as singular value decomposition (SVD) to build an orthonormal basis that spans the snapshots' majority of energy. This oblique subspace, where one seeks a solution, is projected on the original one. Good numerical results have been already reported in the literature [1, 3–7]. But there are still open questions about this procedure as for how and when to choose the snapshots and their number [8, 9]. It is important to emphasize that at every picture it is required solving the FOM regardless of its cost. This chapter thus promotes a new strategy that is snapshot free. The approach originated in signal processing and consists of using the notion of wavelets to compress data in a subspace of smaller dimension which retains the majority of the original energy [10, 11]. The discrete wavelet's low-pass matrix is used as the affine subspace; then, the optimization is performed in this compressed subspace to obtain a cheaper solution that can decompress to its original size.

2. Reduced-order models using wavelet transformations

2.1. Wavelets and data compression

The rationale for using wavelet transformations for model reduction originates from the fields of image processing, image compression, and transform coding. In these fields, massive amounts of information, for example, images, are broadcasted over limited bandwidth communication lines and networks such as the internet. One needs to compress these signals for quick transmission and to diminish storage requirements [10]. In summary, data compression consists of, given a signal $x \in \mathbb{R}^n$ find a lower dimensional signal $\widehat{x} \in \mathbb{R}^r$ with $r \ll n$ to broadcast or store those lower dimensional ones $\widehat{x}$. The most widely used techniques for data compression are based on wavelets transform. The key to using wavelets is to find a lower dimensional signal $\widehat{x}$ that relies on a known subspace properly denoted as energy compaction. It is well comprehended that the wavelets tend to accumulate energy in the low-frequency sub-band of the wavelet decomposition [3, 12–14]. The energy relates to the L^2-norm and is defined as $\epsilon = \|x\|^2 = \sum_{i=1}^{n} x_i^2$. To demonstrate the energy compaction, consider **Figure 1**, in which one has the original image $x \in \mathbb{R}^{512\times512}$. Notice that the upper left quadrant of the wavelet decomposition is a low-dimensional approximation $\widehat{x} \in \mathbb{R}^{256\times256}$ that is one-fourth the size of the original signal and resembles the low-frequency sub-band wavelet coefficients. Using the previously measured energy, the energy enclosed in $\widehat{x}$ is 95.75% using just one-fourth of the coefficients. Since $\widehat{x}$ comprises most of the energy, a simple data compression scheme would execute all of the other wavelet sub-bands to zero and store only the low-frequency information as in **Figure 1** (see in the bottom center). By only employing $\widehat{x}$, one can reproduce an approximation of $x \in \mathbb{R}^n$ by its generalized inverse of the sub-band compression [10]. Next section will provide details.

2.2. Reduced-order models

Let $W \in \mathbb{R}^{n\times n}$ describe an orthonormal wavelet. The transformation encompasses a low-pass and a high-pass submatrix that is given by

$$W_{\mathbf{n}\times\mathbf{n}} = \left[\frac{L_{\mathbf{r}\times\mathbf{n}}}{H_{\mathbf{s}\times\mathbf{n}}}\right] \quad , \quad \mathbf{r}+\mathbf{s}=\mathbf{n}. \tag{1}$$

By orthogonality, $\text{rank}(W) = n$, $\text{rank}(L) = r$, and $\text{rank}(H) = s$. Now, by orthogonality of W, $WW^T = I$, then $LL^T = I_r$, $HH^T = I_s$. The energy of a signal $x \in \mathbb{R}^n$ is

$$\|x\|^2 = \|Lx\|^2 + \|Hx\|^2. \tag{2}$$

Choosing L that contains the majority of the energy such that the energy $\|Hx\| \approx 0$, one has $\|x\| \approx \|Lx\|$. This means that the energy of the original data $x \in \mathbb{R}^n$ is approximately equal to the

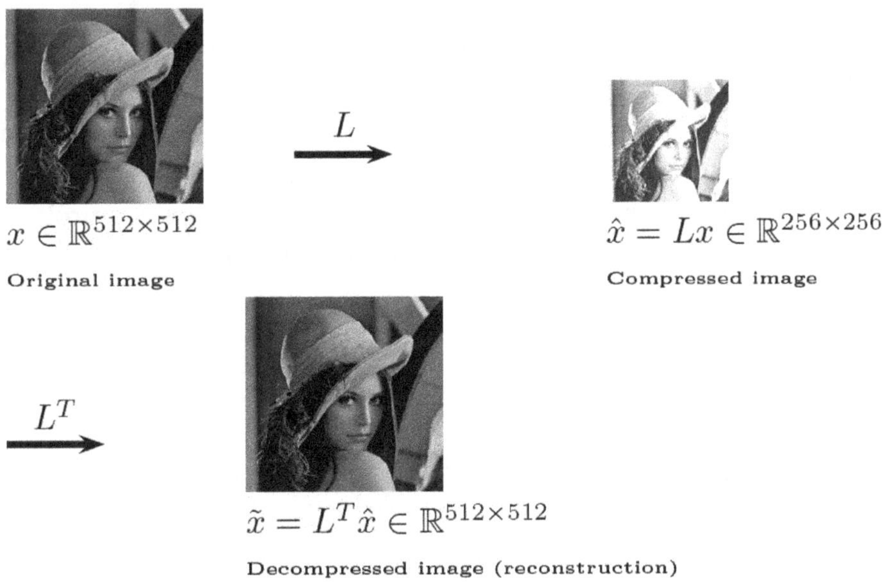

Figure 1. Sample compression and decompression.

energy to the compressed data $\widehat{x} = Lx \in \mathbb{R}^r$. That is: $\|x\| \approx \|\widehat{x}\|$. The decompressed data are $L^T\widehat{x} = \tilde{x} \in \mathbb{R}^{\mathbf{n}}$. On the other hand, the compressed and decompressed energies are equal. That is $\|\widehat{x}\| = \|\tilde{x}\|$ since $LL^T = I_r$. Therefore, the original, compressed, and decompressed data are related as follows

$$\|x\| \approx \|\widehat{x}\| = \|\tilde{x}\|, \tag{3}$$

where $x \in \mathbb{R}^n$ is the original data, $\widehat{x} \in \mathbb{R}^r$ is the compressed data, and $\tilde{x} \in \mathbb{R}^n$ is the decompressed data. Thus, once an appropriate low-pass submatrix L is determined, one proposes solving the corresponding optimization problem in the reduced affine subspace determined by L^T and later coming back to the original size by its generalized inverse.

3. Problem formulation

3.1. Statement of the problem

Given a nonlinear function R from $\mathbb{R}^n$ to $\mathbb{R}^n$, find a solution in an affine subspace determined by an initial displacement point $x_o \in \mathbb{R}^n$ and an orthonormal base $L^T_{\mathbf{n}\times\mathbf{r}}$, $\mathbf{r} < \mathbf{n}$. That is: find $x^* \in \mathbb{R}^n$ with $R(x^*) = 0$ and $x^* \in x_o + \eta(L^T)$, where $\eta(L^T)$ is the subspace generated by the linear combination of the operator L^T.

3.2. Overdetermined problem

This section formulates this problem by using the overdetermined functions H and ϕ from $\mathbb{R}^r$ to $\mathbb{R}^n$

$$H(p) = R(\phi(p)) = 0, \qquad \phi(p) = x_o + L^T p, \qquad and p \in \mathbb{R}^r. \tag{4}$$

The fact that finding a solution p^* of the overdetermined problem draws attention, $H(p^*) = 0$ for $p^* \in \mathbb{R}^r$, is equivalent to finding the solution one is initially seeking. That is: $x^* = \phi(p^*)$ and $R(x^*) = 0$ is a solution on the affine subspace. Therefore, one studies the problem by finding a zero residual of the nonlinear least-squares problem associated to H. Problem (4) is called an overdetermined zero-residual problem.

3.3. Nonlinear least-squares problem

The residual problem (4) is immediately seen to be equivalent to solving the nonlinear zero-residual least-square problem

$$\text{minimize } f(p) = \frac{1}{2}\sum_{i=1}^{n} r_i^2(\phi(p)), p \in \mathbb{R}^r. \tag{5}$$

3.4. First derivatives for the residual functions R and H

The Jacobian of R at $x \in \mathbb{R}^n$ is given by

$$J_R(x) = J(x) = \left[(\nabla r_i(x))^T \right]_{1 \le i \le n}. \tag{6}$$

A direct application of the chain rule, since $J_\phi(p) = L^T$ yields:

$$J_H(p) = J(\phi(p))L^T. \tag{7}$$

3.5. First and second derivatives for problem (5)

The gradient of each term of the problem is $\nabla r_i^2(\phi(p)) = 2L^T r_i(\phi(p)) \nabla r_i(\phi(p))$. Therefore the gradient of $f(p)$ is

$$\nabla f(p) = \left((J(\phi(p))L^T)^T R(\phi(p)) \right). \tag{8}$$

The second-order information is

$$\nabla^2 f(p) = (J(\phi(p))L^T)^T (J\phi(p))L^T) + \sum_{i=1}^{n} h_i(p) \nabla^2 h_i(p)). \tag{9}$$

4. Gauss-Newton method

This section presents a Gauss-Newton method to solve the nonlinear problem (5) which is equivalent to solving the overdetermined nonlinear composite function (4). It describes the standard Newton assumptions for this composite function problems that yield q-quadratic rate of convergence. The inconvenience to use Newton method is that the second-order information associated with the Hessian method is not easily accessible or is impractical for computational time. The latter makes the Newton method impractical for very large-scale problems.

4.1. Model order-reduction-based Gauss-Newton algorithm

This subsection presents a reduced-order Gauss-Newton algorithm for solving problem (5), which is the interest herein.

Algorithm 1. Reduced-order Gauss-Newton (ROGN)

Inputs: Given the compressed base $L^T \in \mathbb{R}^{n \times r}$, and an initial displacement $x_o \in \mathbb{R}^n$.

Output: Approximate solution in the affine subspace $x \in \mathbb{R}^n$.

1: Initial point of the problem. Given $p_o \in \mathbb{R}^r$.

2: Initial point in the affine subspace. $\phi(p_o) = x_o + L^T p_o \in \mathbb{R}^n$.

3: For $k = 0$: until convergence $(\|R(\phi(p_k)\| \leq \epsilon)$.

4: Gauss-Newton direction (compressed direction). Solve for Δp_{k+1}

$$(J(\phi(p_k))L^T)^T (J\phi(p_k))L^T)\Delta p_{k+1} = -(J(\phi(p_k))L^T)^T R(\phi(p_k)). \tag{10}$$

5: Compressed update: $p_{k+1} = p_k + \Delta p_{k+1}$.

6: Decompressed update: $\phi(p_{k+1}) = x_o + L^T p_{k+1}$.

Remarks:

1. The algorithm presents two initial points. The first x_o is the displacement to characterize the affine subspace, and the second one p_o is the initial point for the algorithm.

2. The update $\phi(p_{k+1})$ is the approximation to the solution one is looking for which one denotes by x_{k+1}.

3. Finding Gauss-Newton direction Δp_{k+1} is equivalent to solving the following linear least-squares problem:

$$\min_{\Delta p_{k+1}} \left\{ \frac{1}{2} \| (J\phi(p_k)L^T)\Delta p + R(\phi(p_k)) \|^2 \right\}. \tag{11}$$

4. The Gauss-Newton direction is the Petrov-Galerkin direction obtained by approximating Newton's direction of square nonlinear problems for the following weighted problem:

$$\min_{\Delta p_{k+1}} \left\{ \frac{1}{2} \| \left(L^T \Delta p + R(x) \|^2_{Q^{-1}}, \quad Q = J(x)^T J(x) > 0 \right\}. \tag{12}$$

4.2. Local convergence of reduced Gauss-Newton algorithm

It is known that the Gauss-Newton method retains q-quadratic rate of convergence under standard assumptions for zero-residual single-function problems [15]. The natural question is: What are the standard assumptions that guarantee the Gauss-Newton conditions for the composite function one is working with, that conserve q-quadratic rate of convergence? The next theorem establishes these assumptions.

Theorem: Let H from $\mathbb{R}^r$ to $\mathbb{R}^n$ be defined by $H(p) = R(x_o + L^T p)$, $x_o \in \mathbb{R}^n$, $p \in \mathbb{R}^r$, and $L \in \mathbb{R}^{r \times n}$ are orthonormal operators with $r < n$. Assume there exists a solution $p^* \in \tilde{D} \subset \mathbb{R}^r$, with $\tilde{D}$ convex and open. Define $D = \{x_o\} + L^T(\tilde{D})$, where $L^T(\tilde{D})$ is the image of $\tilde{D}$ under $L^T \in \mathbb{R}^{n \times r}$. Assume that $J_R \in L_\gamma(D)$, J_R is bounded on D, and the minimum eigenvalue of $J(x^*)^T J(x^*)$ is positive. Then, the sequence $\{p_{k+1}\}$ given **ROGN algorithm 1** is well defined, converges, and has q-quadratic rate of convergence. That is:

$$\|p_{k+1} - p^*\| \leq \frac{1}{2}\|p_k - p^*\| \tag{13}$$

$$\|p_{k+1} - p^*\| \leq \frac{\tilde{c}\tilde{\gamma}}{2\tilde{\sigma}}\|p_k - p^*\|^2 \quad with \quad \tilde{c}, \tilde{\gamma}, \tilde{\sigma} \in \mathbb{R}_+. \tag{14}$$

Proof: The residual problem R has solution x^* on D. First, one proves that the Jacobian of H is Lipchitz on $\tilde{D}$. Since $J_H(p) = J(\phi(p))L^T$, then

$$\|J_H(p_1) - J_H(p_2)\| = \|(J(\phi(p_1)) - J(\phi(p_2)))L^T\|, \; for \; p_1, \; p_2 \in \tilde{D}. \tag{15}$$

Since the Jacobian of R is Lipschitz on D, one concludes

$$\|J_H(p_1) - J_H(p_2)\| \leq \tilde{\gamma}\|p_1 - p_2\|, \quad \tilde{\gamma} = \gamma^*\|L^T\|^2, for p_1, p_2 \in \tilde{D}. \tag{16}$$

Second, one proves that the Jacobian of H is bounded on $\tilde{D}$. Since the Jacobian of H at p is $J(\phi(p)L^T$, then

$$\|J_H(p)\| = \|J(\phi(p)L^T\| \quad for p \in \tilde{D}. \tag{17}$$

Now, since the Jacobian of R is bounded on D, one concludes

$$\|J_H(p)\| \leq \tilde{c}, \, \tilde{c} = c^*\|L^T\| for p \in \tilde{D}. \tag{18}$$

Finally, one proves that the smallest eigenvalue of $J_H(p^*)^T J_H(p^*)$ is greater than zero.

$$J_H(p^*)^T J_H(p^*) = \left(J(x^*)L^T\right)^T \left(J(x^*)L^T\right) \quad with \quad x^* = x_o + L^T p^*. \tag{19}$$

Let $p \neq 0 \in \mathbb{R}^k$ and $\sigma \in \mathbb{R}$ be an eigenvector and eigenvalue associated with the last symmetric matrix. Then

$$\|L^T p\|_Q^2 = \sigma\|p\|^2 \;, \;\; Q = J(x^*)^T J(x^*) > 0. \tag{20}$$

Therefore, $\sigma > 0$ since L^T is a full rank and $p \neq 0$. The convergence and its fast rate of convergence given by the last two inequalities follow from the Theorem 10.2.1 in the Dennis and Schnabel book [15].

5. Regularization

Despite the advantages of the Gauss-Newton method, the algorithm will not perform well if either the problem is ill conditioned or in the presence of high nonlinearity of some components of it. The purpose of this section is to introduce two regularizations to overcome these difficulties while retaining the fast rate of convergence of the Gauss-Newton method.

5.1. Levenberg-Marquardt method

To prevent the Gauss-Newton algorithm to preclude in case some eigenvalues are near zero or in case of rank deficiency of the linear systems to solve, the least-squares directions are regularized by

$$\min_{\Delta p}\left\{\frac{1}{2}\|(J(\phi(p))L^T)\Delta p+R(\phi(p))\|^2+\frac{\mu}{2}\|\Delta p\|^2\right\} \tag{21}$$

where $\mu > 0$. The solution is given by

$$\left((J(\phi(p))L^T)^T(J(\phi(x))L^T)+\mu I\right)\Delta p=-(J(\phi(p))L^T)^T R(\phi(p)). \tag{22}$$

Under the standard Gauss-Newton assumptions written before and choosing the regularization parameter as $\mu = O\left(\|(J(\phi(p))L^T)^T R(\phi(p)\|\right)$, the regularized Gauss-Newton algorithm converges and the q-quadratic rate of convergence is retained; see Theorem 10.2.6 [15].

$$\|p_{k+1}-p^*\| \leq \frac{1}{2}\frac{\tilde{\alpha}\tilde{\gamma}}{\tilde{\lambda}}\|p_k-p^*\|^2. \tag{23}$$

5.2. Scaling regularization

To avoid the influence of the high order of magnitude of some components with respect to the rest of the components of the problem, one presents the following regularization:

$$\min_{\Delta p}\left\{\frac{1}{2}\|(J(\phi(p))L^T)\Delta p+R(\phi(p))\|^2+\frac{\sigma}{2}\|L^T\Delta p\|_Q^2\right\}, \tag{24}$$

where $Q=(J(\phi(p))L^T)^T(J(\phi(p))L^T)$. The solution is given by

$$\left((J(\phi(p))L^T)^T(J(\phi(p))L^T)\right)\Delta p=-(J(\phi(p))L^T)^T\frac{R(\phi(p))}{1+\sigma\|R(\phi(p))\|}. \tag{25}$$

This regularization prevents that large components of the problem affect the behavior of the algorithm. It is important to observe the Lipschitz constant of the problem is improved. Considering the preceding two regularizations, one has

$$\left((J(\phi(p))L^T)^T(J(\phi(p))L^T)+\mu I\right)\Delta p=-(J(\phi(p))L^T)^T\frac{R(\phi(p))}{1+\sigma\|R(\phi(p))\|}. \tag{26}$$

This last regularizations prevent the smallest eigenvalue affecting the behavior of the Gauss-Newton algorithm and at the same time, through rescaling, components with small values are not considered by the influence of large components while retain its fast rate of convergence.

6. Globalization strategy

The good performance of the Gauss-Newton algorithm depends on a suitable initial point that must be inside its region of convergence. Rather than absorbing the computational cost associated with choosing an appropriate initial point, the chapter proposes a line-search method that provides convergence for initial points outside of the region of convergence. The goal of this approach is to obtain a sufficient decrease in the merit function. If the direction fails, then a backtracking is used until a sufficient reduction is obtained. A merit function should allow moving toward a solution of the problem.

6.1. Merit function

It is natural to think that the merit function for the unconstrained minimization problem (5) is itself. That is: $M(p) = f(p)$.

6.2. Descent direction

One proves that Gauss-Newton direction is a descent direction for the merit function $M(p) = f(p)$.

Property: The regularized Gauss-Newton direction Δp given by (26) is a descent direction for the merit function $M(p) = f(p)$.

Proof: One proves that the directional derivative of f at the direction Δp is less than zero. The gradient of $f(p)$ is given by $\nabla f(x) = \left(J(\phi(p))L^T\right)^T R(\phi(p))$. Therefore

$$\nabla f(p)^T \Delta p = -\|\left(J(\phi(p))L^T\right)^T R(\phi(p))\|_{Q^{-1}}^2 < 0 \tag{27}$$

since $Q = \left(\left(J(\phi(p))L^T\right)^T \left(J(\phi(p))L^T\right) + \mu\right)$ is positive definite.

Consequently, it is possible to progress toward a solution of the problem in the Δp direction. The purpose is to find a step length $\alpha \in (0, 1]$ that yields a sufficient decrease. To that effect one follows the Armijo-Goldstein conditions given by

$$f(p + \alpha \Delta p) \leq f(p) + \alpha\left(\lambda^* \nabla f(p)^T \Delta p\right) \tag{28}$$

and

$$\nabla f(p + \alpha p)^T \Delta p \geq \beta \nabla f(p)^T \Delta p, \tag{29}$$

for fixed values $\lambda, \beta \in (0, 1)$. The first inequality allows sufficient decrease of the merit function, and the second one avoids step lengths that are very small. It is important to observe that if β is chosen, $\beta \in [\lambda, 1]$, then the two inequalities can be satisfied simultaneously. Wolfe proved that if f is continuously differentiable on $\mathbb{R}^r$, Δp is a descent direction, and assuming the set

$\{f(p+\alpha\Delta p); \alpha \in (0,1]\}$ is bounded below, then there exists an $\alpha^* \in (0,1]$ such that the two inequalities be satisfied simultaneously [16].

It is important to realize that these two inequalities can be reached by using a back-tracking procedure. Therefore, this work uses a line-search strategy to satisfy the inequalities. Next section proposes a line-search regularized Gauss-Newton algorithm for solving the zero-residual composite function problem.

7. A line-search regularized Gauss-Newton method

This section proposes the following regularized Gauss-Newton method with line search to find a solution on the affine subspace $x_o + \eta(L^T)$ for problem (4).

Algorithm 2: A reduced-order regularized Gauss-Newton (RORGN)

Input: Given the compressed base $L^T \in \mathbb{R}^{n\times r}$, and a displacement $x_o \in \mathbb{R}^n$.

Output: The approximate solution in the affine subspace $x \in \mathbb{R}^n$.

1: Initial point of the problem. Given $p_o \in \mathbb{R}^r$.

2: Initial point in affine subspace. $x_1 = x_o + L^T p_o \in \mathbb{R}^n$.

3: For $k = 1$: until convergence ($\|R(x_k)\| \leq \varepsilon$.).

4: Choose $\mu_k = \sigma_k \|R(x_k)\|$, and $\sigma_k \in (0,1]$.

5: Regularized Gauss-Newton direction. Solve for Δp_k

$$\left(\left(\left(J(x_k)L^T\right)^T\left(J(x_k)L^T\right) + \mu_k I\right)\Delta p_k = -\frac{\left(J(x_k)L^T\right)^T R(x_k)}{1+\sigma_k\|R(x_k)\|}. \tag{30}$$

6: Line search (sufficient decrease). Find $\alpha_k \in (0,1]$ such that

$$\|R(x_k + \alpha_k L^T \Delta p_k)\|^2 < \|R(x_k)\|^2 + 2\cdot 10^{-4}\alpha_k \nabla f(p_k)^T \Delta p_k. \tag{31}$$

7: Update. $x_{k+1} = x_k + \alpha_k L^T \Delta p_k$.

Remarks: The algorithm is amenable to the use of any suitable basis, not necessarily a wavelet basis. The algorithm can be tested with different initial displacement points. On the other hand, the election of the initial point of the algorithm is not limited to the origin.

8. Numerical examples

The authors run on a MacBook Pro laptop equipped with an Intel(R) Quad-Core(TM) i7-2720QM CPU @ 2.20GHz and 8 GB of RAM. Section 8.2 presents Bratu's 3D problem. This problem is more challenging since the nonlinear systems become ill conditioned where one approaches the bifurcation point. Therefore, the RORGN algorithm was tested to solve them efficiently. All these Bratu's problems utilized wavelets-based ROM. One takes the regularization by $\mu_k = \sigma_k \|\mathbf{R}(x_k)\|$, with $\sigma_k \in (0,1)$. One employs as stopping criteria for the algorithm; either the norm of the residual, $\epsilon_k = \|\mathbf{R}(x_k)\|$, is less than some small positive real value given, ϵ, or a maximum number of iterations reached, k_{max}.

8.1. Bratu's 2D problem

The Bratu's 1D equation can be generalized by replacing the second derivative by a Laplacian [1, 17]. This section numerically studies the nonlinear diffusion equation with exponential source term in two and three dimensions. Let $\Omega = [0,1]^n, n = 2,3$ be a unitary square or cube, where $x_i \in [0,1], i = 1, \ldots, n$ are the spatial variables while n is the space dimension.

$$\begin{aligned} \Delta u + \zeta \cdot e^u = 0 \ \text{ on } \ \Omega \ ; \ \ u = u(\underline{x}), \\ u = 0 \ \text{ on } \ \partial\Omega, \end{aligned} \tag{32}$$

and $\zeta \in \mathbb{R}$ is a coefficient. The Laplacian is defined by

$$\Delta(\cdot) = \sum_{i=1}^{n} \frac{\partial^2(\cdot)}{\partial x_i^2}. \tag{33}$$

One can discretize (32) by means of central finite differences on regular tensor product meshes. Homogenous Dirichlet boundaries are enforced conditions in all square or cube faces, see [17] for details.

8.2. Bratu's 3D problem

Figures 2 and **3** present results from Bratu's 3D problem. **Figure 2** shows the parameter continuation problem while **Figure 3** compares FOM and WROM results in the whole domain. For visualization purposes, one cuts away the front half of the cube to see inside it. **Figure 2** shows the FOM in continuous line while dashed blue and magenta lines correspond to WROM at 85 and 90%, respectively. The mesh size is $10 \times 10 \times 10$ and $\zeta = 1.5$ is fixed. In the figure, one has from left to right the FOM, the WROM, and the absolute error. Neither of these WROM models could reproduce the FOM behavior beyond $\zeta > 9$. They could not get into neither the second branch nor close to the bifurcation point, where the system becomes highly nonlinear. On the other hand, **Figure 3** compares FOM versus WROM at 90% in order to show that these WROM could properly reproduce the FOM behavior in the whole cube. **Table 1** summarizes the performance of the family of Gauss-Newton algorithms applied to FOM Bratu's 3D problem. One employs these numerical values, $\epsilon_{tol} = 10^{-3}$ and $k_{max} = 32$. Once again, one gets close to the bifurcation point by choosing, $\zeta = 9.9$, to pose a challenging nonlinear system while σ was tuned to achieve performance for a given rank.

One observes for all ranks reported herein that the regularized method provides convergence tolerances likewise but it usually spent two iterations less than standard Newton and hence

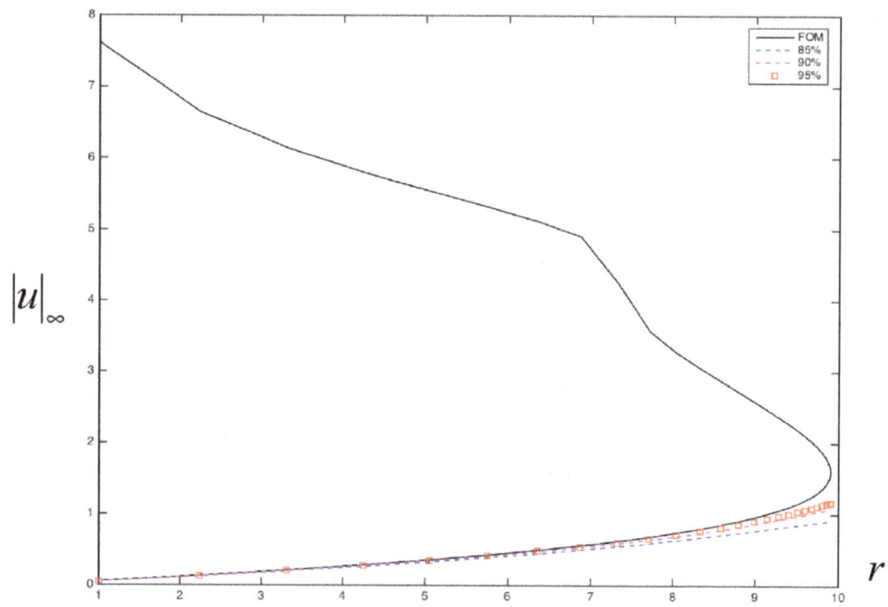

Figure 2. FOM vs. WROM parameter continuation solutions of (32), for $n = 3$, are shown.

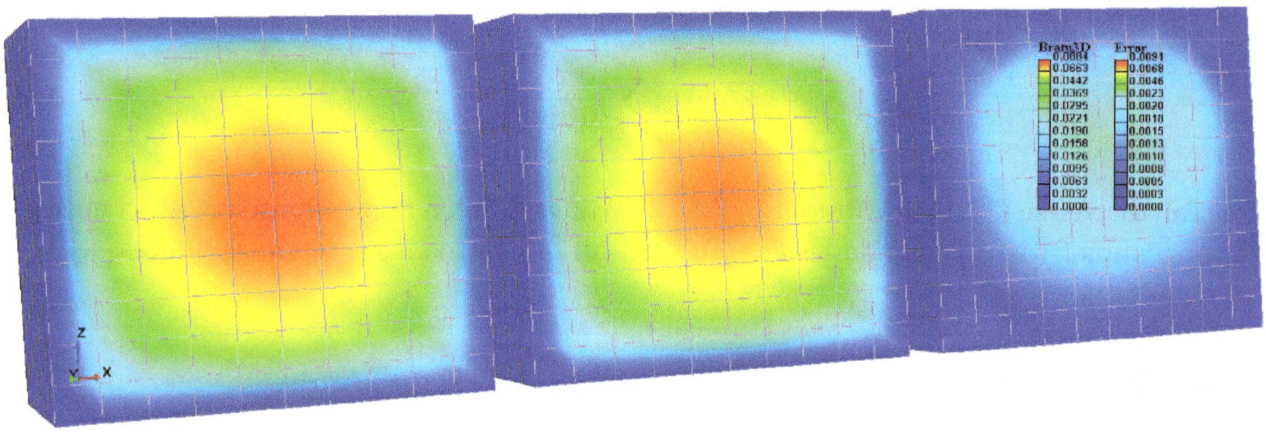

Figure 3. FOM and ROM are compared: FOM (left), WROM at 90% (center), and error (right).

CPU time reduces as well. One notices for this particular problem that line search is equivalent to standard Newton as well as combined matches the performance of the combined plus line-search method.

8.3. Nonlinear benchmark problems

One also considers a benchmark nonlinear problem from the literature in order to challenge the proposed algorithms. The Yamamura [18] problem is a nonlinear system of equations defined by:

$$\mathbf{R} : \mathbb{R}^n \to \mathbb{R}^n, \quad x \in \mathbb{R}^n, \quad R_i(x) = 0, \quad 1 \le i \le n,$$
$$R_i(x) = 2.5x_i^3 - 10.5x_i^2 + 11.8x_i - i + \sum_{i=1}^{n} x_i = 0. \tag{34}$$

Newton method	ϵ_k	δ_k	#Iter	Success
N = 512, $\sigma = 0.03$				
Standard	1.637075E−04	4.497266E−08	7	True
Combined	5.089393E−04	8.391213E−07	5	True
Regularized	5.089393E−04	8.391213E−07	5	True
Scaled	1.637075E−04	4.497266E−08	7	True
Line search	1.637075E−04	4.497266E−08	7	True
Com. and Line search	5.089393E−04	8.391213E−07	5	True
N = 1000, $\sigma = 0.025$				
Standard	3.303599E−04	1.682251E−07	7	True
Combined	1.424983E−05	7.450719E−10	6	True
Regularized	1.424983E−05	7.450719E−10	6	True
Scaled	3.303599E−04	1.682251E−07	7	True
Line search	3.303599E−04	1.682251E−07	7	True
Com. and Line search	1.424983E−05	7.450719E−10	6	True
N = 1728, $\sigma = 0.020$				
Standard	5.792054E−04	4.806531E−07	7	True
Combined	1.197055E−04	6.096723E−08	5	True
Regularized	1.197055E−04	6.096723E−08	5	True
Scaled	5.792054E−04	4.806531E−07	7	True
Line search	5.792054E−04	4.806531E−07	7	True
Com. and Line search	1.197055E−04	6.096723E−08	5	True
N = 2744, $\sigma = 0.015$				
Standard	1.143068E−05	1.756738E−10	8	True
Combined	1.678400E−04	1.052920E−07	6	True
Regularized	1.678400E−04	1.052920E−07	6	True
Scaled	1.143068E−05	1.756738E−10	8	True
Line search	1.143068E−05	1.756738E−10	8	True
Com. and Line search	1.678400E−04	1.052920E−07	6	True

Table 1. Gauss-Newton results for Bratu's 3D problem presented in Section 8.2.

where n is the size of the nonlinear system. One implemented the algorithms with $\epsilon_{tol} = 10^{-6}$, $k_{max} = 128$, and $\sigma = 0.025$.

The objective is to challenge the algorithms presented in this research, and the results are reported in **Table 2**. One can infer from this table that the standard Newton method could not converge in any of these realizations except $n = 32$. Conversely, the regularized method converged for all realizations. This latter method outperformed all others. On the other hand, the

Newton method	ϵ_k	δ_k	#Iter	Success
N = 32				
Standard	6.643960E−06	1.282327E−07	47	True
Regularized	3.547835E−05	8.310947E−07	32	True
Line search	8.835078E−09	2.415845E−13	78	True
Reg. and Line search	4.271577E−06	8.288108E−09	35	True
N = 256				
Standard	2.992259E−01	3.018468E+07	128	False
Regularized	4.260045E−06	5.643326E−08	21	True
Line search	2.569192E−02	2.618397E+03	128	False
Reg. and Line search	2.893999E−06	2.380562E−08	35	True
N = 512				
Standard	2.280299E−02	3.448757E+05	128	False
Regularized	2.384802E−07	2.579578E−10	46	True
Line search	1.127484E−01	1.845466E+07	128	False
Reg. and Line search	3.551655E−08	2.368659E−11	42	True
N = 1024				
Standard	9.540053E−05	4.501800E+01	128	False
Regularized	1.730306E−06	8.414453E−09	41	True
Line search	7.199610E−05	5.095672E+00	128	False
Reg. and Line search	8.857816E−06	4.250476E−07	35	True
N = 2048				
Standard	1.769950E+00	8.504131E+14	128	False
Regularized	7.270952E−08	1.578029E−10	68	True
Line search	6.195869E−02	4.180347E+09	128	False
Reg. and Line search	1.069042E−02	3.229689E−01	128	False

Table 2. Gauss-Newton results for Yamamura problem.

regularized and line-search method could consistently converge for all realizations but $n = 2048$. For larger ranks, that is, 512 and 1024, the latter method is the more efficient bottom line; the regularized method performed well in this example.

9. Hybrid method: HROM

The idea is simple since the wavelet subspace is not a function of a priori known snapshots, it can be determined without executing the so-called, computationally expensive, off-line stage, in which one thoroughly studies the FOM and can sample the input space to record a representative set of snapshots. HROM considers as input snapshots those outputted by a WROM procedure. As shown below, if this WROM happens to reproduce the FOM behavior correctly, then one should expect the resulting PROM, that is, HROM, to replicate the original FOM behavior accurately. At first, glance, depending upon the WROM compression ratio, this procedure reduces the runtime of the off-line stage.

This section conducts a series of preliminary numerical experiments on the well-known Bratu's nonlinear benchmark problem, in particular in one and two dimensions [5] to sell this case. **Figures 4** and **5** depict results for the 1D continuation problem. They utilize the following WROM compression ratios: 10 and 5%, where the compression ratio is constant during the continuation problem. All plots display two distinct HROM compression ratios. The WROM at 20% that is not depicted completely misses the bifurcation zone and the second branch thus the HROM is also way off. However, as the WROM starts to catch up with the FOM then HROM too does. Indeed, it is observed that HROM yields comparable results when comparing it to the version that takes the original snapshots, that is, PROM.

One can repeat a similar experiment with Bratu's 2D continuation problem, which produces the same trend as before. Indeed, if the input WROM is way off targeting then, HROM is off as well. For instance, 27% compression implies that all HROM miss the second branch, but still, the 21% model could slightly reproduce the proper FOM trend. Things significantly improve on models with 21 and 20% as shown in **Figures 6** and **7**. However, these models still miss the bifurcation zone. They just render a flat profile there. These insights suggest that if the WROM can correctly reproduce the FOM behavior, then HROM can do so. One is probably able to improve the accuracy of the WROM by changing the compression ratio during the online

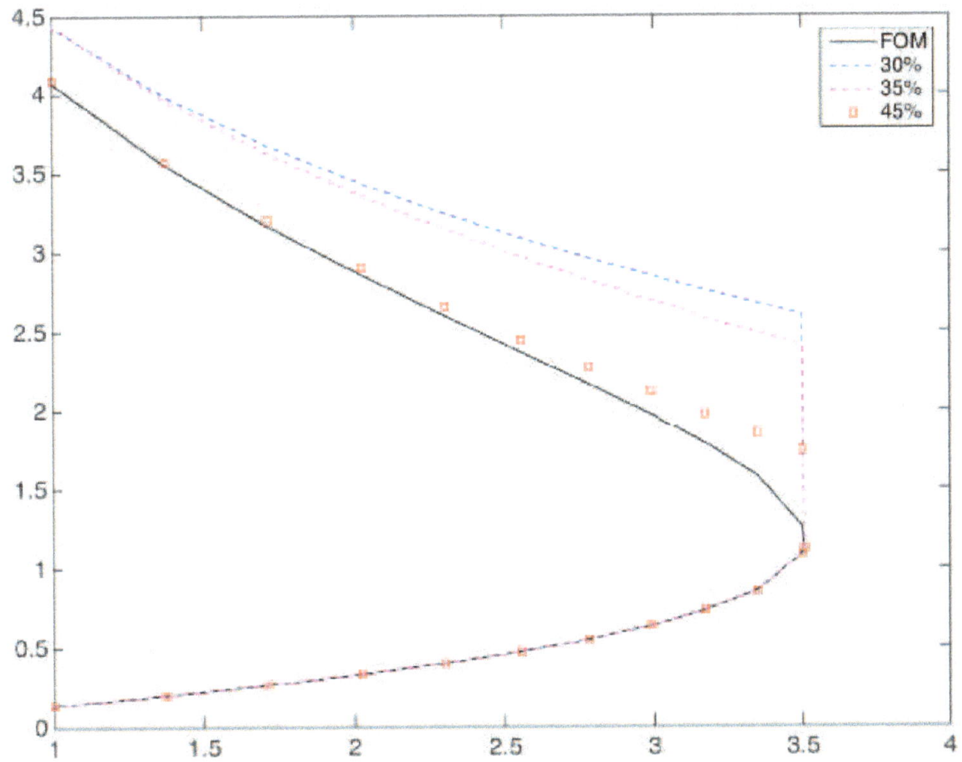

Figure 4. Bratu's 1D, 10% compression.

stage. For Bratu's problem, one can assume a graded energy distribution which provides more of it while approaching the bifurcation point. This approach is referred as "adaptive WROM" or AWROM as shorthand.

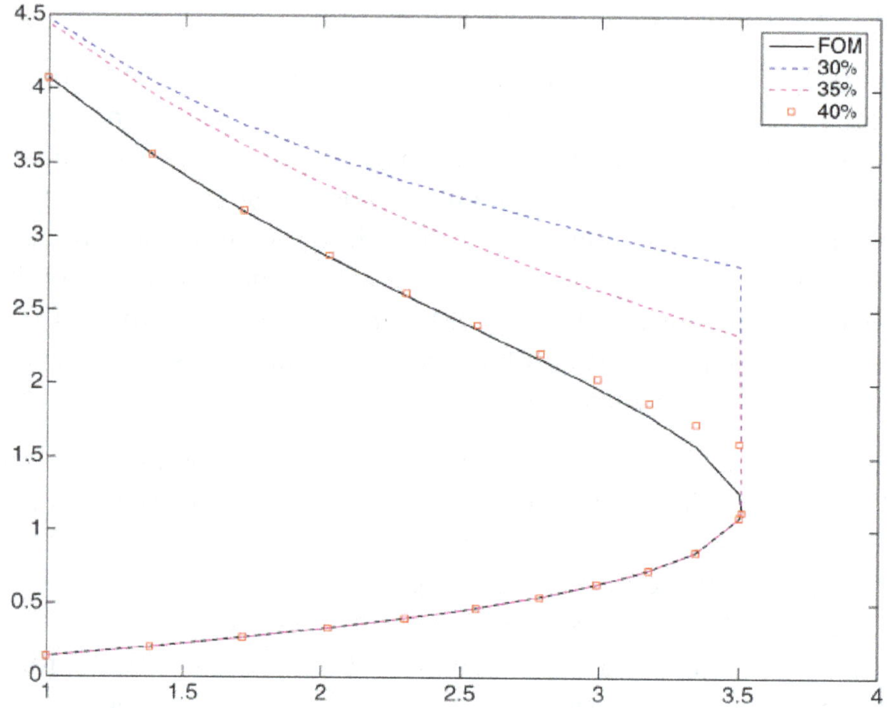

Figure 5. Bratu's 1D, 5% compression.

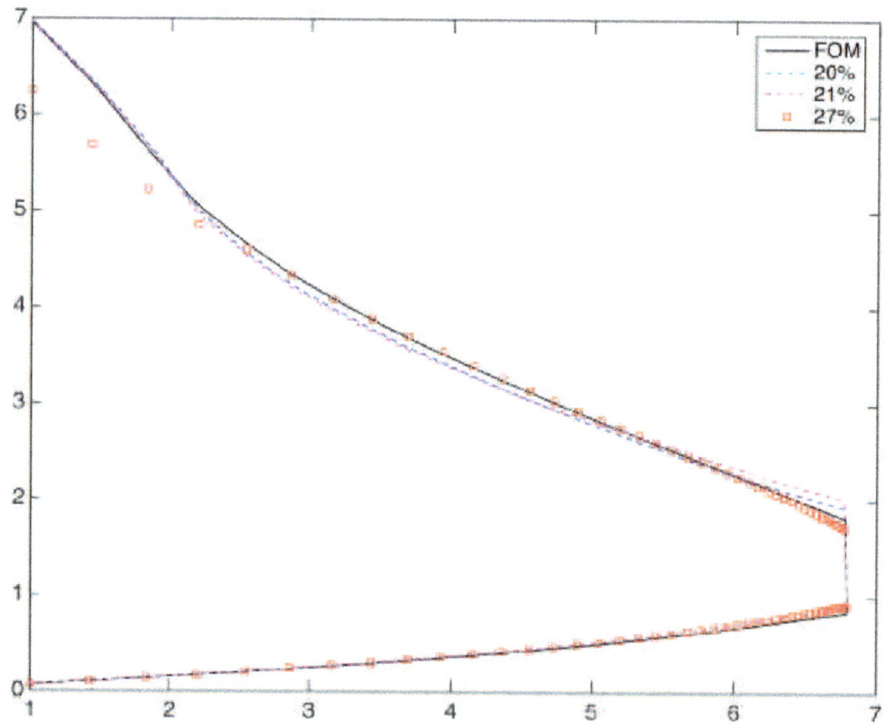

Figure 6. Bratu's 2D, 10% compression.

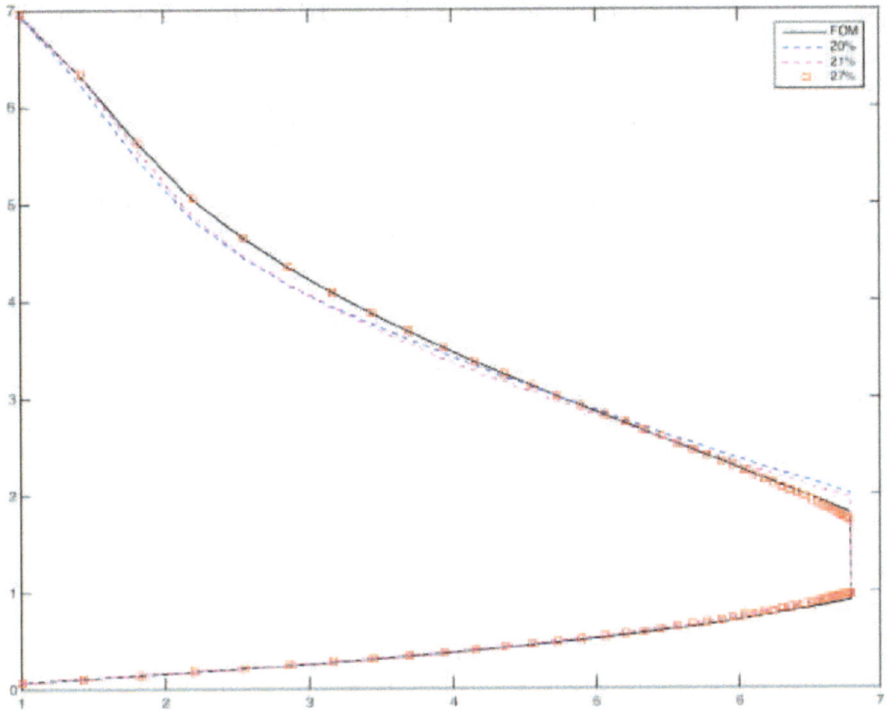

Figure 7. Bratu's 2D, 5% compression.

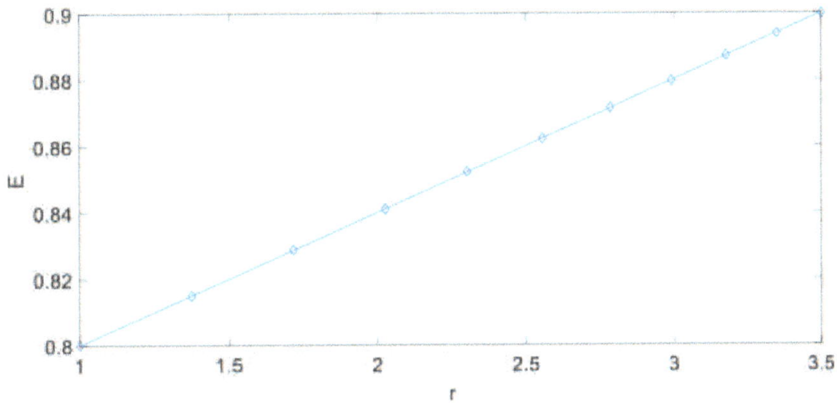

Figure 8. Linear energy distribution.

This section presents an alternative strategy that can rely on insights from the FOM, such as Newton tolerances and number of iterations, which are in turn indirect error estimates. **Figure 8** plots the energy distribution that was utilized as a function of the continuation parameter, ζ, for Bratu's 1D problem. When one approaches the bifurcation point, $\zeta = 3.5$, one should gradually bump up energy as shown. Notice that the distribution tends to concentrate more points toward the bifurcation point. With this energy distribution into account, one obtains the AWROM results in **Figure 9**. This ROM accurately reproduces the FOM behavior as noted. Let now conduct the following experiment. One must run the FOM, and at every snapshot, one needs to store the number of Newton iterations and the resulting error tolerance.

One then computes the energy distribution that **Figure 10** depicts. The following formula was employed to do so:

$$E_i = (1 - \theta) + \theta(nIters_i / nMaxIters), \tag{35}$$

where $nIters_i$ is the number of iterations at the current location, and $nMaxIters$ is the maximum number of iterations reported by the FOM and $\theta \in [0, 1]$. **Figure 11** depicts excellent accordance

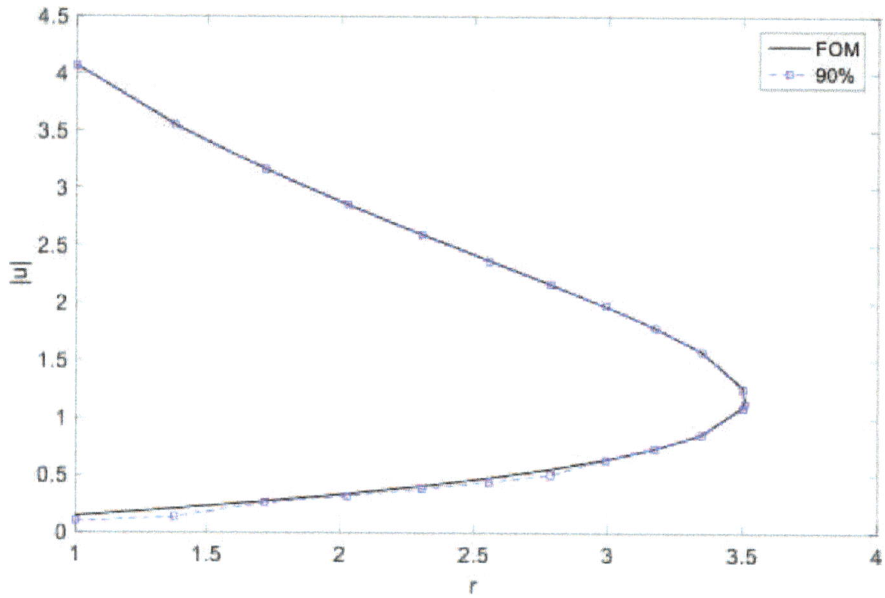

Figure 9. 10–20% variable compression.

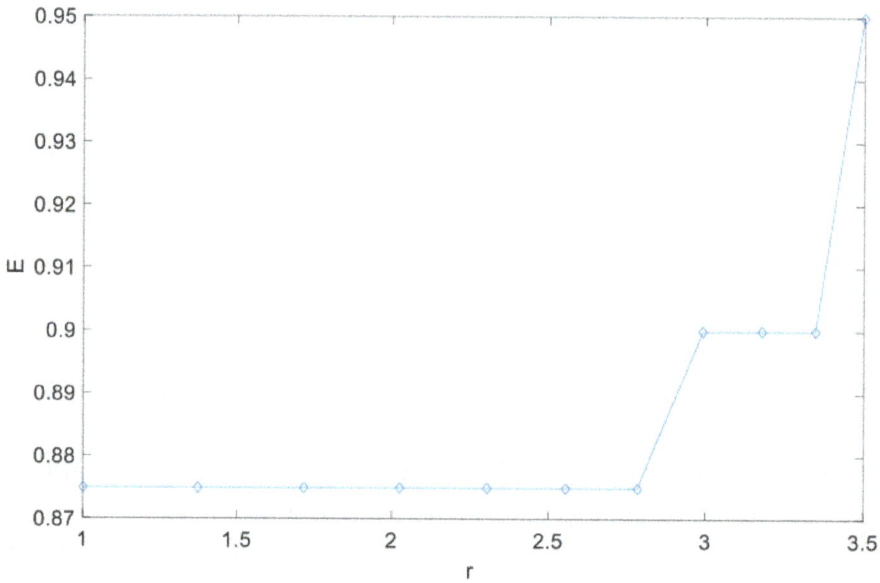

Figure 10. Variable energy distribution.

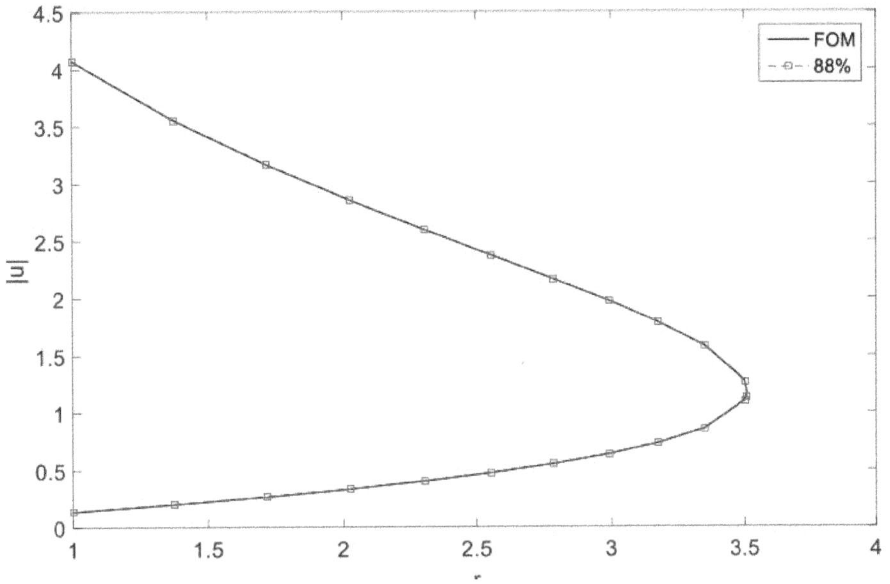

Figure 11. Variable compression.

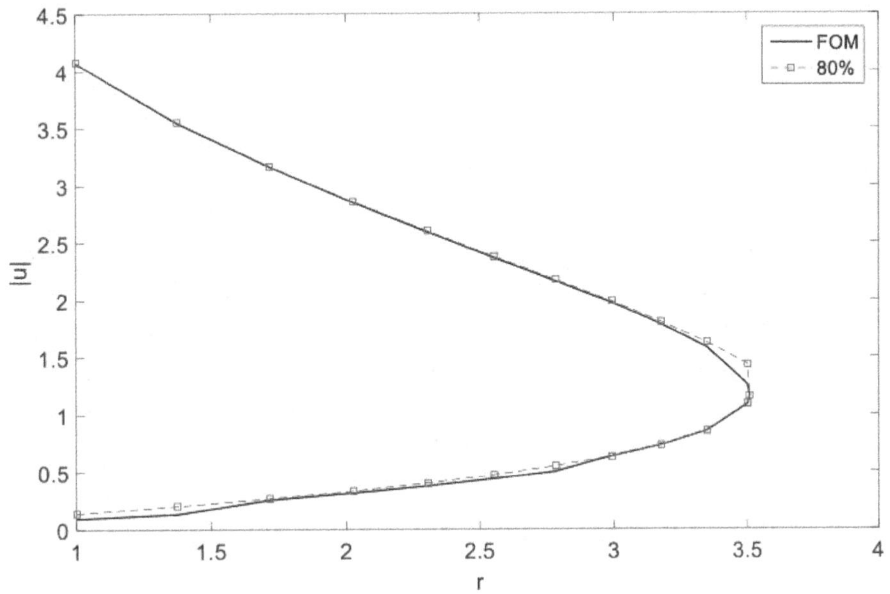

Figure 12. Linear AWROM.

between FOM and AWROM ($\theta = 0.2$). One should expect that if this last AWROM model is inputted for an HROM simulation, HROM should reproduce the original FOM correctly.

Figures 12 and **13** depict preliminary results of HROM applied over a couple of AWROM models whose energy distribution was described in **Figures 8** and **10**, respectively ($\theta = 0.2$). These ROM reproduce the FOM behavior accordingly, which proves that there is potential to study the performance of HROM further. Another important question that arises from further research is how to improve the compression ratio of the AWROM scheme.

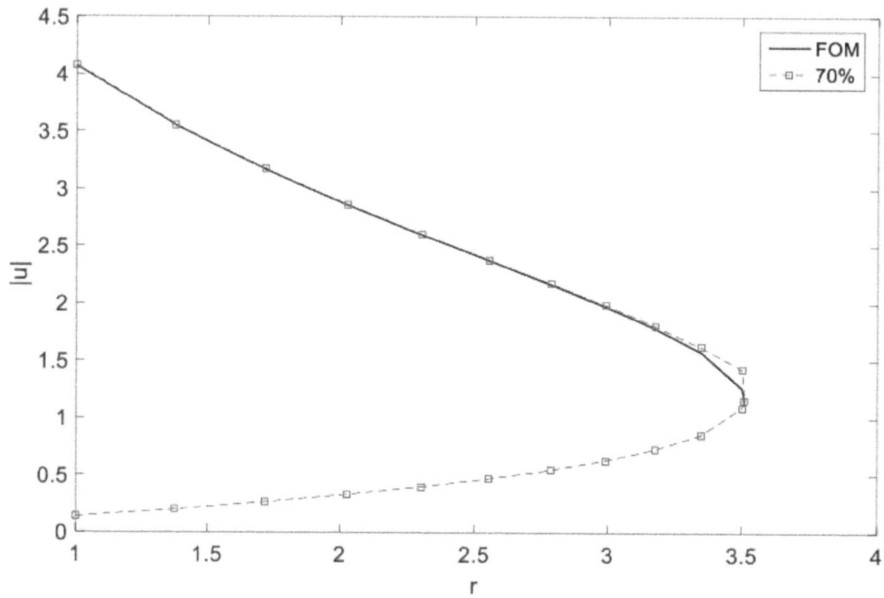

Figure 13. Variable AWROM.

10. Concluding remarks

This chapter introduced a global regularized Gauss-Newton method for resolving square nonlinear algebraic systems in an affine subspace that enable real-time solutions without a priori simulations. Solving a nonlinear least-squares composite function poses the problem where the answer is derived from the inside argument. The authors thus presented the standard Newton assumptions that guarantee a q-quadratic rate of convergence. The findings include that the Petrov-Galerkin projection directions for the Newton method are no other than the Gauss-Newton ones for a composite function. The technique uses two initial points, one that determines the affine subspace and the other is the starting guess for solving the composition mentioned earlier. The notion of compressed sensing with wavelets produces the characterization of an affine subspace that comprises the majority of the energy of the problem. The chapter showed some numerical experimentations that back up the proposed globalization methodology for solving highly nonlinear dynamic systems in real time. These last ones reproduce the principal features of the FOM. Results underline the fact that one does not need to employ information at any particular point. The Bratu's 3D FOM results prove that the proposed RORGN algorithm outperforms the standard GN method while retaining its q-quadratic rate of convergence. This chapter concludes that the regularized and line-search-enabled scheme is the most robust and efficient algorithm for the problems presented herein. The numerical results imply that this approach performs well and it does not significantly increase the CPU time.

From the numerical results presented in Section 9 for Bratu's 1D and 2D problems, the data fusion procedure (HROM) can be used as an alternative procedure when the simulation time for a problem can be limited. This method uses two different sceneries. In the first, one implies that the data come through out of the model, while in the second, the data are obtained independently of the latter. That is, in the first case the model governs the data, and the other the input information rules the model.

Acknowledgements

The authors recognize the financial support of the project "Reduced-Order Parameter Estimation for Underbody Blasts" financed by the Army Research Laboratory, through the Army High-Performance Computing Research Center under Cooperative Agreement W911NF-07-2-0027, and also acknowledge Dr. Martine Ceberio at UTEP for proofreading the manuscript.

Author details

Horacio Florez* and Miguel Argáez

*Address all correspondence to: florezg@gmail.com

Computer Science Department, The University of Texas at El Paso, USA

References

[1] Florez H, Argáez M. A model-order reduction method based on wavelets and POD to solve nonlinear transient and steady-state continuation problems. Applied Mathematical Modelling. 2018;**53**:12-31

[2] Florez H. Linear Thermo-Poroelasticity and Geomechanics. Chapter 10 in Finite Element Method - Simulation, Numerical Analysis and Solution Techniques. In: Pacurar R, editor. Rijeka, Croatia: InTech Open; 2018. pp. 223-242. ISBN: 978-953-51-3849-5. http://dx.doi.org/10.5772/intechopen.71873

[3] Argáez M, Ceberio M, Florez H, Mendez O. A model reduction for highly non-linear problems using wavelets and Gauss-Newton method. In: 2016 Annual Conference of the North American Fuzzy Information Processing Society (NAFIPS); IEEE; 2016. pp. 1-6

[4] Carlberg K, Bou-Mosleh C, Farhat C. Efficient non-linear model reduction via a least-squares Petrov–Galerkin projection and compressive tensor approximations. International Journal for Numerical Methods in Engineering. 2011;**86**:155-181

[5] Willcox K, Peraire J. Balanced model reduction via the proper orthogonal decomposition. AIAA Journal. 2002;**40**(11):2323-2330

[6] Chaturantabut S, Sorensen DC. Application of POD and DEIM on dimension reduction of non-linear miscible viscous fingering in porous media. Mathematical and Computer

Modelling of Dynamical Systems. 2011;**17**(4):337-353. DOI: 10.1080/13873954.2011.547660. URL: http://dx.doi.org/10.1080/13873954.2011.547660

[7] Florez H. Applications of model-order reduction to thermo-poroelasticity. In: 51st US Rock Mechanics/Geomechanics Symposium; American Rock Mechanics Association; 2017

[8] Amsallem D, Zahr MJ, Farhat C. Nonlinear model order reduction based on local reduced-order bases. International Journal for Numerical Methods in Engineering. 2012;**92**:891-916

[9] Kerfriden P, Gosselet P, Adhikari S, Bordas SPA. Bridging proper orthogonal decomposition methods and augmented Newton–Krylov algorithms: An adaptive model order reduction for highly nonlinear mechanical problems. Computer Methods in Applied Mechanics and Engineering. 2011;**200**:850-866

[10] Hernandez M. A reduced order parameter estimation technique using orthonormal wavelets [PhD thesis]. El Paso: ETD Collection for University of Texas; 2011

[11] Le T-H, Caracoglia L. Reduced-order wavelet-Galerkin solution for the coupled, nonlinear stochastic response of slender buildings in transient winds. Journal of Sound and Vibration. 2015;**344**:179-208

[12] Daubechies I. Ten lectures on wavelets. In: CBMS-NSF Conference Series in Applied Mathematics. Philadelphia, PA: SIAM Ed.; 1992

[13] Teolis A. Computational Signal Processing with Wavelets. Boston, MA, USA: Birkhauser; 1998

[14] Mallat S. A Wavelet Tour of Signal Processing. London: Academic Press; 1999

[15] Dennis J, Schnabel R. Numerical Methods for Unconstrained Optimization and Nonlinear Equations. Philadelphia, PA, USA: Society for Industrial and Applied Mathematics; 1996. DOI: 10.1137/1.9781611971200. URL: http://epubs.siam.org/doi/abs/10.1137/1.9781611971200

[16] Wolfe P. Convergence conditions for ascent methods. SIAM Review. 1969;**11**(2):226-235. ISSN: 00361445. URL: http://www.jstor.org/stable/2028111

[17] Florez H, Argáez M. Applications and comparison of model-order reduction methods based on wavelets and POD. In: 2016 Annual Conference of the North American Fuzzy Information Processing Society (NAFIPS); IEEE; 2016. pp. 1-8

[18] Yamamura K, Kawata H, Tokue A. Interval solution of nonlinear equations using linear programming. BIT. 1998;**38**:186-199

2

Invariants of Generalized Fifth Order Non-Linear Partial Differential Equation

Sachin Kumar

Abstract

The fifth order non-linear partial differential equation in generalized form is analyzed for Lie symmetries. The classical Lie group method is performed to derive similarity variables of this equation and the ordinary differential equations (ODEs) are deduced. These ordinary differential equations are further studied and some exact solutions are obtained.

Keywords: generalized fifth order non-linear partial differential equation, lie symmetries, exact solutions

1. Introduction

The theories of modern physics mainly include a mathematical structure, defined by a certain set of differential equations and extended by a set of rules for translating the mathematical results into meaningful statements about the physical work. Theories of non-linear science have been widely developed over the past century. In particular, non-linear systems have fascinated much interest among mathematicians and physicists. A lot of study has been conducted in the area of non-linear partial differential equations (NLPDEs) that arise in various areas of applied mathematics, mathematical physics, and many other areas. Apart from their theoretical importance, they have sensational applications to various physical systems such as hydrodynamics, non-linear optics, continuum mechanics, plasma physics and so on. A large variety of physical, chemical, and biological phenomena is governed by nonlinear partial differential equations (NLPDEs). A number of methods has been introduced for finding solutions of these equations such as Homotopy method [1], G'/G expansion method [2, 3], variational iteration method [4], sub-equation method [5], exp. function method [6], and Lie symmetry method [7–10]. Although solutions of such equations can be obtained easily by numerical computation. However, in order to obtain good understanding of the physical phenomena described by NLPDEs it is important to study the exact solutions of the NLPDEs. Exact solutions of mathematical equations play an major role in the proper understanding of qualitative features of many phenomena and processes in different areas of natural and applied sciences. Exact solutions of non-linear differential equations graphically demonstrate and allow unraveling the mechanisms of many complex non-linear phenomena. However, finding exact solutions of NLPDEs representing some physical phenomena is a tough task. However, because of importance of exact solutions for describing physical phenomena, many powerful methods have been introduced for finding solitons and other type of exact solutions of NLPDEs [2, 11–13]. Comparing to other approximate and numerical methods, which provides approximate solutions [14–16], the Lie group method provides the exact and analytic solutions of the differential structure (**Figures 1–3**).

Lie group method is one of the most effective methods for finding exact solutions of NLPDEs [17, 18]. This method was basically initiated by Norwegian mathematician Sophus Lie [19]. He developed the theory of "Continuous Groups" known as Lie groups. This method is orderly used in various fields of non-linear science. Shopus Lie was the first who arranged differential equations in terms of their symmetry groups, thereby analyzing the set of equations, which could be integrated or reduced to lower order equations by group theoretic algorithms. The Lie group analysis is a mathematical theory that synthesizes symmetry of differential equations. In this method, the differential structure is studied for their invariance by acting one or several parameter continuous group of transformations on the space of dependent and independent variables. We observe a plenty books and research article about Lie group method [17, 18, 20–22].

Wazwaz [23] introduced a fifth order non-linear evolution equation as follows:

$$u_{ttt} - u_{txxxx} - 4(u_x u_t)_{xx} - 4(u_x u_{xt})_x = 0. \tag{1}$$

In this chapter, he obtained multiple soliton solutions of this equation.

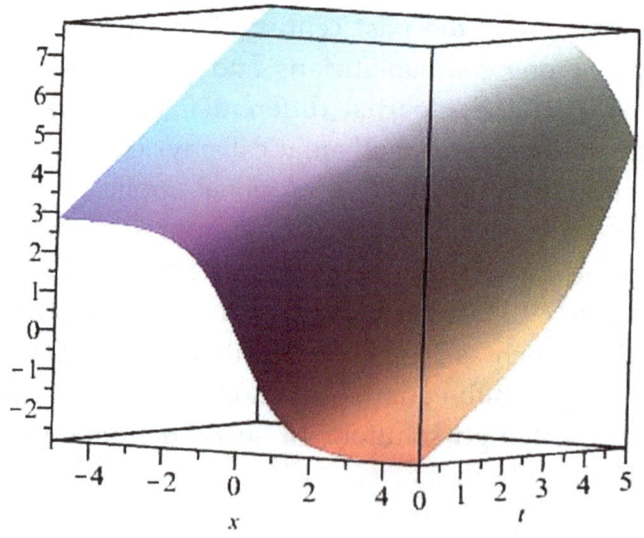

Figure 1. Kink wave solution (17) for $\alpha = \beta = \lambda = \mu = 1, b_1 = b_3 = 0.$

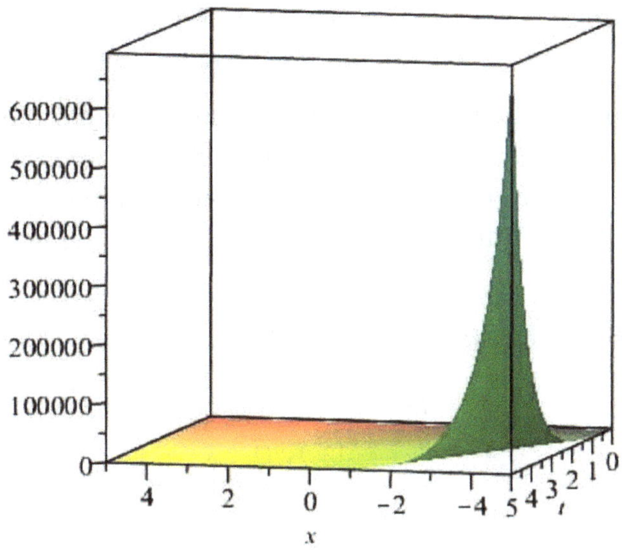

Figure 2. Singularity solution (18) for $\alpha = \lambda = \mu = b_5 = 1, b_2 = b_4 = 0$.

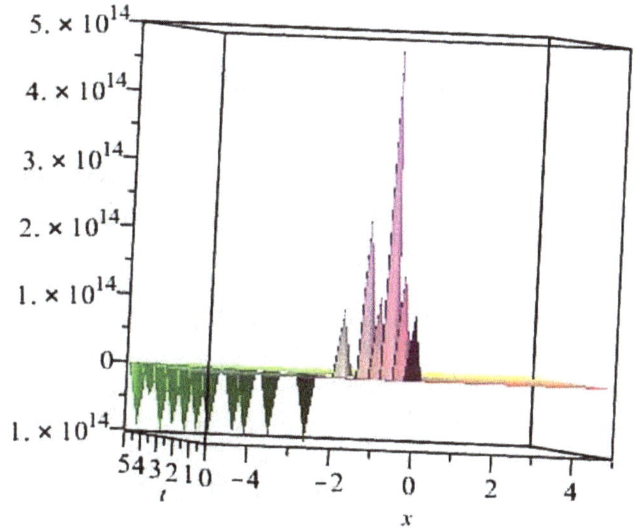

Figure 3. Singularity solution (19) for $\alpha = b_2 = b_4 = 0, b_4 = \lambda = 1, \mu = -1$.

We will consider the generalized fifth order non-linear evolution equation of the form:

$$u_{ttt} - u_{txxxx} - \alpha(u_x u_t)_{xx} - \beta(u_x u_{xt})_x = 0, \tag{2}$$

where α, β are parameters.

In this chapter, we will study the Eq. (2) by the Lie classical method. Firstly, Lie classical method will be used to obtain symmetries of generalized fifth order non-linear evolution Eq. (2). Symmetries will be used to reduce the Eq. (2) to ordinary differential equations (ODEs) and corresponding exact solutions of the generalized fifth order non-linear evolution Eq. (2) will be obtained.

2. Symmetry analysis

Lie classical method of infinitesimal transformation groups reduces the number of independent variables in partial differential equations (PDEs) and reduces the order of ODEs. Lie's method has been widely used in equations of mathematical physics and many other fields [11, 24]. In this chapter, we will perform Lie symmetry analysis [17–19, 24] for the generalized fifth order nonlinear evolution Eq. (2).

Let the group of infinitesimal transformations be defined as:

$$\begin{aligned} t^* &= t + \varepsilon\tau(x,t,u) + O(\epsilon^2) \\ x^* &= x + \varepsilon\xi(x,t,u) + O(\epsilon^2) \\ u^* &= u + \varepsilon\eta(x,t,u) + O(\epsilon^2), \end{aligned} \tag{3}$$

which leaves the Eq. (2) invariant. The infinitesimal transformations (3) are such that if u is solution of Eq. (2), then u^* is also a solution.

Herein, on invoking the invariance criterion as mentioned in [18], the following relation is deduced:

$$\eta^{ttt} - \eta^{txxxx} - \alpha(\eta^{xxx}u_t + \eta^t u_{xxx}) - (2\alpha+\beta)(\eta^{xx}u_{xt} + \eta^{xt}u_{xx}) - (\alpha+\beta)(\eta^x u_{xxt} + \eta^{xxt}u_x) = 0, \tag{4}$$

where $\eta^x, \eta^t, \eta^{xt}, \eta^{xx}, \eta^{xxx}, \eta^{ttt}, \eta^{txxxx}$ and η^{xxt} are extended (prolonged) infinitesimals acting on an enlarged space corresponding to $u_x, u_t, u_{xt}, u_{xx}, u_{xxx}, u_{ttt}, u_{txxxx}$ and u_{xxt}, respectively, given by:

$$\begin{aligned} \eta^x &= D_x\eta - u_x D_x\xi - u_t D_x\tau, \\ \eta^t &= D_t\eta - u_x D_t\xi - u_t D_t\tau, \\ \eta^{xx} &= D_x\eta^x - u_{xx}D_x\xi - u_{xt}D_x\tau, \\ \eta^{xt} &= D_t\eta^x - u_{xx}D_t\xi\_u_{xt}D_t\tau, \\ \eta^{xxx} &= D_x\eta^{xx} - u_{xxx}D_x\xi - u_{xxt}D_x\tau, \\ \eta^{ttt} &= D_t\eta^{tt} - u_{xtt}D_t\xi - u_{ttt}D_t\tau, \\ \eta^{xxxxt} &= D_t\eta^{xxxx} - u_{xxxxx}D_t\xi - u_{xxxxt}D_t\tau, \end{aligned} \tag{5}$$

where D_x and D_t are total derivative operators with respect to x and t respectively given as:

$$\begin{aligned} D_x &= \frac{\partial}{\partial x} + u_x\frac{\partial}{\partial u} + u_{xx}\frac{\partial}{\partial u_x} + \cdots, \\ D_t &= \frac{\partial}{\partial t} + u_t\frac{\partial}{\partial u} + u_{tt}\frac{\partial}{\partial u_t} + \cdots. \end{aligned}$$

Now, after computing (5) we get:

$$
\begin{aligned}
\eta^{x} &= \eta_x + (\eta_u - \xi_x)u_x - \tau_x u_t - \xi_u u_x^2 - \tau_u u_x u_t,\\
\eta^{t} &= \eta_t + (\eta_u - \tau_t)u_t - \xi_t u_x - \tau_u u_t^2 - \xi_u u_x u_t,\\
\eta^{xx} &= \eta_{xx} + u_x(2\eta_{xu} - \xi xx) - u_t\tau_{xx} +_x^2 (\eta_{uu} - 2\xi_{xu}) + u_{xx}(\eta_u - 2\xi_x) - 2u_{xt}\tau_x - 2u_t u_x \tau_{xu}\\
&\quad - u_x^3\xi_{uu} - u_x^2 u_t \tau_{uu} - 2u_x u_{xt\tau_u} - u_{xx}u_t\tau_u - 3u_x u_{xx}\xi_u,\\
\eta^{xt} &= \eta_{xt} + u_x(\eta_{tu} - \xi_{xt}) + \eta_t(\eta_{xu} - \tau_{xt}) - u_x^2\xi_{tu} - u_t^2\tau_{xu} - u_{xx}\xi_t - u_{xt}(\tau_t + \xi_x - \eta_u) - u_{tt}\tau_x\\
&\quad + u_x u_t(\eta_{uu} - \xi_{xu}) - u_x u_{xt}\xi_u - u_{xt}u_x\xi_u - u_{xt}u_t\tau_u - u_{tt}u_x\tau_u - u_x^2 u_t\xi_{uu} - u_t^2 u_x\tau_{uu}\\
&\quad - u_t u_x \tau_{tu} - u_t u_{xt}\tau_u - u_{xx}u_t\xi_u,\\
\eta^{xxx} &= \eta_{xxx} +_x (3\eta_{xxu} - \xi_{xxx}) - u_t\tau_{xxx} + u_{xx}(3\eta_{xu} - 3\xi_{xx}) - 3u_{xt}\tau_{xx} - 3u_x u_t\tau_{xxu} - 3u_{xxt}\tau_x\\
&\quad + u_x^2(3\eta_{xuu} - 3\xi_{xxu}) + u_x u_{xx}(3\eta_{uu} - 9\xi_{xu}) + u_x^3(\eta_{uuu} - 3\xi_{xuu}) + u_{xxx}(\eta_u - 3\xi_x)\\
&\quad - 2u_x u_{xt}\tau_{xu} - u_x^4\xi_{uuu} - 6u_x^2 u_{xx}\xi_{uu} - 3{u_{xx}}^2\xi_u - 4u_x u_{xxx}\xi_u - 3u_t u_x^2\tau_{xuu} - 3u_t u_{xx}\tau_{xu}\\
&\quad - 4u_x u_{tx}\tau_{xu} - u_x^3 u_t\tau_{uuu} - 3u_x u_t u_{xx}\tau_{uu} - 3u_x^2 u_{xt}\tau_{uu} - 3u_{tx}u_{xx}\tau_u - 3u_x u_{xxt}\tau_u\\
&\quad - u_t u_{xxx}\tau_u,\\
\eta^{ttt} &= \eta_{ttt} - u_x\xi_{ttt} + u_t(3\eta_{ttu} - \tau_{ttt}) + u_t^2(3\eta_{tuu} - 3\tau_{ttu}) + u_t^3(\eta_{uuu} - 3\tau_{tuu}) - u_t^4\tau_{uuu} - u_{tt}^2 3\tau_u\\
&\quad - 3u_x u_t\xi_{ttu} - 3u_x^2 u_t\xi_{tuu} - 3u_x u_{tt}\xi_{ut} - 6u_{xt}u_t\xi_{tu} - 3u_x u_{tt}\xi_{tu} + u_{xt}(4\eta_{xxxu} - 3\xi_{tt} - \xi_{xxxx})\\
&\quad + 3u_{tt}(\eta_{tu} - \tau_{tt}) - u_{ttt}\tau_t - 2u_{xtt}(\xi + 2\tau_{xxx})_t + u_{xxx}(\eta_u - 2\tau_t) - u_{xtt}\xi_t - u_{xtt}u_t\xi_u\\
&\quad - u_t^3 u_x\xi_{uuu} - 3u_t^2 u_{xt}\xi_{uu} - 3u_{xt}utt\xi_u - 2u_{xtt}u_t\xi_u - u_t u_{ttt}(\tau_u + \xi_u) - u_t u_{tt}(9\tau_{tu} - 3\eta_{uu})\\
&\quad - 6u_t^2 u_{tt}\tau_{uu} - 3u_t u_{ttt}\xi_u,\\
\eta^{xxxxt} &= \eta_{xxxxt} + u_x(4\eta_{xxxtu} - \xi_{xxxxt}) + u_t\tau_{xxxxt} - u_x^2\tau_{xxxxu} + u_x^3(6\eta_{xxtuu} - 4\xi_{xxxtu})\\
&\quad + u_x^3(4\eta_{xtuuu} - 6\xi_{xxtuu}) + u_x^4(\eta_{tuuuu} - 4\xi_{xtuuuu}) - u_x^5\xi_{tuuuu} - 4u_{xt}\tau_{xxxt}\\
&\quad + u_{xx}(6\eta_{xxtu} - 4\xi_{xxxt}) + 2u_{xxxt}(2\eta_{xu} - 3\xi_{xx} - 2\tau_{xt}) + u_{xxx}(4\eta_{xtu} - 6\xi_{xxt})\\
&\quad + u_{xxxx}(\eta_{tu} - 4\xi_{xt}) + u_{xxt}(6\eta_{xxu} - 6\tau xxt - 4\xi_{xxx}) + u_{xxx}u_{xt}(4\eta_{uu} - 16\xi_{xu})\\
&\quad + 6u_{xx}u_{xxt}(\eta_{uu} - 4\xi_{xu}) + u_{xxxx}u_t(\eta_{uu} - 4\xi_{xu}) + 4u_x u_{xxxt}(\eta_{uu} - 4\xi_{xu})\\
&\quad - 6u_x^2 u_{tt}\tau_{xxuu} - 24u_x u_t u_{xt}\tau_{xxuu} - 6u_{xx}u_t^2\tau_{xxuu} - 4u_x u_{tt}\tau_{xxxu} - 8u_{xt}u_t\tau_{xxxu}\\
&\quad - 10u_x^2 u_{xxxt}\xi_{uu} - 5u_x u_t u_{xxxx}\xi_{uu} - 30u_{xx}u_{xxt}u_x\xi_{uu} - 20u_{xxx}u_{xt}u_x\xi_{uu}\\
&\quad - 15u_{xx}^2 u_{xt}\xi_{uu} - 5u_x u_{xxxxt}\xi_u - 10u_{xx}u_{xxxt}\xi_u - 5u_{xt}u_{xxxx}\xi_u - 10u_{xxt}u_{xxx}\xi_u\\
&\quad - 5u_x u_{xxxx}\xi_{tu} - 10u_{xx}u_{xxx}\xi_{tu} - 6u_x^2 u_{xxt}\tau_{tuu} - 12u_{xx}u_{xt}u_x\tau_{tuu} - 4u_x u_t u_{xxx}\tau_{tuu}\\
&\quad - 3u_{xx}^2 u_t\tau_{tuu} - 12u_x u_{xxt}\tau_{xtu} - 4u_t u_{xxx}\tau_{xtu} - u_{xx}u_{xt}(12\tau_{xtu} + 18\xi_{xxu}) - 18u_x u_{xxt}\xi_{xxu}\\
&\quad - 6u_t u_{xxx}\xi_{xxu} - 4u_{xxx}u_{xtt}\tau_u - 6u_{xxt}^2\tau u - u_{tt}u_{xxxx}\tau u - 8u_{xt}u_{xxxt}\tau u - 6u_{xx}u_{xxtt}\tau u\\
&\quad - u_t u_{xxxxt}\tau u - 4u_x u_{xxxtt}\tau u - 12u_x^2 u_{xtt}\tau xuu - 24u_x u_t u_{xxt}\tau xuu - 12u_x u_{xx}u_{tt}\tau xuu\\
&\quad - 24u_{xt}^2 u_x\tau xuu - 4u_t^2 u_{xxx}\tau xuu - 24u_t u_{xx}u_{xt}\tau xuu - 4u_x^3 u_{xtt}\tau uuu - 12u_t u_x^2 u_{xxt}\tau uuu
\end{aligned}
$$

$$\begin{aligned}
&- 6u_{xx}u_{tt}u_x^2\tau uuu - 12u_{xt}^2u_x^2\tau uuu - 4u_t^2u_{xxx}\tau uuu - 24u_xu_tu_{xx}u_{xt}\tau uuu - 3u_t^2u_{xx}^2\tau uuu \\
&- 12u_xu_{xxtt}\tau xu - 8u_tu_{xxxt}\tau xu - 12u_{xx}u_{xtt}\tau xu - 24u_{xxt}u_{xt}\tau xu - 4u_{xxx}u_{tt}\tau xu \\
&- u_{tt}\tau_{xxxx} + u_t\eta_{xxxxu} - 10u_x^3u_{xx}\xi_{tuuu} - 24u_x^2u_{xx}\xi_{xtuu} + u_xu_{xx}(12\eta_{xtuu} - 18\xi_{xxtu}) \\
&- 6u_x^2u_t^2\tau_{xxuuu} - u_x^2u_t(4\xi_{xxxuu} - 6\tau_{xxtuu}) + u_x^3u_t(4\eta_{xuuuu} - 6\xi_{xxuuu} - 4\tau_{xtuuu}) \\
&+ u_x^4u_t(\eta_{uuuuu} - 4\xi_{xuuuu} - \tau_{tuuuu}) + 6u_x^2u_t\eta_{xxuuu} - 4u_x^3u_t^2\tau_{xuuuu} - 10u_x^3u_{xx}u_t\xi_{uuuu} \\
&- 5u_x^4u_{xt}\xi_{uuuu} + u_x^2u_{xx}u_t(6\eta_{uuuu} - 24\xi_{xuuu} - 6\tau_{tuuu}) + u_x^3u_{xt}(4\eta_{uuuu} - 16\xi_{xuuu} - 4\tau_{tuuu}) \\
&- 4u_x^3u_{tt}\tau_{xuuu} - 24u_x^2u_tu_{xt}\tau_{xuuu} - 12u_{xx}u_t^2\tau_{xuuu} - 10u_x^2u_{xxx}\xi_{xuu} - 15u_xu_{xx}^2\xi_{xuu} \\
&+ 6u_x^2u_{xt}(2\eta_{xuuu} - 2\tau_{xtuu} - 3\xi_{xxuu}) + 6u_xu_tu_{xx}(2\eta_{xuuu} - 2\tau_{xtuu} - 3\xi_{xxuu}) - 10u_x^3u_{xxt}\xi_{uuu} \\
&- 10u_tu_x^2u_{xxx}\xi_{uuu} - 30u_{xx}u_{xt}uu_x^2\xi_{uuu} - 15u_{xx}^2u_xu_t\xi_{uuu} - u_x^4u_t^2\tau_{uuuuu} - 8u_x^3u_tu_{xt}\tau_{uuuu} \\
&+ u_xu_t(4\eta_{xxxuu} - \xi_{xxxxu} - 4\tau_{xxxtu}) - u_x^5u_t\xi_{uuuuu} - u_x^4u_{xt}\tau_{uuuu} - 6u_x^2u_t^2u_{xx}\tau_{uuuu} \\
&- 4u_xu_t^2\tau_{xxxuu} + 6u_x^2u_{tt}\eta_{tuuu} - 24u_x^2u_{xxt}\xi_{xuu} + 12u_{xx}u_{xt}u_x(\eta_{uuu} - 4\xi_{xuu}) - 16u_xu_tu_{xxx}\xi_{xuu} \\
&- 12u_{xx}^2u_t\xi_{xuu} + u_xu_{xxt}12\eta_{xuu} + u_{xxx}u_t4\eta_{xuu} + u_{xx}u_{xt}\eta_{xuu} - 6u_x^2u_{xxxtt}\tau_{uu} - 8u_xu_tu_{xxxt}\tau_{uu} \\
&- 12u_xu_{xx}u_{xtt}\tau_{uu} - 24u_xu_{xt}u_{xxt}\tau_{uu} - 4u_xu_{tt}u_{xxx}\tau_{uu} - u_t^2u_{xxxx}\tau_{uu} - 12u_{xx}u_{xxt}u_t\tau_{uu} \\
&- 8u_{xxx}u_{xt}u_t\tau_{uu} - 3u_{xx}^2u_{tt}\tau_{uu} - 12u_{xx}u_{xt}^2\tau_{uu} + 6u_x^2u_{xxt}\eta_{uuu} + 4u_xu_tu_{xxx}\eta_{uuu} + 3u_{xx}^2u_t\eta_{uuu} \\
&- 12u_xu_{xt}\tau_{xxtu} - 6u_{xx}u_t\tau_{xxtu} + u_xu_{xxx}(\eta_{tuu} - 16\xi_{xtu}) + u_{xx}^2(\eta_{tuu} - 12\xi_{xtu}) - 8u_xu_{xt}\xi_{xxxu} \\
&- 4u_{xx}u_t\xi_{xxxu} + 12u_xu_{xt}\eta_{xxuu} + 6u_{xx}u_t\eta_{xxuu} - 12u_{xxt}u_t\tau_{xxu} - 12u_xu_{xtt}\tau_{xxu} - 6u_{xx}u_{tt}\tau_{xxu} \\
&- 12u_{xt}^2\tau_{xxu} - 4u_{xxx}u_{xt}\tau_{xu} - 6u_{xx}u_{xxt}\tau_{tu} - u_tu_{xxxx}\tau_{tu} - 4u_xu_{xxxt}\tau_{tu} - 6u_{xxtt}\tau_{xx} \\
&- 4u_{xxxtt} - u_{xxxxx}\xi_t + u_{xxxxt}(\eta_u - \tau_t - 4\xi_x) - u_{xxxxt}u_t\tau_u - u_{xxxxx}u_t\xi_u.\tau_x
\end{aligned} \tag{6}$$

The Lie classical method for determining the similarity variables of (2) is mainly consists of finding the infinitesimals $\tau, \xi,$ and $\eta,$ which are functions of x, t, u. After substituting the values of $\eta^x, \eta^t, \eta^{xt}, \eta^{xx}, \eta^{xxx}, \eta^{ttt}, \eta^{txxxx}$ and η^{xxt} from (5) to (4) and equating the coefficients of different differentials of u to zero, we get a number of PDEs in $\tau, \xi,$ and η, that need to be satisfied. Solving these system of PDEs, we obtain the infinitesimals $\tau, \xi,$ and η as follows:

$$\begin{aligned}
\tau &= C_1 + tC_4 \\
\xi &= C_2 + \frac{x}{2}C_4 \\
\eta &= C_3 - \frac{u}{2}C_4,
\end{aligned} \tag{7}$$

where $C_1, C_2, C_3,$ and C_4 are arbitrary constants.

Corresponding vector fields can be written as:

$$V_1 = \frac{\partial}{\partial t}, V_2 = \frac{\partial}{\partial x}, V_3 = \frac{\partial}{\partial u}, V_4 = \frac{x}{2}\frac{\partial}{\partial x} + t\frac{\partial}{\partial t} - \frac{u}{2}\frac{\partial}{\partial u}. \tag{8}$$

3. Symmetry reductions and invariant solutions

To obtain the symmetry reductions of Eq. (2), we have to solve the characteristic equation:

$$\frac{dx}{\xi} = \frac{dt}{\tau} = \frac{du}{\eta}, \tag{9}$$

where ξ, τ and η are given by Eq. (7).

To solve Eq. (9), following cases will be considered: (*i*) $V_1 + \mu V_2 + \lambda V_3$ and (*ii*) V_4, where μ, λ are arbitrary constants.

Case (i) $V_1 + \mu V_2 + \lambda V_3$

On solving Eqs. (9) we have,

$$\begin{aligned} \rho &= x - \mu t \\ u &= \lambda t + F(\rho), \end{aligned} \tag{10}$$

where ρ is new independent variables and $F(\rho)$ is new dependent variable. Substituting (10) into Eq. (2), we obtain the reduced ODE which reads:

$$\left[\mu(2\alpha+\beta)F' - \mu^3 - \alpha\lambda\right]F''' + \mu\left[(2\alpha+\beta)F''^2 + F'''''\right] = 0, \tag{11}$$

where primes ($'$) denotes derivative with respect to ρ.

Let assume the solution of ODE (11) in following form:

$$F = a_0 + a_1\rho + \frac{a_2}{\rho}, \tag{12}$$

where $a_0, a_1,$ and a_2 needs to be determined. Substituting (12) into ODE (11) and equating coefficients of the different powers of ρ equal to zero, we obtain:

$$\begin{aligned} a_0 &= arbitrary \\ a_1 &= \frac{\mu^3 + \alpha\lambda}{\mu(2\alpha+\beta)} \\ a_2 &= \frac{12}{2\alpha+\beta}. \end{aligned} \tag{13}$$

Corresponding solution of ODE (11) can be written as:

$$F = a_0 + \left(\frac{\mu^3 + \alpha\lambda}{\mu(2\alpha+\beta)}\right)\rho + \frac{12}{(2\alpha+\beta)\rho}, \tag{14}$$

where $\beta \neq -2\alpha$.

Corresponding solution of main Eq. (2) is given by:

$$u(x,t) = \lambda t + a_0 + \left(\frac{\mu^3 + \alpha\lambda}{\mu(2\alpha+\beta)}\right)(x-\mu t) + \frac{12}{(2\alpha+\beta)(x-\mu t)}, \tag{15}$$

with $\beta \neq -2\alpha$.

Some more solutions of ODE (11) are given by:

$$(i)\, F(\rho) = b_3 \pm \frac{6\sqrt{\mu(\alpha\lambda+\mu^3)}}{\mu(2\alpha+\beta)} \tanh\left(b_1 - \frac{\sqrt{\mu(\alpha\lambda+\mu^3)}\rho}{2\mu}\right) \text{with } \beta \neq -2\alpha,$$

$$(ii)\, F(\rho) = b_4 + b_5 \cosh\left(b_2 \pm \frac{\sqrt{\mu(\alpha\lambda+\mu^3)}\rho}{\mu}\right) \text{with } \beta = -2\alpha, \tag{16}$$

$$(iii)\, F(\rho) = b_3 + b_4 \coth\left(b_1 + \frac{1}{2}\frac{\sqrt{\mu(\alpha\lambda+\mu^3)}\rho}{\mu}\right) \text{with } \beta = \frac{2}{b_4}\left(-b_4\alpha + 3\frac{\sqrt{\mu(\alpha\lambda+\mu^3)}}{\mu}\right),$$

where b_1, b_2, b_3, b_4 and b_5 are arbitrary constants.

Corresponding solutions of main Eq. (2) are given by:

$$(i)\, u(x,t) = \lambda t + b_3 \pm \frac{6\sqrt{\mu(\alpha\lambda+\mu^3)}}{\mu(2\alpha+\beta)} \tanh\left(b_1 - \frac{\sqrt{\mu(\alpha\lambda+\mu^3)}(x-\mu t)}{2\mu}\right) \text{with } \beta \neq -2\alpha, \tag{17}$$

$$(ii)\, u(x,t) = \lambda t + b_4 + b_5 \cosh\left(b_2 \pm \frac{\sqrt{\mu(\alpha\lambda+\mu^3)}(x-\mu t)}{\mu}\right) \text{with } \beta = -2\alpha, \tag{18}$$

$$\begin{aligned}(iii)\, u(x,t) &= \lambda t + b_3 + b_4 \coth\left(b_1 + \frac{1}{2}\frac{\sqrt{\mu(\alpha\lambda+\mu^3)}(x-\mu t)}{\mu}\right) \\ &\text{with } \beta = \frac{2}{b_4}\left(-b_4\alpha + 3\frac{\sqrt{\mu(\alpha\lambda+\mu^3)}}{\mu}\right),\end{aligned} \tag{19}$$

where b_1, b_2, b_3, b_4, and b_5 are arbitrary constants.

Case (ii) V_4

On solving Eq. (9) for vector field V_4, we have:

$$\begin{aligned}\phi &= \frac{t}{x^2} \\ u &= \frac{G(\phi)}{x},\end{aligned} \tag{20}$$

where ϕ is new independent variables and $G(\phi)$ is new dependent variable. Substituting (20) into Eq. (2), we obtain the reduced ODE which reads

$$\begin{aligned}&(138\alpha+54\beta)\phi G'^2+(8(beta+2\alpha)\phi^3G'''+(30\alpha+18\beta)G+(148\alpha+68\beta)\phi^2G''-360)G'\\&+8(\beta+2\alpha)\phi^3G''^2+4(\alpha+\beta)\phi^2GG'''+(26\alpha+22\beta)\phi GG''-16\phi^4G'''''+(1-1020\phi^2)G'''\\&-1320\phi G''-240\phi^3G''''=0,\end{aligned} \tag{21}$$

where primes ($'$) denotes derivatives with respect to ϕ.

Let assume the solution of ODE (21) in following form:

$$G=\frac{b_2}{\phi^2}+\frac{b_1}{\phi}+a_0+a_1\phi+a_2\phi^2, \tag{22}$$

where b_1, b_2, a_0, a_1 and a_2 needs to be determined.

Substituting (22) into ODE (21) and equating coefficients of the different powers of ϕ equal to zero, we obtain:

$$\begin{aligned}&(i)\ a_0=arbitrary, a_1=a_2=b_1=0, b_2=\frac{1}{5\alpha+3\beta}\\&(ii)\ a_0=a_1=a_2=0, b_1=arbitrary, b_2=\frac{1}{5\alpha+3\beta}\end{aligned} \tag{23}$$

Corresponding solution of ODE (21) can be written as:

$$\begin{aligned}&(i)\ G=\frac{1}{(5\alpha+3\beta)\phi^2}+a_0,\\&(ii)\ G=\frac{1}{(5\alpha+3\beta)\phi^2}+\frac{b_1}{\phi},\end{aligned} \tag{24}$$

where b_1 is arbitrary constant.

Corresponding solution of main Eq. (2) can be written as:

$$\begin{aligned}&(i)\ u(x,t)=\frac{1}{x}\left(\frac{x^4}{(5\alpha+3\beta)t^2}+a_0\right),\\&(ii)\ u(x,t)=\frac{x^3}{(5\alpha+3\beta)t^2}+\frac{b_1x}{t},\end{aligned} \tag{25}$$

where b_1 is arbitrary constant.

4. Conclusion

In this chapter, we derived the symmetry variables and symmetry transformations of the generalized fifth order non-linear partial differential equation. We applied Lie symmetry analysis for investigating considered nonlinear partial differential equation and using similarity variables, given equation is reduced into ordinary differential equations. We derived explicit exact solutions of considered partial differential equation corresponding to each ordinary differential equation obtained by reduction.

Author details

Sachin Kumar

Address all correspondence to: sachin1jan@yahoo.com

Department of Mathematics, Central University of Punjab, Bathinda, Punjab, India

References

[1] Dehghan M, Manafian J, Saadatmandi A. Solving nonlinear fractional partial differential equations using the homotopy analysis method. Numerical Methods for Partial Differential Equations. 2010;**26**(2):448-479

[2] Naher H. New approach of (G/G)-expansion method and new approach of generalized (Ga/G)-expansion method for ZKBBM equation. Journal of the Egyptian Mathematical Society. 2015;**23**(1):42-48

[3] Zheng B. (G'/G)-expansion method for solving fractional partial differential equations in the theory of mathematical physics. Communications in Theoretical Physics (Beijing). 2012;**58**(5):623-630

[4] Fatoorehchi H, Abolghasemi H. The variational iteration method for theoretical investigation of falling film absorbers. National Academy Science Letters. 2015;**38**(1):67-70

[5] Zhang S, Zhang H-Q. Fractional sub-equation method and its applications to nonlinear fractional PDEs. Physics Letters A. 2011;**375**(7):1069-1073

[6] He J-H, Wu X-H. Exp-function method for nonlinear wave equations. Chaos Solitons Fractals. 2006;**30**(3):700-708

[7] Liu H, Li J, Liu L. Lie symmetry analysis, optimal systems and exact solutions to the fifth-order kdv types of equations. Journal of Mathematical Analysis and Applications. 2010; **368**(2):551-558

[8] Sahoo S, Garai G, Ray SS. Lie symmetry analysis for similarity reduction and exact solutions of modified kdv–zakharov–kuznetsov equation. Nonlinear Dynamics. 2017; **87**(3):1995-2000

[9] Zhang Y, Zhao Z. Lie symmetry analysis, lie-bäcklund symmetries, explicit solutions, and conservation laws of drinfeld-sokolov-wilson system. Boundary Value Problems. 2017; **1**(2017):154

[10] Zhao Z, Han B. Lie symmetry analysis of the heisenberg equation. Communications in Nonlinear Science and Numerical Simulation. 2017;**45**:220-234

[11] Khalique C, Biswas A. A Lie symmetry approach to nonlinear Schrödinger's equation with non-kerr law nonlinearity. Communications in Nonlinear Science and Numerical Simulation. 2009;**14**(12):4033-4040

[12] Martnez HY, Gmez-Aguilar JF, Baleanu D. Beta-derivative and sub-equation method applied to the optical solitons in medium with parabolic law nonlinearity and higher order dispersion. Optik–International Journal for Light and Electron Optics. February 2018;**155**:357–365

[13] Zhoua Q. Optical solitons in the parabolic law media with high-order dispersion. Optik. 2014;**125**:5432-5435

[14] Alshaery A, Ebaid A. Accurate analytical periodic solution of the elliptical Kepler equation using the Adomian decomposition method. Acta Astronautica. 2017;**140**:27-33

[15] Sami Bataineh A, Noorani MSM, Hashim I. Approximate analytical solutions of systems of pdes by homotopy analysis method. Computers & Mathematics with Applications. 2008;**55**(12):2913-2923

[16] David UU, Qamar S, Morgenstern AS. Analytical and numerical solutions of two-dimensional general rate models for liquid chromatographic columns packed with coreshell particles. Chemical Engineering Research and Design. 2018;**130**:295-320

[17] Bluman G, Anco S. Symmetry and Integration Methods for Differential Equations. Vol. 154. New York: Springer-Verlag Inc; 2002 http://www.springer.com/us/book/9780387986548

[18] Olver P. Applications of Lie Groups to Differential Equations, Vol. 107. New York: Springer-Verlag Inc.; 1986. http://www.springer.com/us/book/9780387950006

[19] Lie S. Uber die Integration durch bestimmte integrale von einer Klasse linear partieller Differntialgleichungen. Archiv for Matematik. 1881;**6**:328-368

[20] Bluman GW, Cheviakov AF, Anco SC. Applications of symmetry methods to partial differential equations. Vol. 168. New York: Springer; 2010

[21] Lekalakala SL, Motsepa T, Khalique CM. Lie symmetry reductions and exact solutions of an option-pricing equation for large agents. Mediterranean Journal of Mathematics. 2016; **13**(4):1753-1763

[22] Liu H, Li J, Zhang Q. Lie symmetry analysis and exact explicit solutions for general burgers equation. Journal of Computational and Applied Mathematics. 2009;**228**(1):1-9

[23] Wazwaz AM. A new fifth-order nonlinear integrable equation: Multiple soliton solutions. Physica Scripta. 2011;**83**:015012

[24] Kumar S, Singh K, Gupta R. Painlevé analysis, Lie symmetries and exact solutions for $(2+1)$-dimensional variable coefficients Broer-Kaup equations. Communications in Nonlinear Science and Numerical Simulation. 2012;**17**(4):1529-1541

3

Monte Carlo Set-Membership Filtering for Nonlinear Dynamic Systems

Zhiguo Wang, Xiaojing Shen and Yunmin Zhu

Abstract

This chapter considers the nonlinear filtering problem involving noises that are unknown and bounded. We propose a new filtering method via set-membership theory and boundary sampling technique to determine a state estimation ellipsoid. In order to guarantee the online usage, the nonlinear dynamics are linearized about the current estimate, and the remainder term is then bounded by an optimization ellipsoid, which can be described as the solution of a semi-infinite optimization problem. It is an analytically intractable problem for general nonlinear dynamic systems. Nevertheless, for a typical nonlinear dynamic system in target tracking, some certain regular properties for the remainder are analytically derived; then, we use a randomized method to approximate the semi-infinite optimization problem efficiently. Moreover, for some quadratic nonlinear dynamic systems, the semi-infinite optimization problem is equivalent to solving a semi-definite program problem. Finally, the set-membership prediction and measurement update are derived based on the recent optimization method and the online bounding ellipsoid of the remainder other than a priori bound. Numerical example shows that the proposed method performs better than the extended set-membership filter, especially in the situation of the larger noise.

Keywords: nonlinear dynamic systems, set-membership filter, randomization, semi-definite optimization, target tracking

1. Introduction

Filtering techniques for dynamic systems are widely used in practiced fields such as target tracking, signal processing, automatic control, and computer vision. The Kalman filter is a fundamental tool for solving a broad class of filtering problems with linear dynamic systems. When dynamic systems are nonlinear, some well-known generalizations include the extended

Kalman filter (EKF) and unscented Kalman filtering (UKF) [1]. These methods are based on local linear approximations of the nonlinear system where the higher order terms are ignored. Most recently, [2] proposes box particle filter to handle interval data based on interval analysis and constraint satisfaction techniques. The advantage of the box particle filter against the standard particle filter is its reduced computational complexity [3–5]. However, most of Monte Carlo filtering techniques are based on the assumptions that probability density functions of the state noise and measurement noise are known.

Actually, when the underlying probabilistic assumptions are not realistic (e.g., the main perturbation may be deterministic), it seems more natural to assume that the state noise and measurement noise are unknown but bounded [6]; then, [7] proposed set-membership estimation technique. The idea of propagating bounding ellipsoids (or boxes, polytopes, simplexes, parallelotopes, and polytopes) for systems with bounded noises has also been extensively investigated (e.g., see recent papers [6, 8–12] and references therein). Most of these methods concentrate on the linear dynamic systems.

The set-membership filtering for nonlinear dynamic systems is known to be a challenging problem. Based on ellipsoid-bounded, fuzzy-approximated, or Lipschitz-like nonlinearities, several results have been made [13–15]. These results assume that the ellipsoid bounds, the coefficients of fuzzy-approximation, or Lipschitz constants are known before filtering, which limits them in real time implementation. For example, for a typical nonlinear dynamic system in a radar, the bounds of the remainder depend on the past estimates so that they cannot be obtained before filtering. Recently, the paper [16] gives an overview of recent developments in set-theoretic methods for nonlinear systems, with a particular focus on a two-reaction model of anaerobic digestion, and the key idea of [17, 18] consists in a combination of Bayesian and set-valued estimation concepts. To our knowledge, [19, 20] develop nonlinear set-membership filters which can estimate the bounding ellipsoid of nonlinearities in real time, and the filters are called the extended set-membership filter (ESMF) and set-valued nonlinear filter (SVNF), respectively. Both [19, 20] derive the bounds of the remainder by an outer bounding box. Actually, if the remainder can be bounded by a tighter ellipsoid and using some recent advanced optimization techniques for filtering, it should be able to derive a tighter set-membership filtering for the nonlinear dynamic system.

In this chapter, when the underlying state noises and measurement noises are unknown but bounded, we propose a tighter set-membership filtering methods via set-membership estimation theory and boundary sampling technique. In order to guarantee the online usage, the nonlinear dynamics are linearized about the current estimate, and the remainder terms are then bounded by an ellipsoid, which can be formulated as the solution of a semi-infinite optimization problem. In general, it is an analytically intractable problem when dynamic systems are nonlinear. The main contributions of the paper are summarized as follows:

- For a typical nonlinear dynamic system in target tracking, we can analytically derive some regular properties for the remainder. Then, the semi-infinite optimization problem can be efficiently solved by using boundary sampling technique.

- For some quadratic nonlinear dynamic systems, using the samples on all vertices of a polyhedron, we obtain a tight bounding ellipsoid, which can cover the remainder by solving a semi-definite program (SDP) problem.
- The set-membership prediction and measurement update are derived based on the recent optimization method and the online bounding ellipsoid of the remainder other than a priori bound.

The rest of the paper is organized as follows. Preliminaries are given in Section 2. In Section 3, the bounding ellipsoid of the remainder set is calculated. In Section 4, the prediction step and the measurement update step of the set-membership filtering for nonlinear dynamic systems are derived, respectively. Examples and conclusions are given in Section 5 and Section 6, respectively.

2. Preliminaries

2.1. Problem formulation

We consider a nonlinear dynamic system:

$$\mathbf{x}_{k+1} = f_k(\mathbf{x}_k) + \mathbf{w}_k, \quad (1)$$

$$\mathbf{y}_k = h_k(\mathbf{x}_k) + \mathbf{v}_k, \quad (2)$$

where $\mathbf{x}_k \in \mathcal{R}^n$ is the state of system at time k and $\mathbf{y}_k \in \mathcal{R}^{n_1}$ is the measurement. $f_k(\mathbf{x}_k)$ and $h_k(\mathbf{x}_k)$ are nonlinear functions of $\mathbf{x}_k$, $\mathbf{w}_k \in \mathcal{R}^n$ is the uncertainty of process noises or system biases, and $\mathbf{v}_k \in \mathcal{R}^{n_1}$ is the uncertainty of measurement noises or system biases. They are assumed to be confined to the specified ellipsoidal sets:

$$\mathbf{W}_k = \{\mathbf{w}_k : \mathbf{w}_k^T \mathbf{Q}_k^{-1} \mathbf{w}_k \le 1\}$$
$$\mathbf{V}_k = \{\mathbf{v}_k : \mathbf{v}_k^T \mathbf{R}_k^{-1} \mathbf{v}_k \le 1\},$$

where $\mathbf{Q}_k$ and $\mathbf{R}_k$ are the *shape matrices* of the ellipsoids $\mathbf{W}_k$ and $\mathbf{V}_k$, respectively, which are known as symmetric positive-definite matrices. At time k given that $\mathbf{x}_k$ belongs to a current bounding ellipsoid:

$$\begin{aligned}\mathcal{E}_k &= \left\{\mathbf{x} \in R^n : (\mathbf{x} - \widehat{\mathbf{x}}_k)^T (\mathbf{P}_k)^{-1} (\mathbf{x} - \widehat{\mathbf{x}}_k) \le 1\right\} \\ &= \left\{\mathbf{x} \in R^n : \mathbf{x} = \widehat{\mathbf{x}}_k + \mathbf{E}_k \mathbf{u}_k, \mathbf{P}_k = \mathbf{E}_k \mathbf{E}_k^T, \|\mathbf{u}_k\| \le 1\right\}\end{aligned} \quad (3)$$

where $\widehat{\mathbf{x}}_k$ is the center of ellipsoid $\mathcal{E}_k$ and $\mathbf{P}_k$ is a known symmetric positive-definite matrix. Moreover, we assume that when the nonlinear functions are linearized, the remainder terms can be bounded by an ellipsoid. Specifically, by Taylor's theorem, f_k and h_k can be linearized to

$$f_k(\widehat{\mathbf{x}}_k + \mathbf{E}_k\mathbf{u}_k) = f_k(\widehat{\mathbf{x}}_k) + \mathbf{J}_{f_k}\mathbf{E}_k\mathbf{u}_k + \Delta f_k(\mathbf{u}_k), \tag{4}$$

$$h_k(\widehat{\mathbf{x}}_k + \mathbf{E}_k\mathbf{u}_k) = h_k(\widehat{\mathbf{x}}_k) + \mathbf{J}_{h_k}\mathbf{E}_k\mathbf{u}_k + \Delta h_k(\mathbf{u}_k), \tag{5}$$

where $\mathbf{E}_k$ and $\mathbf{u}_k$ are defined in (3), $\mathbf{J}_{f_k} = \frac{\partial f_k(\mathbf{x}_k)}{\partial \mathbf{x}}\Big|_{\widehat{\mathbf{x}}_k}$ and $\mathbf{J}_{h_k} = \frac{\partial h_k(\mathbf{x}_k)}{\partial \mathbf{x}}\Big|_{\widehat{\mathbf{x}}_k}$ are Jacobian matrices, and $\Delta f_k(\mathbf{u}_k)$ and $\Delta h_k(\mathbf{u}_k)$ are high-order remainders, which can be bounded in an ellipsoid for all $\|\mathbf{u}_k\| \le 1$, respectively, i.e.,

$$\Delta f_k(\mathbf{u}_k) \in \mathcal{E}_{f_k} = \left\{\mathbf{x} \in R^n : (\mathbf{x} - \mathbf{e}_{f_k})^T (\mathbf{P}_{f_k})^{-1} (\mathbf{x} - \mathbf{e}_{f_k}) \le 1\right\}, \tag{6}$$

$$= \left\{\mathbf{x} \in R^n : \mathbf{x} = \mathbf{e}_{f_k} + \mathbf{B}_{f_k}\Delta_{f_k}, \mathbf{P}_{f_k} = \mathbf{B}_{f_k}\mathbf{B}_{f_k}^T, \|\Delta_{f_k}\| \le 1\right\}, \tag{7}$$

$$\Delta h_k(\mathbf{u}_k) \in \mathcal{E}_{h_k} = \left\{\mathbf{x} \in R^{n_1} : (\mathbf{x} - \mathbf{e}_{h_k})^T (\mathbf{P}_{h_k})^{-1} (\mathbf{x} - \mathbf{e}_{h_k}) \le 1\right\}, \tag{8}$$

$$= \left\{\mathbf{x} \in R^{n_1} : \mathbf{x} = \mathbf{e}_{h_k} + \mathbf{B}_{h_k}\Delta_{h_k}, \mathbf{P}_{h_k} = \mathbf{B}_{h_k}\mathbf{B}_{h_k}^T, \|\Delta_{h_k}\| \le 1\right\}, \tag{9}$$

where $\mathbf{e}_{f_k}$ and $\mathbf{e}_{h_k}$ are the centers of the ellipsoids $\mathcal{E}_{f_k}$ and $\mathcal{E}_{h_k}$, respectively, and $\mathbf{P}_{f_k}$ and $\mathbf{P}_{h_k}$ are the shape matrices of the ellipsoids $\mathcal{E}_{f_k}$ and $\mathcal{E}_{h_k}$, respectively. Note that we do not assume that the ellipsoids $\mathcal{E}_{f_k}$ and $\mathcal{E}_{h_k}$ are given before filtering, and we will compute these ellipsoids online.

Suppose that the initial state $\mathbf{x}_0$ belongs to a given bounding ellipsoid:

$$\mathcal{E}_0 = \left\{\mathbf{x} \in R^n : (\mathbf{x} - \widehat{\mathbf{x}}_0)^T (\mathbf{P}_0)^{-1} (\mathbf{x} - \widehat{\mathbf{x}}_0) \le 1\right\}, \tag{10}$$

where $\widehat{\mathbf{x}}_0$ is the center of ellipsoid $\mathcal{E}_0$ and $\mathbf{P}_0$ is the shape matrix of the ellipsoid $\mathcal{E}_0$ which is a known symmetric positive-definite matrix.

The proposed set-membership filter mainly contains two steps: prediction step and measurement update step. The goal of prediction step is to determine a bounding ellipsoid $\mathcal{E}_{k+1|k}$ based on the measurement $\mathbf{y}_k$ at time k, i.e., look for $\widehat{\mathbf{x}}_{k+1|k}, \mathbf{P}_{k+1|k}$ such that the state $\mathbf{x}_{k+1}$ belongs to

$$\mathcal{E}_{k+1|k} = \left\{\mathbf{x} \in R^n : (\mathbf{x} - \widehat{\mathbf{x}}_{k+1|k})^T (\mathbf{P}_{k+1|k})^{-1} (\mathbf{x} - \widehat{\mathbf{x}}_{k+1|k}) \le 1\right\},$$

whenever (i) $\mathbf{x}_k$ is in $\mathcal{E}_k$; (ii) the processes $\mathbf{w}_k, \mathbf{v}_k$ are bounded in ellipsoids, i.e., $\mathbf{w}_k \in \mathbf{W}_k$, $\mathbf{v}_k \in \mathbf{V}_k$; and (iii) the remainders $\Delta f_k(\mathbf{u}_k) \in \mathcal{E}_{f_k}$ and $\Delta h_k(\mathbf{u}_k) \in \mathcal{E}_{h_k}$. The robust measurement update step is aimed to determine a bounding ellipsoid $\mathcal{E}_{k+1}$ based on the measurement $\mathbf{y}_{k+1}$ at time $k+1$, i.e., look for $\widehat{\mathbf{x}}_{k+1}, \mathbf{P}_{k+1}$ such that the state $\mathbf{x}_{k+1}$ belongs to

$$\mathcal{E}_{k+1} = \left\{\mathbf{x} \in R^n : (\mathbf{x} - \widehat{\mathbf{x}}_{k+1})^T (\mathbf{P}_{k+1})^{-1} (\mathbf{x} - \widehat{\mathbf{x}}_{k+1}) \le 1\right\},$$

whenever (i) $\mathbf{x}_{k+1}$ is in $\mathcal{E}_{k+1|k}$; (ii) the measurement noises $\mathbf{v}_{k+1}$ is bounded in ellipsoid, i.e., $\mathbf{v}_{k+1} \in \mathbf{V}_{k+1}$; and (iii) the remainders $\Delta h_{k+1}(\mathbf{u}_{k+1}) \in \mathcal{E}_{h_{k+1}}$. The **key issue** is to determine two tight bounding ellipsoids $\mathcal{E}_{f_k}$ and $\mathcal{E}_{h_k}$ in real time so that the filtering algorithm can be implemented online.

3. Ellipsoidal remainder bounding

In this section, we discuss the key problem on how to adaptively determine a tighter bounding ellipsoid to cover the high-order remainders from the optimization point of view.

3.1. Ellipsoidal remainder bounding by sampling

By (4)–(5), the high-order remainders are

$$\Delta f_k(\mathbf{u}_k) = f_k(\widehat{\mathbf{x}}_k + \mathbf{E}_k\mathbf{u}_k) - f_k(\widehat{\mathbf{x}}_k) - \mathbf{J}_{f_k}\mathbf{E}_k\mathbf{u}_k,$$
$$\Delta h_k(\mathbf{u}_k) = h_k(\widehat{\mathbf{x}}_k + \mathbf{E}_k\mathbf{u}_k) - h_k(\widehat{\mathbf{x}}_k) - \mathbf{J}_{h_k}\mathbf{E}_k\mathbf{u}_k,$$

whenever $\|\mathbf{u}_k\| \leq 1$. Obviously, it is a hard problem to cover a remainder by an ellipsoid since f_k and h_k are generally nonlinear functions. The outer bounding ellipsoid for $\Delta f_k(\mathbf{u}_k)$ is not uniquely defined, which can be optimized by minimizing the size $f(P)$ of the bounding ellipsoid. Thus, the optimization problem for the bounding ellipsoid of $\Delta f_k(\mathbf{u}_k)$ defined in (6) can be written as

$$\min f(\mathbf{P}_{f_k}) \tag{11}$$

$$\text{subject to } \left(\Delta f_k(\mathbf{u}_k) - \mathbf{e}_{f_k}\right)^T \left(\mathbf{P}_{f_k}\right)^{-1} \left(\Delta f_k(\mathbf{u}_k) - \mathbf{e}_{f_k}\right) \leq 1, \text{ for all } \|\mathbf{u}_k\| \leq 1. \tag{12}$$

where $\mathbf{P}_{f_k} = \mathbf{B}_{f_k}\mathbf{B}_{f_k}^T$ and $\mathbf{e}_{f_k}$ and $\mathbf{P}_{f_k}$ are decision variables. Since the optimization problem (11)–(12) has an infinite number of constraints, it is called a semi-infinite optimization problem in [21]. In general, it is a NP-hard problem.

Remark 1. *In practice, we want to achieve a state estimation ellipsoid by minimizing its "size" at each time; it is a function of the shape matrix* $\mathbf{P}$ *denoted by* $f(\mathbf{P})$*. If we choose trace function, i.e.,* $f(\mathbf{P}) = tr(\mathbf{P})$*, it means the sum of squares of semiaxes lengths of the ellipsoid* $\mathcal{E}$*. The other common "size" of the ellipsoid is* $logdet(\mathbf{P})$*, which corresponds to the volume of the ellipsoid* $\mathcal{E}$*.*

For a general nonlinear dynamic system, it is hard to solve the problem (11)–(12) [22]. It is this reason that the literatures [23, 24] are sought to find the particular relaxations of the original optimization problem (11)–(12). One of the typical methods is based on randomization of the parameter $\mathbf{u}_k$. Specifically, to solve the problem (11)–(12), we may take some samples from the boundary and interior points of the sphere $\|\mathbf{u}_k\| \leq 1$ so that we can get a finite set of $\mathbf{u}_k^1, \ldots, \mathbf{u}_k^N$, and then the infinite constraint (12) can be approximated by N constraints based on $\mathbf{u}_k^1, \ldots, \mathbf{u}_k^N$. Moreover, by Schur complement, an approximate bounding ellipsoid for $\Delta f_k(\mathbf{u}_k)$ can be derived by solving the following SDP optimization problem:

$$\min\ f(\mathbf{P}_{f_k}) \tag{13}$$

$$\text{subject to} \begin{bmatrix} -1 & (\Delta f_k(\mathbf{u}_k^i) - \mathbf{e}_{f_k})^T \\ \Delta f_k(\mathbf{u}_k^i) - \mathbf{e}_{f_k} & -\mathbf{P}_{f_k} \end{bmatrix} \preccurlyeq 0 \quad i = 1, \ldots, N. \tag{14}$$

Although the randomized solution may not be feasible for all $\|\mathbf{u}_k\| \leq 1$, [24] has used statistical learning techniques to provide an explicit bound on the measure of the set of original constraints that are possibly violated by the randomized solution, and they prove this measure rapidly decreases to zero as N is increasing. Therefore, the obtained randomized solution of the optimization problem (13)–(14) can be made approximately feasible for the semi-infinite optimization (11)–(12) by sampling a sufficient number of constraints.

Similarly, the outer bounding ellipsoid for $\Delta h_k(\mathbf{u}_k)$ can be derived by solving

$$\min\ f(\mathbf{P}_{h_k}) \tag{15}$$

$$\text{subject to} \begin{bmatrix} -1 & (\Delta h_k(\mathbf{u}_k^i) - \mathbf{e}_{h_k})^T \\ \Delta h_k(\mathbf{u}_k^i) - \mathbf{e}_{h_k} & -\mathbf{P}_{h_k} \end{bmatrix} \preccurlyeq 0, \quad i = 1, \ldots, N. \tag{16}$$

Remark 2. *Note that the bounding ellipsoid of* [19] *is derived by interval mathematics. We derive the bounding ellipsoid by solving a semi-infinite optimization problem.* **Figure 1** *illustrates the difference of two methods. It is obvious to see that the bounding ellipsoid derived by solving the SDP (13) is tighter than that obtained by interval mathematics. The cumulative effect of the conservative bounding ellipsoid at each time step may yield divergence of a filtering.*

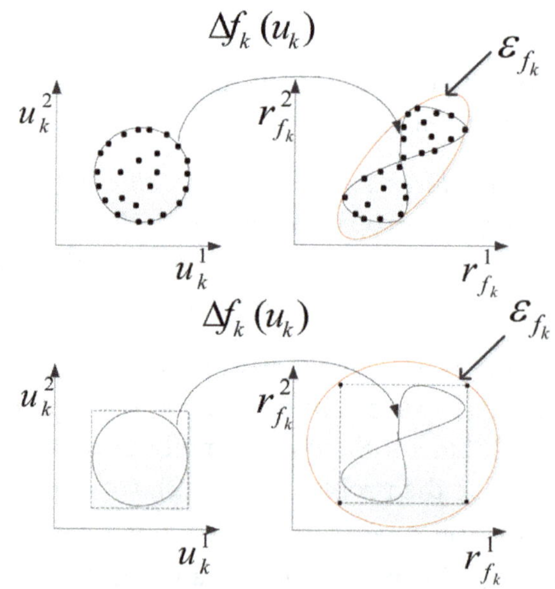

Figure 1. (Top) The bounding ellipsoid is derived by covering the solid points of the remainder which are obtained by Monte Carlo sampling. (Bottom) The bounding ellipsoid is derived by covering the vertices of the rectangle obtained by interval mathematics [19].

3.2. Ellipsoidal remainder bounding by boundary sampling

In this subsection, for a typical nonlinear dynamic system in target tracking, we discuss that the remainder can be bounded by an ellipsoid via *boundary* sampling for target tracking. Thus, the new method can reduce the computation complexity efficiently in the bounding step.

Let us consider the following nonlinear measurement Eq. [1]:

$$h(\mathbf{x}) = \begin{bmatrix} \sqrt{(\mathbf{x}(1)-a)^2 + (\mathbf{x}(2)-b)^2} \\ arctan\left(\frac{\mathbf{x}(2)-b}{\mathbf{x}(1)-a}\right) \end{bmatrix}, a, b \in \mathcal{R} \tag{17}$$

where $\mathbf{x}$ is a four-dimensional state variable that includes position and velocity $[x, y, \dot{x}, \dot{y}]^T$. Note that $h(\mathbf{x})$ only depends on the first two dimensions $\mathbf{x}(1)$ and $\mathbf{x}(2)$.

We discuss the relationship between the set $\{\|\mathbf{u}_k\| \leq 1, \mathbf{u}_k = [\mathbf{u}_k(1)\ \mathbf{u}_k(2)]\}$ and the remainder set $\{\Delta h_{k+1}(\mathbf{u}_k) : \|\mathbf{u}_k\| \leq 1\}$.

Proposition 1. *If we let the remainder* $g(\mathbf{u}) = h(\widehat{\mathbf{x}} + \mathbf{E}\mathbf{u}) - h(\widehat{\mathbf{x}}) - \mathbf{J}_h\mathbf{E}\mathbf{u}$ *where* $h(\widehat{\mathbf{x}})$ *is defined in (17),* $\mathbf{E}$ *is a Cholesky factorization of a positive-definite* $\mathbf{P}$ *such that ellipsoid* $\{\mathbf{x} = \widehat{\mathbf{x}} + \mathbf{E}\mathbf{u} : \|u\| \leq 1\}$ *does not intersect with the radial* $\{\mathbf{x} : \mathbf{x}(1) <= a, \mathbf{x}(2) = b\}$, *and then the boundary of the remainder set* $\mathbf{S} = \{g(\mathbf{u}) : \|\mathbf{u}\| \leq 1\}$ *belongs to the set* $\{g(\mathbf{u}) : \|\mathbf{u}\| = 1\}$.

Remark 3. *Note that the ellipsoid* $\{\mathbf{x} = \widehat{\mathbf{x}} + \mathbf{E}\mathbf{u} : \|u\| \leq 1\}$ *does not intersect with the radial.* $\{\mathbf{x} : \mathbf{x}(1) <= a, \mathbf{x}(2) = b\}$ *is a weak condition, which is in order to satisfy the continuity of* $g(\mathbf{u})$, *and we can verify the condition by using the distance from the ellipsoid center* $\widehat{\mathbf{x}}$ *to the radial. Moreover, if the condition is violated, i.e., the true target is near the radial, we can transform the data to a new coordinate system where the target is far way the radial, and then the assumption can be satisfied.*

The proof of Proposition 1 relies on the following three lemmas:

Lemma 1. (Remainder lemma). *The determinant of the derivative of the remainder* $g(\mathbf{u})$ *is not less than 0, and the equality holds if and only if* $c\mathbf{u}(1) + d\mathbf{u}(2) = 0$, *where* $c = \mathbf{E}_{11}(\widehat{\mathbf{x}}(2) - b) - \mathbf{E}_{21}(\widehat{\mathbf{x}}(1) - a)$, $d = \mathbf{E}_{12}(\widehat{\mathbf{x}}(2) - b) - \mathbf{E}_{22}(\widehat{\mathbf{x}}(1) - a)$, *and* $\mathbf{E}_{ij}$ *are the entries of the ith row and the jth column of the matrix* $\mathbf{E}$. *Meanwhile, if* $c\mathbf{u}(1) + d\mathbf{u}(2) = 0$, *then* $g(\mathbf{u}) = 0$.

Lemma 2. *If the sets* $\mathbf{S}^1 \cup \mathbf{S}^2 = \mathbf{S}^3 \cup \mathbf{S}^4$, $\mathbf{S}^3 \cap \mathbf{S}^4 = \emptyset$, *and* $\mathbf{S}^1 \subset \mathbf{S}^3$, *then* $\mathbf{S}^4 \subset \mathbf{S}^2$.

Lemma 3. (Inverse function theorem [25]) *Suppose that* $\varphi : \mathbf{R}^n \to \mathbf{R}^n$ *is continuously differentiable in an open set containing* $\mathbf{u}$ *and* $\det(\varphi'(\mathbf{u})) \neq 0$, *then there is an open set* $\mathbf{V}$ *containing* $\mathbf{u}$ *and open set* $\mathbf{W}$ *containing* $\varphi(\mathbf{u})$ *such that* $\varphi : \mathbf{V} \to \mathbf{W}$ *has a continuous inverse* $\varphi^{-1} : \mathbf{W} \to \mathbf{V}$ *which is differentiable and for all* $\mathbf{y} \in \mathbf{W}$ *satisfies* $(\varphi^{-1})'(\mathbf{y}) = [\varphi'(\varphi^{-1}(\mathbf{y}))]^{-1}$.

Example 1. *To illustrate Proposition 1, we give an example as follow: if* $a = 50$, $b = 100$, $\widehat{x} = [80\ 130]^T$, *and* $\mathbf{P} = \text{diag}(500, 1000)$, *it is easy to check that* $g(\mathbf{u})$ *is continuously differentiable in set* $\mathbf{S}_1 = \{\mathbf{u} : \|\mathbf{u}\| \leq 1\}$. *We divide* $\mathbf{S}_1$ *into three parts, i.e.,* $\mathbf{S}_1 = \mathbf{A}^1 \cup \mathbf{B}^1 \cup \mathbf{C}^1$, *where* $\mathbf{A}^1 = \{\mathbf{u} : c\mathbf{u}(1) + d\mathbf{u}(2) < 0, \|\mathbf{u}\| \leq 1\}$, $\mathbf{B}^1 = \{\mathbf{u} : c\mathbf{u}(1) + d\mathbf{u}(2) > 0, \|\mathbf{u}\| \leq 1\}$, *and* $\mathbf{C}^1 = \{\mathbf{u} : c\mathbf{u}(1) + d\mathbf{u}(2) = 0,$

$\|\mathbf{u}\| \leq 1\}$. *Meanwhile, we can also divide* **S** *into the corresponding parts, such that* $\mathbf{A} = \{g(\mathbf{u}) : \mathbf{u} \in \mathbf{A}^1\}$, $\mathbf{B} = \{g(\mathbf{u}) : \mathbf{u} \in \mathbf{B}^1\}$, $\mathbf{C} = \{g(\mathbf{u}) : \mathbf{u} \in \mathbf{C}^1\}$, *and then* $\mathbf{S} = \mathbf{A} \cup \mathbf{B} \cup \mathbf{C}$.

Figure 2 *shows the separation area of the circle and their corresponding area of* $g(\mathbf{u})$. *Three observations can be made as follows:*

- *The remainder set is the union of two sets.*
- *The (red) line* $\mathbf{C}^1$ *is mapped to the point 0.*
- *The boundary of* **S** *belongs to the set* $\{g(\mathbf{u}) : \|\mathbf{u}\| = 1\}$.

In summary, when we take samples from the boundary, they are sufficient to derive the outer bounding ellipsoids of the remainder set. Therefore, based on Proposition 1, *the computation complexity in the bounding step of the new method can be reduced much more.*

Remark 4. *In order to further reduce the samples and cover the remainder set at the same time, we can heuristically enlarge the sampling area, such as* $\{\|\mathbf{u}_k\| = 1.1\}$; *then, the remainder set becomes a little larger. If we derive an ellipsoid to cover the little larger remainder, then this ellipsoid can cover the original remainder set* $\{\Delta h_{k+1}(\mathbf{u}_k) : \|\mathbf{u}_k\| \leq 1\}$.

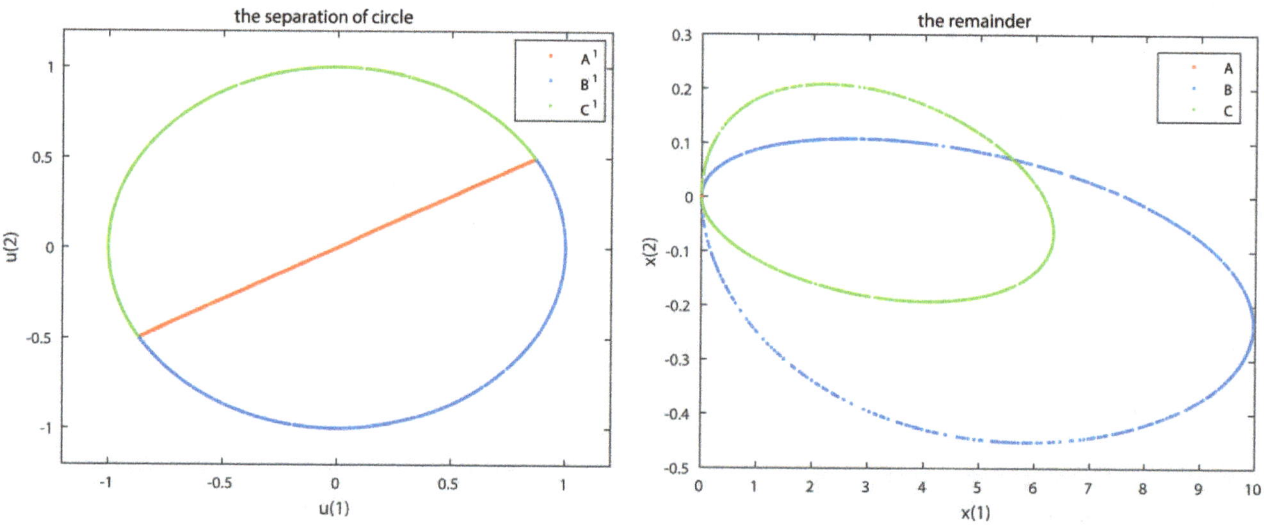

Figure 2. (Left) The separation of circle. (Right) The corresponding area of $g(u)$.

3.3. A tight solution

In this subsection, for some quadratic nonlinear dynamic systems, the semi-infinite optimization problem (11)–(12) may be equivalent to solving an SDP problem via sampling on all vertices of a polyhedron. Thus, we can obtain a tight bounding ellipsoid to cover the remainder.

We consider a quadratic nonlinear state equation:

$$f_k(\mathbf{x}_k) = \begin{bmatrix} \alpha_1 & 0 & 0 & 0 \\ 0 & \alpha_2 & 0 & 0 \\ 0 & 0 & \ddots & 0 \\ 0 & 0 & 0 & \alpha_n \end{bmatrix} \begin{bmatrix} (\mathbf{x}_k^1)^2 \\ (\mathbf{x}_k^2)^2 \\ \vdots \\ (\mathbf{x}_k^n)^2 \end{bmatrix} \tag{18}$$

where $\mathbf{x}_k \in \mathcal{R}^n$ is the state of system at time k, and $\mathbf{x}_k^i$ denoted the *ith* component of $\mathbf{x}_k$; α_i are known parameters, $i = 1, \ldots, n$.

Proposition 2. *If we let the high-order remainder* $\Delta f_k(\mathbf{u}_k) = f(\widehat{\mathbf{x}}_k + \mathbf{E}_k\mathbf{u}_k) - f(\widehat{\mathbf{x}}_k) - \mathbf{J}_{f_k}\mathbf{E}_k\mathbf{u}_k$ *where* $\|\mathbf{u}_k\| \leq 1$, *assume that* $f_k(\widehat{\mathbf{x}}_k)$ *is a quadratic function defined in* (18) *and* $\mathbf{E}_k$ *is a diagonal matrix; then, a tight bounding ellipsoid can be derived to cover the high-order remainder* $\Delta f_k(\mathbf{u}_k)$ *by solving the following optimization problem:*

$$\min f\left(\mathbf{P}_{f_k}\right) \tag{19}$$

$$\textit{subject to} \quad \left[\begin{array}{c:ccc} -1 & -\mathbf{e}_{f_k}^1 & \cdots & -\mathbf{e}_{f_k}^n \\ \hdashline -\mathbf{e}_{f_k}^1 & & & \\ \vdots & & -\mathbf{P}_{f_k} & \\ -\mathbf{e}_{f_k}^2 & & & \end{array}\right] \preceq 0,$$

$$\left[\begin{array}{c:c} -1 & -\mathbf{e}_{f_k}^1 \cdots \alpha_i(\mathbf{E}_k^{ii})^2 - \mathbf{e}_{f_k}^i \cdots - \mathbf{e}_{f_k}^n \\ \hdashline -\mathbf{e}_{f_k}^1 & \\ \vdots & \\ \alpha_i(\mathbf{E}_k^{ii})^2 - \mathbf{e}_{f_k}^i & \\ \vdots & -\mathbf{P}_{f_k} \\ -\mathbf{e}_{f_k}^n & \end{array}\right] \preceq 0, \quad \textit{for } i = 1, 2, \ldots, n. \tag{20}$$

where $\mathbf{E}_k^{ii}$ is the *ith* row and *jth* column of $\mathbf{E}_k$.

In summary, we can determine the remainder bounding ellipsoid by sampling as follows.

1 For general nonlinear functions, samples may be taken from the sphere $\|\mathbf{u}_k\| \leq 1$.

2 For a typical nonlinear dynamic system in target tracking or nonlinear functions, samples may be taken on the boundary of the sphere $\|\mathbf{u}_k\| \leq 1$.

3 For some quadratic nonlinear functions, samples only need vertices of a polyhedron.

4. Ellipsoidal state bounding via SDP

In this section, we present the prediction step and the measurement step of the set-membership filtering by extending El Ghaoui and Calafiore's optimization method [26]. The point is that when the nonlinear dynamics are linearized on the current estimate, the uncertainties of the new linearized dynamic system include the uncertain bounding ellipsoids of the remainder terms and the noises.

4.1. Prediction step

Proposition 3. *At time $k+1$, based on measurements $\mathbf{y}_k$, the bounding ellipsoids of the state, and the remainders $\mathcal{E}_k$, $\mathcal{E}_{f_k}$, and $\mathcal{E}_{h_k}$, a predicted bounding ellipsoid $\mathcal{E}_{k+1|k} = \left\{\mathbf{x} : \left(\mathbf{x} - \widehat{\mathbf{x}}_{k+1|k}\right)^T \left(\mathbf{P}_{k+1|k}\right)^{-1} \left(\mathbf{x} - \widehat{\mathbf{x}}_{k+1|k}\right) \leq 1\right\}$ can be obtained by solving the optimization problem in the variables $\mathbf{P}_{k+1|k}$ and $\widehat{\mathbf{x}}_{k+1|k}$ and nonnegative scalars $\tau^u \geq 0, \tau^w \geq 0, \tau^v \geq 0, \tau^f \geq 0, \tau^h \geq 0$:*

$$\min f\left(\mathbf{P}_{k+1|k}\right) \tag{21}$$

$$\textit{subject to} \quad -\tau^u \leq 0, \; -\tau^w \leq 0, \; -\tau^v \leq 0, \tag{22}$$

$$-\tau^f \leq 0, \; -\tau^h \leq 0, \; -\mathbf{P}_{k+1|k} \prec 0, \tag{23}$$

$$\begin{bmatrix} -\mathbf{P}_{k+1|k} & \Phi_{k+1|k}\left(\Psi_{k+1|k}\right)_\perp \\ \left(\Phi_{k+1|k}\left(\Psi_{k+1|k}\right)_\perp\right)^T & -\left(\Psi_{k+1|k}\right)_\perp^T \Xi \left(\Psi_{k+1|k}\right)_\perp \end{bmatrix} \preccurlyeq 0, \tag{24}$$

where

$$\Phi_{k+1|k} = \left[f_k(\widehat{\mathbf{x}}_k) + \mathbf{e}_{f_k} - \widehat{\mathbf{x}}_{k+1|k}, \mathbf{J}_{f_k}\mathbf{E}_k, \mathbf{I}, 0, \mathbf{B}_{f_k}, 0\right], 0 \in \mathcal{R}^{n, n_1}, \tag{25}$$

$$\Psi_{k+1|k} = \left[h_k(\widehat{\mathbf{x}}_k) + \mathbf{e}_{h_k} - \mathbf{y}_k, \mathbf{J}_{h_k}\mathbf{E}_k, 0, \mathbf{I}, 0, \mathbf{B}_{h_k}\right]. \tag{26}$$

$\left(\Psi_{k+1|k}\right)_\perp$ is the orthogonal complement of $\Psi_{k+1|k}$. $\mathbf{E}_k$ is the Cholesky factorization of $\mathbf{P}_k$, i.e., $\mathbf{P}_k = \mathbf{E}_k(\mathbf{E}_k)^T$. $\mathbf{e}_{f_k}$, $\mathbf{e}_{h_k}$, $\mathbf{B}_{f_k}$, and $\mathbf{B}_{h_k}$ are denoted by (7) and (9), respectively. $\mathbf{J}_{f_k} = \frac{\partial f_k(\mathbf{x}_k)}{\partial \mathbf{x}}\Big|_{\widehat{\mathbf{x}}_k}$ and $\mathbf{J}_{h_k} = \frac{\partial h_k(\mathbf{x}_k)}{\partial \mathbf{x}}\Big|_{\widehat{\mathbf{x}}_k}$:

$$\Xi = \mathrm{diag}\left(1 - \tau^u - \tau^w - \tau^v - \tau^f - \tau^h, \tau^u I, \tau^w \mathbf{Q}_k^{-1}, \tau^v \mathbf{R}_k^{-1}, \tau^f I, \tau^h I\right). \tag{27}$$

Remark 5. *The objective function (21) is aimed at minimizing the shape matrix of the predicted ellipsoid, and the constraints (22)–(24) ensure that the true state is contained in predicted bounding ellipsoid $\mathcal{E}_{k+1|k}$. Notice that if $f(P) = tr(P)$, the optimization problems* (13)–(16), (21)–(24), and (28)–(31) *are SDP problems, which can be efficiently solved by modern interior-point methods* [27]. *According to the guidelines in* [28], *the computational complexity of solving an SDP problem is $O\left(\max(m,n)^4 n^{1/2} \log 1/\epsilon\right)$, where n is the number of the states. With the development of convex optimization technology technique, one can also use first-order optimizing algorithm. The computational complexity may be reduced further* (see [29]).

4.2. Measurement update step

Similarly, we can derive the measurement update step of the nonlinear filtering.

Proposition 4. *At time $k+1$, based on measurement $\mathbf{y}_{k+1}$, the predicted bounding ellipsoid $\mathcal{E}_{k+1|k}$, and the bounding ellipsoid of the remainder $\mathcal{E}_{h_{k+1}}$, an estimated bounding ellipsoid*

$\mathcal{E}_{k+1} = \left\{\mathbf{x} : (\mathbf{x} - \widehat{\mathbf{x}}_{k+1})^T (\mathbf{P}_{k+1})^{-1} (\mathbf{x} - \widehat{\mathbf{x}}_{k+1}) \le 1\right\}$ *can be obtained by solving the optimization problem in the variables* $\mathbf{P}_{k+1}$ *and* $\widehat{\mathbf{x}}_{k+1}$ *and nonnegative scalars* $\tau^u \ge 0$, $\tau^v \ge 0$, $\tau^h \ge 0$:

$$\min f(\mathbf{P}_{k+1}) \tag{28}$$

$$\textit{subject to} \quad -\tau^u \le 0,\ -\tau^v \le 0,\ -\tau^h \le 0, \tag{29}$$

$$-\mathbf{P}_{k+1} \prec 0, \tag{30}$$

$$\begin{bmatrix} -\mathbf{P}_{k+1} & \Phi_{k+1}(\Psi_{k+1})_\perp \\ (\Phi_{k+1}(\Psi_{k+1})_\perp)^T & -(\Psi_{k+1})_\perp^T \Xi (\Psi_{k+1})_\perp \end{bmatrix} \preccurlyeq 0, \tag{31}$$

where

$$\Phi_{k+1} = \left[\widehat{\mathbf{x}}_{k+1|k} - \widehat{\mathbf{x}}_{k+1}, \mathbf{E}_{k+1|k}, 0, 0\right], \quad 0 \in \mathcal{R}^{n, n_1}, \tag{32}$$

$$\Psi_{k+1} = \left[h_{k+1}\left(\widehat{\mathbf{x}}_{k+1|k}\right) + \mathbf{e}_{h_{k+1}} - \mathbf{y}_{k+1}, \mathbf{J}_{h_{k+1|k}} \mathbf{E}_{k+1|k}, \mathbf{I}, \mathbf{B}_{h_{k+1}}\right]. \tag{33}$$

$(\Psi_{k+1})_\perp$ *is the orthogonal complement of* Ψ_{k+1}. $\mathbf{E}_{k+1|k}$ *is the Cholesky factorization of* $\mathbf{P}_{k+1|k}$, *i.e.,* $\mathbf{P}_{k+1|k} = \mathbf{E}_{k+1|k}\left(\mathbf{E}_{k+1|k}\right)^T$. $\widehat{\mathbf{x}}_{k+1|k}$ *is the center of the predicted bounding ellipsoid* $\mathcal{E}_{k+1|k}$. $\mathbf{e}_{h_{k+1}}$ *and* $\mathbf{B}_{h_{k+1}}$ *are denoted by* (9) *at the time step* $k+1$. $\mathbf{J}_{h_{k+1|k}} = \left.\frac{\partial h_{k+1}(\mathbf{x}_k)}{\partial \mathbf{x}}\right|_{\widehat{\mathbf{x}}_{k+1|k}}$:

$$\Xi = \mathrm{diag}(1 - \tau^u - \tau^v - \tau^h, \tau^u \mathbf{I}, \tau^v \mathbf{R}_{k+1}^{-1}, \tau^h I). \tag{34}$$

4.3. Sampling-based ellipsoidal bounding filter algorithm

- Step 1: (Initialization step) Set $k = 0$ and initial values $(\widehat{\mathbf{x}}_0, \mathbf{P}_0)$ such that $\mathbf{x}_0 \in \mathcal{E}_0$.
- Step 2: (Bounding step) Take samples $\mathbf{u}_k^1, \ldots, \mathbf{u}_k^N$ from the sphere $\|\mathbf{u}_k\| \le 1$, and then determine two bounding ellipsoids to cover the remainders Δf_k and Δh_k by (13)–(14) and (15)–(16), respectively.
- Step 3: (Prediction step) Optimize the center and shape matrix of the state prediction ellipsoid $(\widehat{\mathbf{x}}_{k+1|k}, \mathbf{P}_{k+1|k})$ such that $\mathbf{x}_{k+1|k} \in \mathcal{E}_{k+1|k}$ by solving the optimization problem (21)–(24).
- Step 4: (Bounding step) Take samples $\mathbf{u}_{k+1|k}^1, \ldots, \mathbf{u}_{k+1|k}^N$ from the sphere $\|\mathbf{u}_{k+1|k}\| \le 1$, and then determine one bounding ellipsoid to cover the remainder Δh_{k+1} by (15)–(16).
- Step 5: (Measurement update step) Optimize the center and shape matrix of the state estimation ellipsoid $(\widehat{\mathbf{x}}_{k+1}, \mathbf{P}_{k+1})$ such that $\mathbf{x}_{k+1} \in \mathcal{E}_{k+1}$ by solving the optimization problem (28)–(31).
- Step 6: Set $k = k+1$ and go to Step 2.

5. Numerical example in target tracking

In this section, we compare the performance between the proposed set-membership filter and extended set-membership filter (ESMF) [19], which can also be implemented online for target tracking with a nonlinear dynamic system, when the state noises and measurement noises are unknown but bounded.

By considering a two-dimensional Cartesian, coordinate system as follows [1]:

$$\mathbf{x}_{k+1} = \begin{bmatrix} 1 & 0 & T & 0 \\ 0 & 1 & 0 & T \\ 0 & 0 & 1 & 0 \\ 0 & 0 & 0 & 1 \end{bmatrix} \mathbf{x}_k + \mathbf{w}_k, \tag{35}$$

$$\mathbf{y}_k = \begin{bmatrix} \sqrt{(\mathbf{x}_k(1))^2 + (\mathbf{x}_k(2))^2} \\ \arctan\left(\dfrac{\mathbf{x}_k(2)}{\mathbf{x}_k(1)}\right) \end{bmatrix} + \mathbf{v}_k, \tag{36}$$

$\mathbf{x}$ is a four-dimensional state variable that includes position and velocity $[x, y, \dot{x}, \dot{y}]^T$, and $T = 1$ s is the time sampling interval. The process noise and measurement noise assumed to be confined to specified ellipsoidal sets:

$$\mathbf{W}_k = \left\{\mathbf{w}_k : \mathbf{w}_k^T \mathbf{Q}_k^{-1} \mathbf{w}_k \leq 1\right\}$$
$$\mathbf{V}_k = \left\{\mathbf{v}_k : \mathbf{v}_k^T \mathbf{R}_k^{-1} \mathbf{v}_k \leq 1\right\}.$$

where

$$\mathbf{Q}_k = \sigma^2 \begin{bmatrix} \frac{T^3}{3} & 0 & \frac{T^2}{2} & 0 \\ 0 & \frac{T^3}{3} & 0 & \frac{T^2}{2} \\ \frac{T^2}{2} & 0 & T & 0 \\ 0 & \frac{T^2}{2} & 0 & T \end{bmatrix}, \mathbf{R}_k = q \cdot \begin{bmatrix} 3^2 & 0 \\ 0 & 1^2 \end{bmatrix}.$$

The target acceleration σ^2 equals to 10. The parameter q is used to control the uncertainty of the measurement noise. In the example, the target starts at the point $(50, 30)$ with a velocity of $(5, 5)$. The center and the shape matrix of the initial bounding ellipsoid are $\widehat{\mathbf{x}}_0 = [49.5\ \ 29.5\ \ 5\ \ 5]^T$ and $\mathbf{P}_0 = \text{diag}([5, 5, 2, 2])$, respectively.

The following simulation results include three parts: the first part is about the size of the remainder bounding ellipsoid, the second part is about the root-mean-square error (RMSE) of the state estimation, and the third part is about the computation time. They are illustrated and discussed by the number of samples, the time steps, and the uncertain parameter q of the measurement noise, respectively:

- In **Figure 3**, the log size of the remainder bounding ellipsoid is plotted as a function of the time steps with the uncertain parameter $q = 0.01$. It shows that the size of the new method is much smaller than that of the ESMF, i.e., the new method derives a tighter ellipsoid to cover the remainder. Moreover, we use 30 samples to calculate a remainder bounding ellipsoid on a time step based on solving the optimization problem (15)–(16). The corresponding bounding ellipsoid is presented in **Figure 4**. It shows that the bounding ellipsoid can cover all points of the remainder set with a very small size. In **Figures 5** and **6**, the average size of the remainder bounding ellipsoid through the time steps 1–20 is plotted as a function of the uncertain parameter q of the measurement noise. The larger q means that the measurement noise is more uncertain. Thus, **Figures 5** and **6** show that when the uncertainty of the measurement noise is increasing, the size of the remainder bounding ellipsoid of ESMF is quickly increasing; however, that of the new method is slowly increasing and relatively stable.

- RMSE of the state estimation along the position direction is plotted as a function of the time steps in **Figure 7**. It shows that RMSE of the new method is less than that of ESMF. The reason may be that the new method derives a tighter ellipsoid to cover the remainder, which can be seen in the **Figure 4**. **Figure 7** also shows that RMSEs of the proposed filter based on 30 and 40 samples are almost same. The reason may be that the remainder bounding ellipsoid is same when the number of samples is more than 30. In **Figure 8**, the average RMSE of the state estimation through the time steps 1 to 20 is plotted as a function of the uncertain parameter q. It

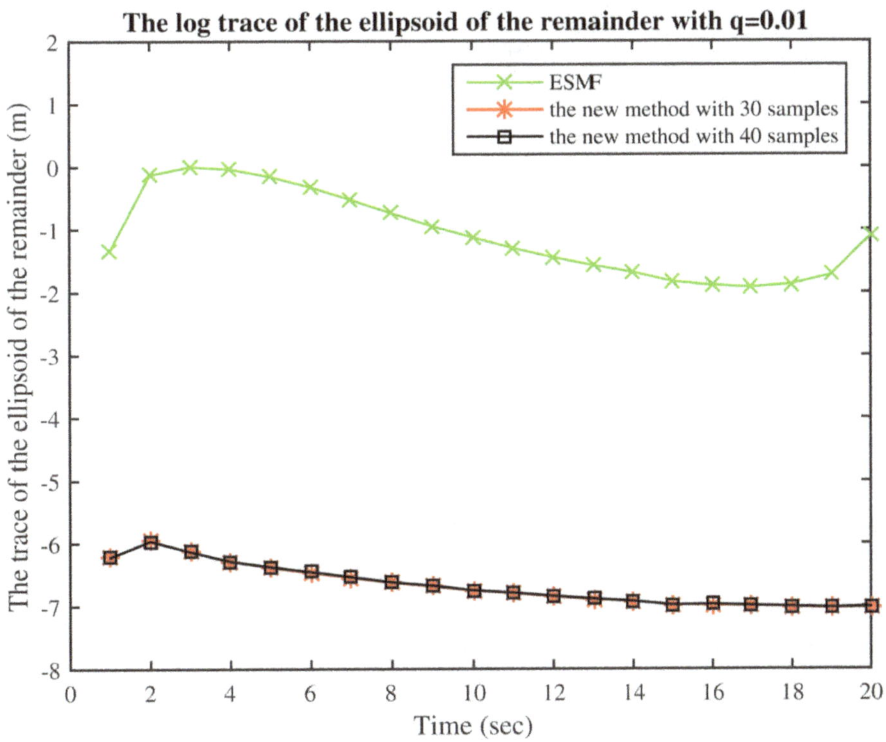

Figure 3. The log size of the remainder bounding ellipsoid is plotted as a function of time steps with $q = 0.01$.

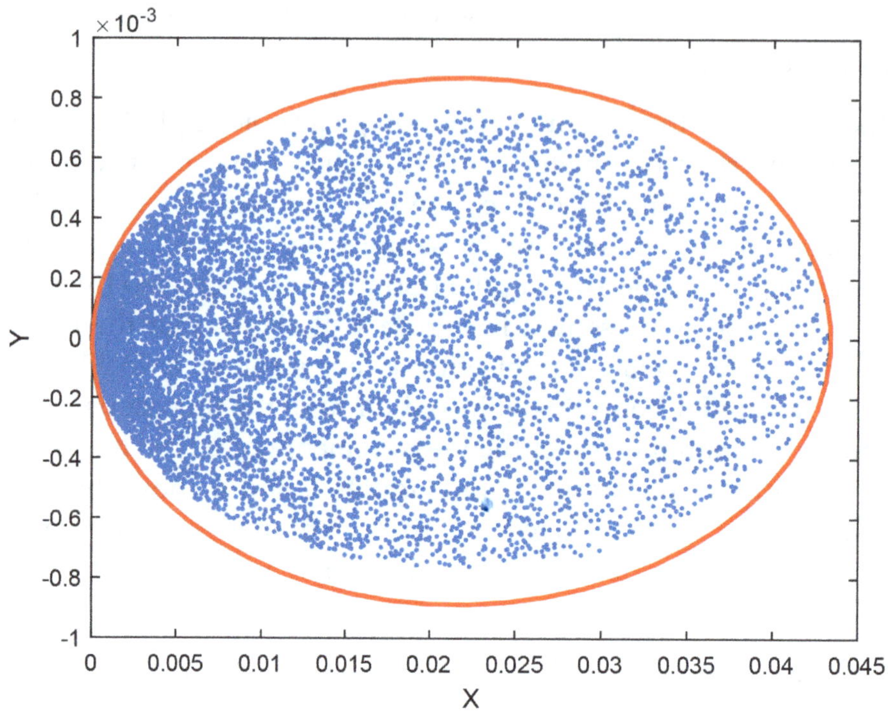

Figure 4. The remainder bounding ellipsoid on a time step based on 30 samples from the boundary.

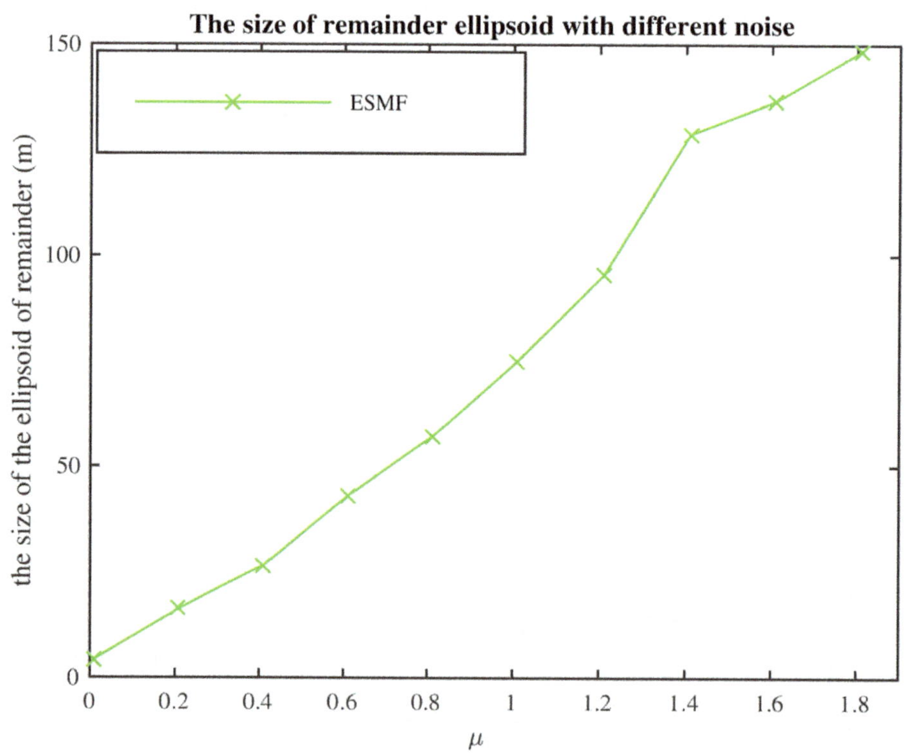

Figure 5. The average size of the remainder bounding ellipsoid is plotted as a function of the uncertain parameter q by ESMF.

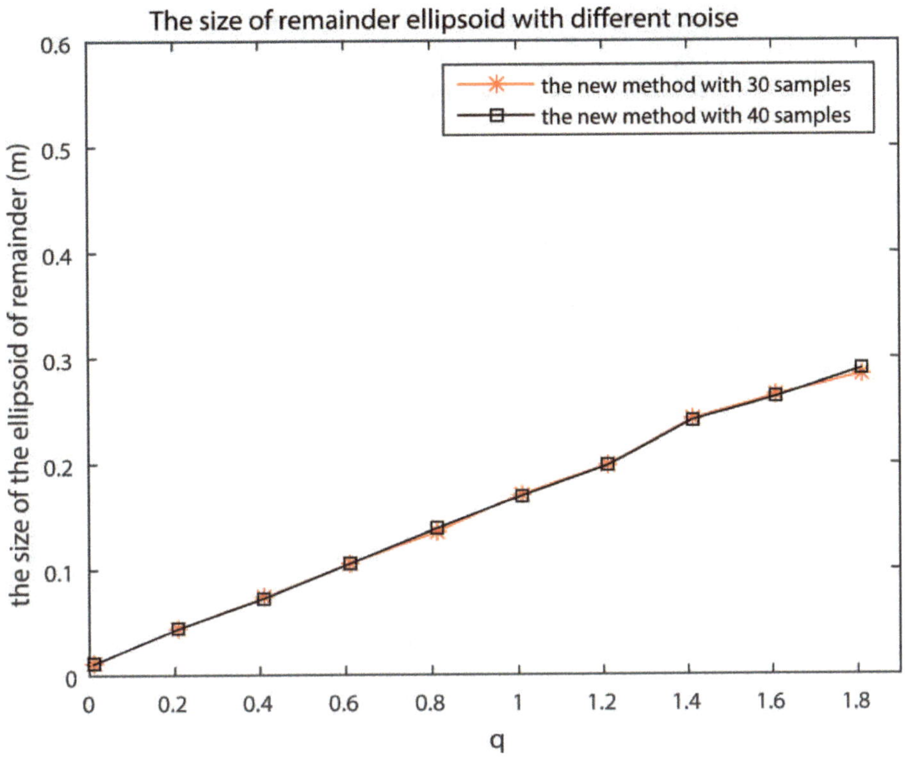

Figure 6. The average size of the remainder bounding ellipsoid is plotted as a function of the uncertain parameter q by the new method.

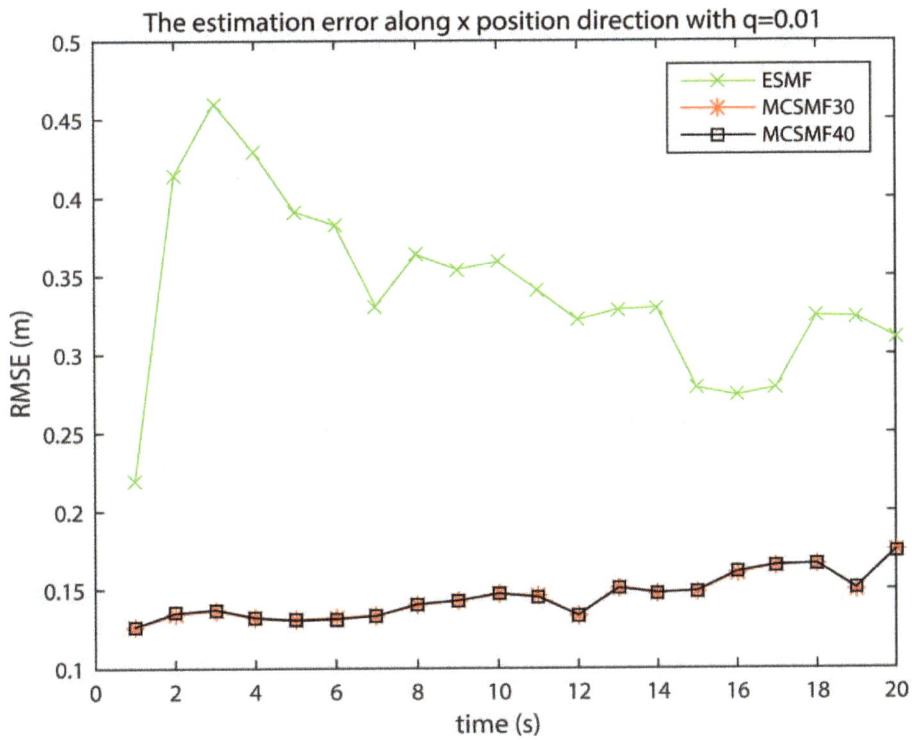

Figure 7. The RMSE of the state estimation is plotted as a function of time steps with $q = 0.01$.

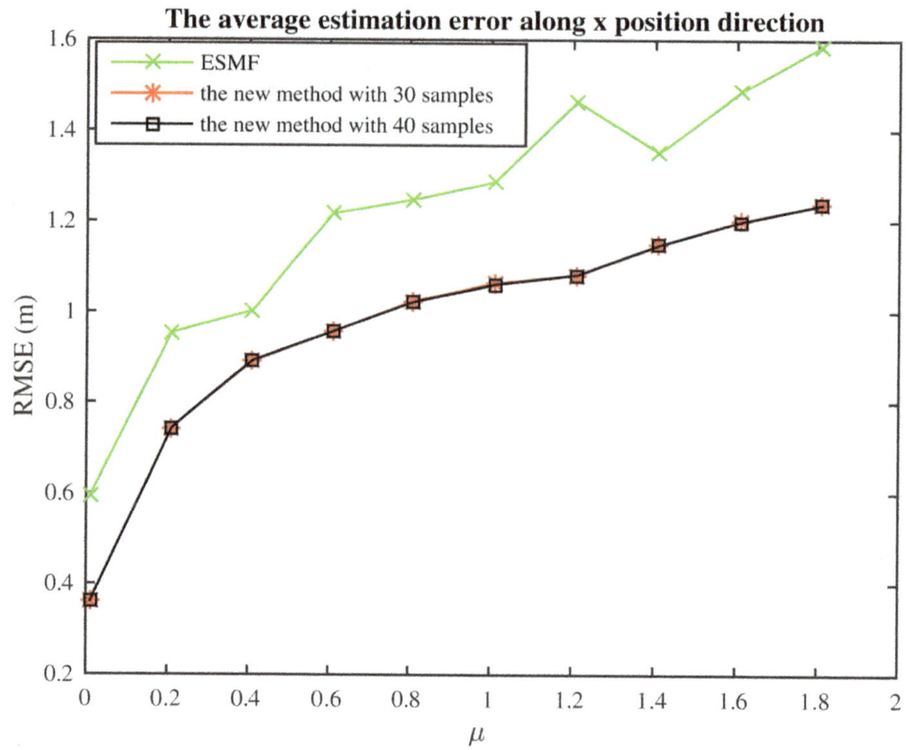

Figure 8. The average RMSE of the state estimation through the time steps 1 to 20 is plotted as a function of the uncertain parameter q.

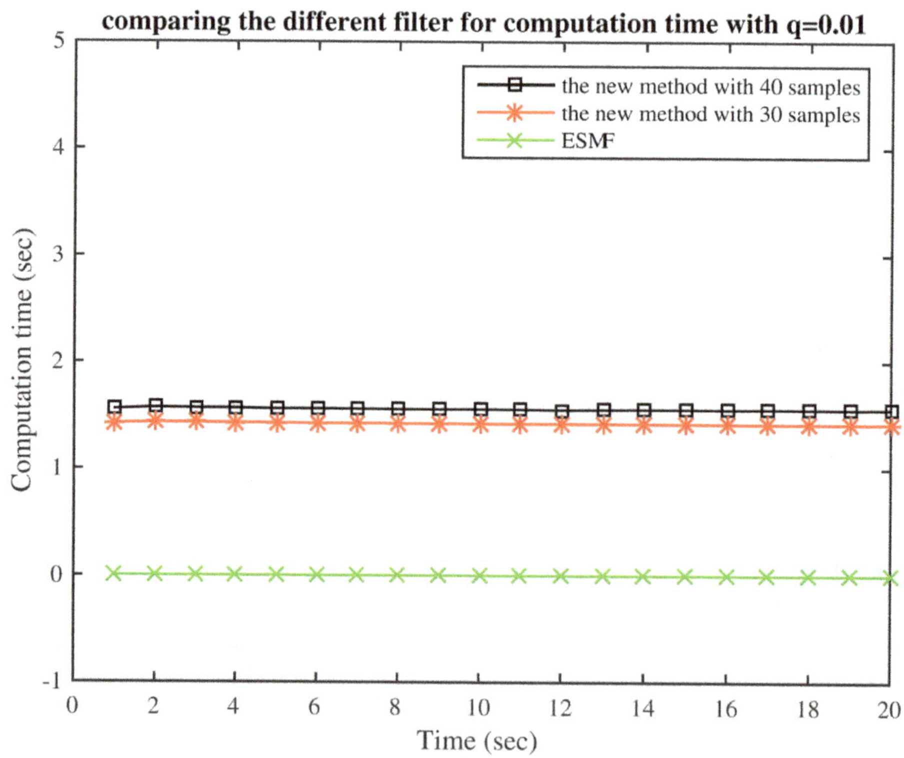

Figure 9. The computation time of the proposed state bounding filter and ESMF at each time step.

shows that the average RMSE of the state estimation based on the new method is also less than that of ESMF. The larger uncertain parameter q is a better performance of the new method than that of ESMF. In summary, **Figures 5–8** indicate that the new method performs much better than ESMF, especially in the situation of the larger noise.

• Since the predictive step and measurement update step of the new method are calculated by solving an SDP, the computation time of the new method is greater than that of ESMF, which can be seen in the right of **Figure 9**, but it may be tolerated and be done in polynomial time.

6. Conclusion

In order to deal with the nonlinear dynamic systems with unknown but bounded noise, we have proposed a new filtering method via set-membership theory and boundary sampling technique to determine a state estimation ellipsoid. To guarantee the online usage, the nonlinear dynamics are linearized about the current estimate, and the remainder terms are then bounded by an ellipsoid, which can be written as the solution of a semi-infinite optimization problem. For a typical nonlinear dynamic system in target tracking, the semi-infinite optimization problem can be efficiently approximated by a randomized method. Moreover, for some quadratic nonlinear dynamic systems, using the samples on all vertices of a polyhedron, we obtain a tight bounding ellipsoid, which covers the remainder by solving an SDP problem. Finally, the set-membership prediction and measurement update are derived based on the recent optimization method and the online bounding ellipsoid of the remainder other than a priori bound, so that a tighter set-membership filter can be achieved. Numerical example shows that the proposed method performs much better than ESMF, especially in the situation of the larger noise. Future work will include that the multisensor fusion, multiple target tracking, and various applications such as sensor management and placement for structures and different types of wireless networks.

Acknowledgements

This work was supported in part by the NSFC No. 61673282, the open research funds of BACC-STAFDL of China under Grant No. 2015afdl010, and the PCSIRT16R53.

Author details

Zhiguo Wang, Xiaojing Shen* and Yunmin Zhu

*Address all correspondence to: shenxj@scu.edu.cn

Department of Mathematics, Sichuan University, Chengdu, Sichuan, China

References

[1] Bar-Shalom Y, Li X, Kirubarajan T. Estimation with Applications to Tracking and Navigation. New York: Wiley; 2001

[2] Abdallah F, Gning A, Bonnifait P. Box particle filtering for nonlinear state estimation using interval analysis. Automatica. 2008;**44**:807-815

[3] Gning A, Ristic B, Mihaylova L, Abdallah F. An introduction to box particle filtering. IEEE Signal Processing Magazine. 2013;**30**:166-171

[4] Freitas AD, Mihaylova L, Gning A, Angelova D, Kadirkamanathan V. Autonomous crowds tracking with box particle filtering and convolution particle filtering. Automatica. 2016;**69**: 380-394

[5] Merlinge N, Dahia K, Piet-Lahanier H. A box regularized particle filter for terrain navigation with highly non-linear measurements. IFAC-Papers OnLine. 2016;**49**:361-366

[6] Polyak BT, Nazin SA, Durieu C, Walter E. Ellipsoidal parameter or state estimation under model uncertainty. Automatica. 2004;**40**:1171-1179

[7] Schweppe FC. Recursive state estimation: Unknown but bounded errors and system inputs. IEEE Transactions on Automatic Control. 1968;**13**:22-28

[8] Jaulin L. Nonlinear bounded-error state estimation of continuous-time systems. Automatica. 2002;**38**:1079-1082

[9] Rohou S, Jaulin L, Mihaylova L, Bars FL. Guaranteed computation of robot trajectories. Robotics and Autonomous Systems. 2017;**97**:76-84

[10] Calafiore G, El Ghaoui L. Ellipsoidal bounds for uncertain equations and dynamical systems. Automatica. 2004;**40**:773-787

[11] Shen X, Zhu Y, Song E, Luo Y. Minimizing Euclidian state estimation error for linear uncertain dynamic systems based on multisensor and multi-algorithm fusion. IEEE Transactions on Information Theroy. 2011;**57**:7131-7146

[12] Durieu C, Walter E, Polyak BT. Multi-input multi-output ellipsoidal state bounding. Journal of Optimization Theory and Applications. 2001;**111**:273-303

[13] Shamma JS, Tu K. Approximate set-valued observers for nonlinear systems. IEEE Transactions on Automatic Control. 1997;**42**:648-658

[14] Morrell DR, Stirlling WC. An extended set-valued Kalman filter. Proceeding of ISIPTA. 2003:396-407

[15] Wei G, Wang Z, Shen B. Error-constrained filtering for a class of nonlinear time-varying delay systems with non-Gaussian noises. IEEE Transaction on Automatic Control. 2010; **55**:2876-2882

[16] Chachuat B, Houska B, Paulen R, Perić N, Rajyaguru J, Villanueva ME. Set-theoretic approaches in analysis, estimation and control of nonlinear systems. IFAC-PapersOnLine. 2015;**48**:981-998

[17] Noack B, Klumpp V, Hanebeck UD. State estimation with sets of densities considering stochastic and systematic errors. In: Proceedings of the IEEE International Conference on Information Fusion; 2009

[18] Noack B, Klumpp V, Petkov N, Hanebeck UD. Bounding linearization errors with sets of densities in approximate Kalman filtering. Proceedings of the IEEE International Conference on Information Fusion. 2010

[19] Scholte E, Campbell ME. A nonlinear set-membership filter for on-line applications. International Journal of Robust and Nonlinear Control. 2003;**13**:1337-1358

[20] Calafiore G. Reliable localization using set-valued nonlinear filters. IEEE Transactions on Systems, Man, and Cybernetics - Part A: Systems and Humans. 2005;**35**:189-197

[21] Boyd S, Vandenberghe L. Convex Optimization. Cambridge, United Kingdom: Cambridge University Press; 2004

[22] Ben-Tal A, Nemirovski A. Robust convex optimization. Mathematics of Operations Research. 1998;**23**:769-805

[23] Still G. Discretization in semi-infinite programming: The rate of convergence. Mathematical programming. 2001;**91**:53-69

[24] Calafiore G, Campi M. Uncertain convex programs: Randomized solutions and confidence levels. Mathematical Programming. 2005;**102**:25-46

[25] Spivak M. Calculus on manifolds. New York: Benjamin; 1965

[26] El Ghaoui L, Calafiore G. Robust filtering for discrete-time systems with bounded noise and parametric uncertainty. IEEE Transactions on Automatic Control. 2001;**36**:1084-1089

[27] Nesterov Y, Nemirovski A. Interior point polynomial methods in convex programming: Theorey and applications. Philadelphia, PA: SIAM;1994

[28] Vandenberghe L, Boyd S. Semidefinite programming. SIAM Review. 1996;**38**:49-95

[29] Lan G, Lu Z, Monteiro RD. Primal-dual first-order methods with $\mathcal{O}(1/\epsilon)$ iteration-complexity for cone programming. Mathematical Programming. 2011;**126**:1-29

4

Canonical Generalized Inversion Form of Kane's Equations of Motion for Constrained Mechanical Systems

Abdulrahman H. Bajodah and Ye-Hwa Chen

Abstract

The canonical generalized inversion dynamical equations of motion for ideally constrained discrete mechanical systems are introduced in the framework of Kane's method. The canonical equations of motion employ the acceleration form of constraints and the Moore-Penrose generalized inversion-based Greville formula for general solutions of linear systems of algebraic equations. Moreover, the canonical equations of motion are explicit and nonminimal (full order) in the acceleration variables, and their derivation is made without appealing to the principle of virtual work or to Lagrange multipliers. The geometry of constrained motion is revealed by the canonical equations of motion in a clear and intuitive manner by partitioning the canonical accelerations' column matrix into two portions: a portion that drives the mechanical system to abide by the constraints and a portion that generates the momentum balance dynamics of the mechanical system. Some geometrical perspectives of the canonical equations of motion are illustrated via vectorial geometric visualization, which leads to verifying the Gauss' principle of least constraints and its Udwadia-Kalaba interpretation.

Keywords: canonical equations of motion, discrete mechanical systems, Kane's method, Gauss' principle of least constraints, cononical generalized speeds, Greville formula

1. Introduction

Deriving mathematical models for dynamical systems is in the core of the discipline of analytical dynamics, and it is the step that precedes dynamical system's analysis, design, and control synthesis. For discrete mechanical systems, i.e., those composed of particles and rigid bodies, the mathematical models are in the forms of differential equations or differential/algebraic equations that are derived by using fundamental laws of motion or energy principles. Because many mechanical systems nowadays are multi-bodied with numerous degrees of freedom and large numbers of *holonomic* and *nonholonomic* constraints, simplicity of the derived equations of motion is important for facilitating studying the mechanical system's characteristics and for extracting useful information out of its mathematical model. Hence, deriving the simplest possible form of the equations of motion that govern the dynamics of the mechanical system is crucial. Moreover, because the mechanical system's equations of motion are simulated on digital computers, computational efficiency of the derived differential equations of motion when numerically integrated is another factor by which the quality of the mathematical model is judged on.

It has been a general trend for over two centuries to employ d'Alembert's *principle of virtual work* [1] to derive equations of motion that involve no constraint forces. The principle was implemented by Lagrange [2] for deriving the first set of such equations, which constituted the first paradigm shift from the Newton-Euler's approach. The only other alternative to employing d'Alembert's principle has been to augment the equations with undetermined multipliers, an approach that was initiated by Lagrange himself. Other formulations that followed the trend include the Maggi [3] and Boltzmann-Hamel [4] formulations. A remarkable contribution of the Lagrangian approach to analytical dynamics is utilizing the concept of *generalized coordinates* instead of the Cartesian coordinate concept. The choice of generalized coordinates greatly affects simplicity of the derived equations of motion.

Another paradigm shift in the subject took place when Gibbs [5] and Appell [6] independently derived their equations of motion. For the first time, formulating the dynamical equations involved neither invoking d'Alembert's principle nor augmenting undetermined multipliers. Because d'Alembert's principle was to many analytical dynamics practitioners, "an ill-defined, nebulous, and hence objectionable principle," [7] the Gibbs-Appell model was widely accepted within the analytical dynamics community. Moreover, the absence of undetermined multipliers from the Gibbs-Appell equations contributed to maintaining simplicity and practicality of the equations for large constrained mechanical systems. Another feature of the Gibbs-Appell approach was initiating the concept of *quasi-velocities*, which equal in their number to the number of the degrees of freedom of the mechanical system. Similar to the advantage of generalized coordinates, carefully chosen quasi-velocities can lead to dramatic simplifications of the dynamical equations of motion.

One feature that is associated with the Gibbs-Appell's approach is that it is based on the differential *Gauss' principle of least constraints* [8] as was shown by Appell, [9] in contrast to the Lagrange's approach that is based on the variational *Hamilton's principle of least action* [10] as opposed. Another feature is adopting the *acceleration form of constraints* to model a mechanical system's constraints. Although easy by itself, employing the acceleration form eased the historical hurdle of modeling nonholonomic constraints that used to obstruct variational-based formulations, and it is a consequence of the differential theme that is based on Gauss' principle. In particular, the acceleration form bypassed d'Alembert's principle and the undetermined multiplier augmentation practices that produce false equations of nonholonomically constrained motion, and it unified the treatments of holonomic and nonholonomic constraints.

A key developments in the arena of analytical dynamics is the Kane's method for modeling constrained discrete mechanical systems [11–13]. Kane's method adopts a vector approach that inspired useful geometric features of the derived equations of motion [14]. The *generalized active forces* and *generalized inertia forces* are obtained by scalar (dot) multiplications of the active and inertia forces, respectively, with the vector entities *partial angular velocities* and *partial velocities*. This process delicately eliminates the contribution of constraint forces without invoking the principle of virtual work. The resulting equations are simple and effective in describing the motion of nonconservative and nonholonomic systems within the same framework, requiring neither energy methods nor Lagrange multipliers.

The standard Kane's equations of motion for nonholonomic systems are minimal in generalized speeds, i.e., their number is equal to the number of degrees of freedom of the dynamical system, and only the independent portion of generalized speeds and their time derivatives appear in the equations. Nevertheless, information about dependent generalized speeds can be practically important, e.g., for the purpose of obtaining stability information about a dependent dynamics or when it desired to target a dependent dynamics with a control system design by using state space control methodologies.

On the other hand, generalized inversion and the Greville formula for general solutions of linear systems of algebraic equations were introduced to the subject of analytical dynamics by Udwadia and Kalaba [15, 16] as tools for deriving equations of constrained motion for discrete mechanical systems. The success that the formula met in modeling ideally constrained motion is due to its geometrical structure that captures orthogonality of ideal constraint forces on active and inertia forces, which is the essence of the principle of virtual work.

Inspired by the Udwadia-Kalaba equations of motion and the Greville formula, this chapter introduces a new form of Kane's equations of motion. The introduced equations of motion employ the acceleration form of constraints, and therefore holonomic and nonholonomic constraints are augmented within the momentum balance formulation in a unified manner and irrespective of being linear or nonlinear in generalized coordinates and generalized speeds. The equations of motion are nonminimal, i.e., no reduction of generalized speed's space dimensionality takes place from the number of generalized coordinates to the number of degrees of freedom. Furthermore, the new equations of motion are explicit, i.e., are separated in the generalized acceleration variables, and only one generalized acceleration variable appears in each equation.

The main feature of the derived equations of motion is the explicit algebraic and geometric partitioning of the generalized acceleration vector at every instant of time into two portions: one portion drives the mechanical system to abide by the constraint dynamics, and the other portion generates the momentum balance of the mechanical system as to follow Newton-Euler's laws of motion.

2. Kane's equations of motion for holonomic systems

Consider a set of ν particles and μ rigid bodies that form a *holonomic system* S_h possessing n degrees of freedom in an inertial reference frame J. Assume that n *generalized coordinates* $q_1, \ldots, q_n$ are used to describe the configuration of the system. Then a corresponding set of n

holonomic generalized speeds $u_{h_1}, \ldots, u_{h_n}$ is used to model the kinematics of the system. The two sets are related by the kinematical differential equations [12, 13]:

$$\dot{q} = C(q,t)u_h + D(q,t), \tag{1}$$

where $q \in \mathbb{R}^n$ is a column matrix containing the generalized coordinates; $u_h \in \mathbb{R}^n$ is a column matrix containing the generalized speeds, $\dot{q} = dq/dt$, $C \in \mathbb{R}^{n \times n}$, $D \in \mathbb{R}^n$; and C^{-1} exists for all $q \in \mathbb{R}^n$ and all $t \in \mathbb{R}$ [12, 13]. Kane's dynamical equations of motion for S_h are given by [12, 13]

$$F_r(q,u_h,t) + F_r^\star(q,u_h,\dot{u}_h,t) = 0 \ , \qquad r = 1, \ldots, n, \tag{2}$$

where F_r and $F_r^\star$ are the rth holonomic generalized active force and the rth holonomic generalized inertia force on the system, respectively, and $\dot{u}_h = du_h/dt \in \mathbb{R}^n$ is a column matrix containing the *generalized accelerations*. Furthermore, the velocities and angular velocities of the particles and bodies comprising a mechanical system are linear in the generalized speeds u_{h_r}. Hence, the accelerations, angular accelerations, and consequently the generalized inertia forces are linear in the generalized accelerations $\dot{u}_{h_r}$. Therefore, a column matrix $F^\star \in \mathbb{R}^n$ containing $F_r^\star, r = 1, \ldots, n$ can be written in the following form [17]:

$$F^\star(q,u_h,\dot{u}_h,t) = -Q(q,t)\dot{u}_h - L(q,u_h,t), \tag{3}$$

where the generalized inertia matrix $Q \in \mathbb{R}^{n \times n}$ is assumed symmetric and positive definite and $L \in \mathbb{R}^n$. Hence, a matrix form of (2) is written as [17]

$$Q(q,t)\dot{u}_h = -L(q,u_h,t) + F(q,u_h,t). \tag{4}$$

3. Kane's equations of motion for nonholonomic systems

Let us now consider a modification of the kinematics of S_h that is made by imposing the following simple nonholonomic constraints on the generalized speeds [12, 13]:

$$u_{p+r} = \sum_{s=1}^{p} A_{rs}(q,t)u_s + B_r(q,t), \qquad r = 1, \ldots, m, \tag{5}$$

where $u_1, \ldots, u_n$ are the generalized speeds of the *nonholonomic system* S that is resulting from constraining S_h according to (5), $m = n - p$, and A_{rs} and B_r are scalar functions of the generalized coordinates $q_1, \ldots, q_n,$ and t. The *nonholonomic generalized speeds* are considered to satisfy the same kinematical relations with generalized coordinates as their holonomic counterparts, i.e.,

$$\dot{q} = C(q,t)u + D(q,t). \tag{6}$$

The system dynamics of S changes from that given by (2) accordingly. Let the generalized speed column matrix be partitioned as

$$u = \begin{bmatrix} u_I^T & u_D^T \end{bmatrix}^T, \tag{7}$$

where $u_I = \begin{bmatrix} u_1 & \cdots & u_p \end{bmatrix}^T$ and $u_D = \begin{bmatrix} u_{p+1} & \cdots & u_n \end{bmatrix}^T$. Kane's dynamical equations of motion for S are given by [12, 13]

$$\widetilde{F}_r(q, u_I, t) + \widetilde{F}_r^\star(q, u_I, \dot{u}_I, t) = 0 \ , \qquad r = 1, \ldots, p, \tag{8}$$

where $\widetilde{F}_r$ and $\widetilde{F}_r^\star$ are the rth nonholonomic generalized active force and the rth nonholonomic generalized inertia force on S, respectively. The relationships between holonomic generalized active forces on S_h and nonholonomic generalized active forces on S are given by [12, 13]

$$\widetilde{F}_r(q, u_I, t) = F_r(q, u, t) + \sum_{s=1}^{m} F_{p+s}(q, u, t)A_{sr}, \qquad r = 1, \ldots, p. \tag{9}$$

In a similar manner, the relationships between holonomic generalized inertia forces on S_h and nonholonomic generalized inertia forces on S are given by [12, 13]

$$\widetilde{F}_r^\star(q, u_I, \dot{u}_I, t) = F_r^\star(q, u, \dot{u}, t) + \sum_{s=1}^{m} F_{p+s}^\star(q, u, \dot{u}, t)A_{sr}(q, t), \qquad r = 1, \ldots, p. \tag{10}$$

Substituting (9) and (10) in (8) yields the unreduced form of Kane's equations of motion for S [12, 13, 17]:

$$F_r(q, u, t) + F_r^\star(q, u, \dot{u}, t) + \sum_{s=1}^{m} \Big(F_{p+s}(q, u, t) + F_{p+s}^\star(q, u, \dot{u}, t)\Big)A_{sr}(q, t) = 0, \qquad r = 1, \ldots, p. \tag{11}$$

The simple nonholonomic constraint equations given by (5) can be rewritten in the following matrix representation [17]:

$$u_D = A(q, t)u_I + B(q, t), \tag{12}$$

where $A \in \mathbb{R}^{m \times p}$ and $B \in \mathbb{R}^m$. Furthermore, (12) can be rewritten as [17]

$$A_1(q, t)u = B(q, t), \tag{13}$$

where $A_1 \in \mathbb{R}^{m \times n}$ is given by

$$A_1(q, t) = \begin{bmatrix} -A(q, t) & I_{m \times m} \end{bmatrix}. \tag{14}$$

Also, (11) can be rewritten in the matrix form [17]:

$$A_2(q, t)F^\star(q, u, \dot{u}, t) = -A_2(q, t)F(q, u, t), \tag{15}$$

where $A_2 \in \mathbb{R}^{p \times n}$ is given by

$$A_2(q,t) = \begin{bmatrix} I_{p\times p} & A^T(q,t) \end{bmatrix}. \tag{16}$$

Hence, (15) becomes [17]

$$A_2(q,t)Q(q,t)\dot{u} = -A_2(q,t)L(q,u,t) + A_2(q,t)F(q,u,t). \tag{17}$$

Notice that (17) is obtained by multiplying both sides of (4) by $A_2(q,t)$. Therefore, the unique holonomic generalized acceleration vector $\dot{u}_h$ that solves the fully determined system given by (4) solves the underdetermined system given by (17) also, among an infinite number of generalized acceleration vectors that satisfy (17), each of which preserves a constrained momentum balance dynamics of the mechanical system.

4. Canonical generalized speeds

Choosing the set of generalized speeds is a crucial step in formulating Kane's dynamical Eqs. (2) and (8) because the extent of how complex these equations appear is affected by this choice. For every choice of nonholonomic generalized speeds $u_1, \ldots, u_n$, we define the *canonical set of nonholonomic generalized speeds* $w_1, \ldots, w_n$ such that

$$w = Q^{1/2}(q,t)u, \tag{18}$$

where w is the column matrix containing $w_1, \ldots, w_n$ and $Q^{1/2}$ is the square root matrix of Q. With this choice of generalized speeds, (13) becomes

$$\mathscr{A}_1(q,t)w = B(q,t), \tag{19}$$

where $\mathscr{A}_1(q,t) = A_1(q,t)Q^{-1/2}(q,t)$. The time derivative of (18) is

$$\dot{w} = \dot{Q}^{1/2}(q,u,t)u + Q^{1/2}(q,t)\dot{u} \tag{20}$$

where $\dot{Q}^{1/2}$ is the element-wise time derivative of $Q^{1/2}$ along the trajectory solutions of the kinematical differential Eqs. (6). Therefore, (17) becomes

$$A_2(q,t)Q^{1/2}(q,t)\dot{w} = -A_2(q,t)L(q,u,t) + A_2(q,t)F(q,u,t) + A_2(q,t)Q^{1/2}(q,t)\dot{Q}^{1/2}(q,u,t)u \tag{21}$$

and can be simplified further to the following form:

$$\mathscr{A}_2(q,t)\dot{w} = \mathscr{A}_2(q,t)Q^{-1/2}(q,t)(F(q,u,t) - L(q,u,t)) + \mathscr{A}_2(q,t)\dot{Q}^{1/2}(q,u,t)u \tag{22}$$

where $\mathscr{A}_2(q,t) = A_2(q,t)Q^{1/2}(q,t)$. We view the nonholonomic mechanical system dynamics as being composed of two parts: a constraint dynamics that is modeled by (19) and a momentum balance dynamics that is modeled by (22). Scaling velocity variables and constraint matrices by square roots of the inertia matrices for the purpose of characterizing constrained motion is implicit in Gauss' principle of least constraints [8] as will be shown later in this paper.

Moreover, deriving explicit equations of motion for constrained mechanical systems by utilizing this type of scaling is first due to Udwadia and Kalaba. [15, 16] The arguments of the functions are omitted in the remaining sections for brevity, unless necessary to clarify concepts.

5. Generalized accelerations from the acceleration form of constraints

Time differentiating the constraint dynamics given by (19) yields [17]

$$\mathscr{A}_1 \dot{w} = V_1, \tag{23}$$

where $V_1 \in \mathbb{R}^m$ is given by

$$V_1 = \dot{B}(q, u, t) - \dot{\mathscr{A}}_1(q, u, t) w, \tag{24}$$

where $\dot{B}$ and $\dot{\mathscr{A}}_1$ are the element-wise time derivatives of B and $\mathscr{A}_1$ along the trajectory solutions of the kinematical differential Eqs. (6). The general solution of the above-written acceleration form of constraint equations for $\dot{w}$ is given by the Greville formula as [18–20]

$$\dot{w} = \mathscr{A}_1^+ V_1 + \mathscr{P}_1 y_1, \tag{25}$$

where A_1^+ is the Moore-Penrose generalized inverse (MPGI) [21, 22] ofA_1 and

$$\mathscr{P}_1 = I_{n \times n} - \mathscr{A}_1^+ \mathscr{A}_1 \tag{26}$$

is the projection matrix on the nullspace of $\mathscr{A}_1$ and $y_1 \in \mathbb{R}^n$ is an arbitrary vector as for satisfying the acceleration form given by (25) but is yet to be determined to obtain the unique natural generalized acceleration. Because $Q^{-1/2}$ is of full rank, it follows that $\mathscr{A}_1$ retains the full row rank of A_1 and hence that $\mathscr{A}_1^+ \in \mathbb{R}^{n \times m}$ is given by the closed form expression:

$$\mathscr{A}_1^+ = \mathscr{A}_1^T \left(\mathscr{A}_1 \mathscr{A}_1^T \right)^{-1}. \tag{27}$$

In (25), the following holds

$$\mathscr{A}_1^+ V_1 \in \mathscr{R}\left(\mathscr{A}_1^T\right), \qquad \mathscr{P}_1 y_1 \in \mathscr{N}(\mathscr{A}_1) \tag{28}$$

where $\mathscr{R}(\cdot)$ and $\mathscr{N}(\cdot)$ refer to the range space and the nullspace, respectively. The term $\mathscr{A}_1^+ V_1$ in (25) is the minimum norm solution of (23) for $\dot{w}$ among infinitely many solutions that are parameterized by y_1.

6. Generalized accelerations from the momentum balance dynamics

Let $V_2 \in \mathbb{R}^n$ be given by

$$V_2 = Q^{-1/2}(F - L) + \dot{Q}^{1/2}u. \tag{29}$$

Then the momentum balance Eq. (22) takes the following compact form:

$$\mathscr{A}_2\dot{w} = \mathscr{A}_2 V_2 \tag{30}$$

where A_2 retains the full row rank of A_2 because $Q^{1/2}$ is of full row rank. Hence, another expression for the general solution of $\dot{w}$ is obtained by utilizing the Greville formula to solve (30) and is given by

$$\dot{w} = \mathscr{A}_2^+\mathscr{A}_2 V_2 + \mathscr{P}_2 y_2 \tag{31}$$

where $A_2^+ \in \mathbb{R}^{n\times p}$ is given by the closed form expression:

$$\mathscr{A}_2^+ = \mathscr{A}_2^T\left(\mathscr{A}_2\mathscr{A}_2^T\right)^{-1} \tag{32}$$

and

$$\mathscr{P}_2 = I_{n\times n} - \mathscr{A}_2^+\mathscr{A}_2 \tag{33}$$

and $y_2 \in \mathbb{R}^n$ is an arbitrary vector as for satisfying the momentum balance dynamics given by (30), but its unique value that solves for the natural generalized acceleration vector $\dot{w}$ is yet to be determined, and

$$\mathscr{A}_2^+\mathscr{A}_2 V_2 \in \mathscr{R}\left(\mathscr{A}_2^+\mathscr{A}_2\right) = \mathscr{R}\left(\mathscr{A}_2^T\right), \qquad \mathscr{P}_2 y_2 \in \mathscr{N}\left(\mathscr{A}_2\right). \tag{34}$$

The term $\mathscr{A}_2^+\mathscr{A}_2 V_2$ in (31) is the minimum norm solution of (30) for $\dot{w}$ among infinitely many solutions that are parameterized by y_2.

7. Canonical generalized inversion Kane's equations of motion

Since A_1 and A_2 are full row rank matrices and their numbers of rows m and p sum up to the full space dimension n and since

$$\begin{aligned}\mathscr{A}_1\mathscr{A}_2^T &= \left(A_1Q^{-1/2}\right)\left(A_2Q^{1/2}\right)^T = \left(A_1Q^{-1/2}\right)\left(A_2Q^{1/2}\right)^T \\ &= A_1Q^{-1/2}Q^{1/2}A_2^T = A_1A_2^T = -A + A = \mathbf{0}_{m\times m}\end{aligned} \tag{35}$$

it follows that the row spaces of $\mathscr{A}_1$ and $\mathscr{A}_2$ are orthogonal complements, i.e.,

$$\mathscr{R}\left(\mathscr{A}_1^T\right) = \left[\mathscr{R}\left(\mathscr{A}_2^T\right)\right]^\perp. \tag{36}$$

Nevertheless, since [23]

$$\mathscr{R}(\mathscr{A}_1^T) = [\mathscr{N}(\mathscr{A}_1)]^{\perp} \tag{37}$$

then it follows from (36) that

$$\mathscr{R}(\mathscr{A}_2^T) = \mathscr{N}(\mathscr{A}_1). \tag{38}$$

Since the only part in the expression of $\dot{w}$ given by (25) that is in $\mathscr{N}(\mathscr{A}_1)$ is the second term $\mathscr{P}_1 y_1$, and since the ony part in the equivalent expression of $\dot{w}$ given by (31) that is in $\mathscr{R}(\mathscr{A}_2^T)$ is the first term $\mathscr{A}_2^+ \mathscr{A}_2 V_2$, it follows from (38) that

$$\mathscr{P}_1 y_1 = \mathscr{A}_2^+ \mathscr{A}_2 V_2. \tag{39}$$

Substituting (39) in (25) yields the canonical generalized inversion form of Kane's equations for nonholonomic systems:

$$\dot{w} = \mathscr{A}_1^+ V_1 + \mathscr{A}_2^+ \mathscr{A}_2 V_2. \tag{40}$$

The same result is obtained by using the fact:

$$\mathscr{N}(\mathscr{A}_2) = [\mathscr{R}(\mathscr{A}_2^T)]^{\perp} \tag{41}$$

which implies by using (36) that

$$\mathscr{R}(\mathscr{A}_1^T) = \mathscr{N}(\mathscr{A}_2) \tag{42}$$

Since the only part in the expression of $\dot{w}$ given by (25) that is in $\mathscr{R}(\mathscr{A}_1^T)$ is the first term $\mathscr{A}_1^+ V_1$, and since the only part in the equivalent expression of $\dot{w}$ given by (31) that is in $\mathscr{N}(\mathscr{A}_2)$ is the second term $\mathscr{P}_2 y_2$, it follows from (42) that

$$\mathscr{P}_2 y_2 = \mathscr{A}_1^+ V_1. \tag{43}$$

(Substituting (43) in (31) yields Eq. (40). Eq. (20) can be used to express (40) in terms of the original generalized acceleration vector $\dot{u}$, resulting in

$$\dot{u} = Q^{-1/2}\left(\mathscr{A}_1^+ V_1 + \mathscr{A}_2^+ \mathscr{A}_2 V_2 - \dot{Q}^{1/2} u\right). \tag{44}$$

8. Geometric interpretation of the canonical generalized inversion form

Adopting the canonical set $w_1, \cdots, w_n$ of generalized speeds in deriving the dynamical equations for a mechanical system reveals the geometry of its constrained motion. **Figure 1** depicts a geometrical visualization of the n dimensional Euclidian space at an arbitrary time instant t. The vertical and the horizontal axes resemble the orthogonally complements m dimensional and p dimensional subspaces $\mathscr{R}(\mathscr{A}_1^T)$ and $\mathscr{R}(\mathscr{A}_2^T)$, respectively.

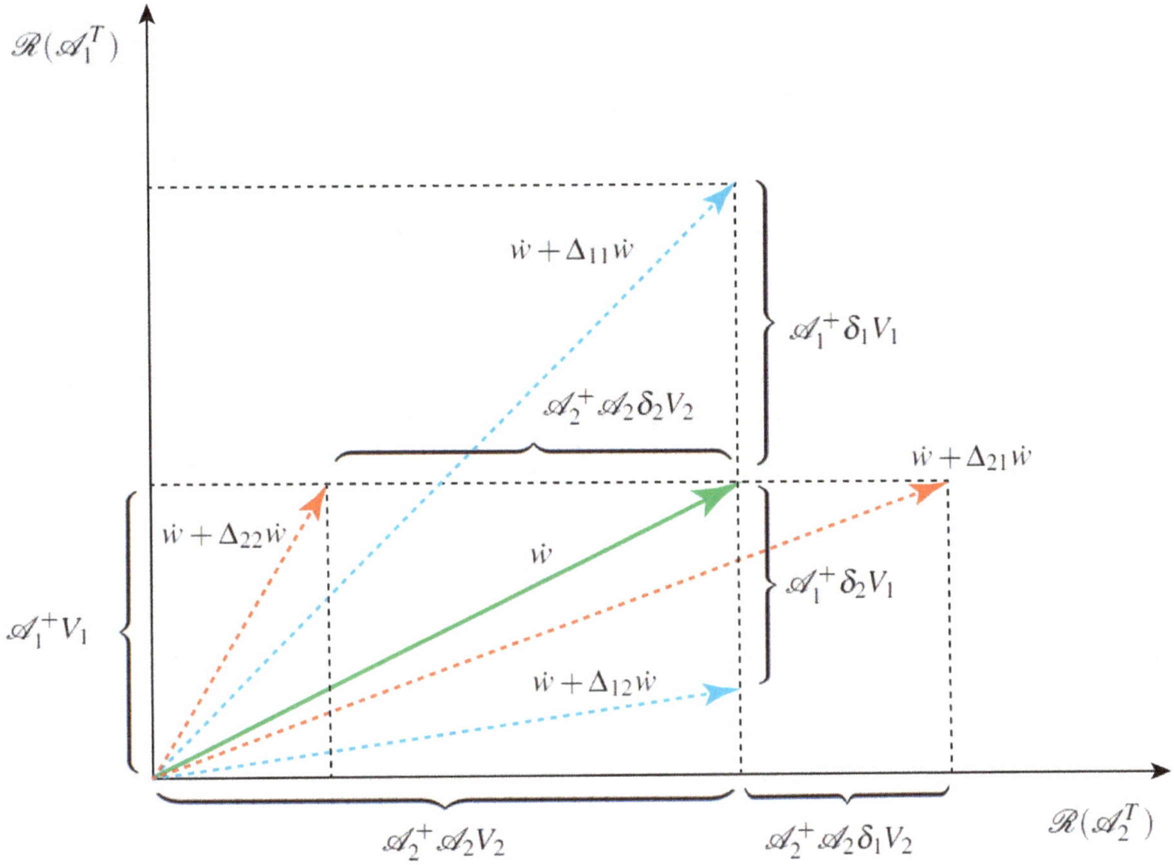

Figure 1. Geometric visualization of the constrained generalized acceleration vector $\dot{w}$.

In viewing the canonical generalized acceleration $\dot{w}$ given by (40) as the geometrical vector shown in **Figure 1**, it is shown to be composed of two components that are orthogonal to each other: The vertical component $\mathscr{A}_1^+ V_1$ resides in $\mathscr{R}(\mathscr{A}_1^T)$, and it enforces the constraint dynamics given by (23), and the horizontal component $\mathscr{A}_2^+\mathscr{A}_2 V_2$ resides in $\mathscr{R}(\mathscr{A}_2^T)$, and it generates the momentum balance dynamics given by (30).

Moreover, the vertical component of $\dot{w}$ is the shortest in length "minimum norm" solution among infinitely many solutions of (23) that are parameterized by y_1 according to (25). These solutions can also be represented by arbitrary horizontal deviation vectors: $\Delta_{2i}\dot{w} = \mathscr{A}_2^+\mathscr{A}_2\delta_i V_2 \in \mathscr{R}(\mathscr{A}_2^T), i = 1, 2, \ldots$ as

$$\dot{w} + \Delta_{2i}\dot{w} = \mathscr{A}_1^+ V_1 + \mathscr{A}_2^+\mathscr{A}_2(V_2 + \delta_i V_2) \tag{45}$$

and are shown to solve (23) by direct substitution and noticing that $\mathscr{A}_1\mathscr{A}_1^+ = I_{m\times m}$ and $\mathscr{A}_1\mathscr{A}_2^+ = \mathbf{0}_{m\times p}$. Two of these solutions are plotted (in dotted red) in **Figure 1** for arbitrary vectors $\delta_1 V_2$ and $\delta_2 V_2$ in $\mathbb{R}^n$, in addition to the natural generalized acceleration vector $\dot{w}$ that is obtained by setting $\delta_i V_2 = \mathbf{0}_n$.

Similarly, the horizontal component $\mathscr{A}_2\mathscr{A}_2^+ V_2$ of $\dot{w}$ is the shortest solution among infinitely many solutions of (30) that are parameterized by y_2 according to (31). These solutions can also be represented by arbitrary vertical deviation vectors: $\Delta_{1i}\dot{w} = \mathscr{A}_1^+\delta_i V_1 \in \mathscr{R}(\mathscr{A}_1^T), i = 1, 2, \ldots$ as

$$\dot{w} + \Delta_{1i}\dot{w} = \mathscr{A}_2^+\mathscr{A}_2 V_2 + \mathscr{A}_1^+(V_1 + \delta_i V_1) \tag{46}$$

and are shown to solve (30) by direct substitution and noticing that $\mathscr{A}_2\mathscr{A}_2^+ = I_{p\times p}$ and $\mathscr{A}_2\mathscr{A}_1^+ = \mathbf{0}_{p\times m}$. Two of these solutions are plotted (in dotted blue) in **Figure 1** for arbitrary vectors $\delta_1 V_1$ and $\delta_2 V_1$ in $\mathbb{R}^m$, in addition to the natural generalized acceleration vector $\dot{w}$ that is obtained by setting $\delta_i V_1 = \mathbf{0}_m$. Notice that the canonical generalized acceleration vector $\dot{w}$ is the only solution that solves (45, 46) simultaneously and is obtained by setting $\delta_i V_2 = \mathbf{0}_n$ and $\delta_i V_1 = \mathbf{0}_m$.

Now consider a general deviation vector $\Delta\dot{w}$ that is composed of arbitrary vertical and horizontal deviation components from $\dot{w}$ as shown in **Figure 2**. The vertical component $\mathscr{A}_1^+\delta V_1$ abides by (46) but violates (45), and the horizontal component $\mathscr{A}_2^+\mathscr{A}_2\delta V_2$ abides by (45) but violates (46). Hence:

$$\Delta\dot{w} = \mathscr{A}_1^+\delta V_1 + \mathscr{A}_2^+\mathscr{A}_2\delta V_2. \tag{47}$$

The deviated canonical generalized acceleration vector $\dot{w} + \Delta\dot{w}$ is obtained by summing (40) and (47) as

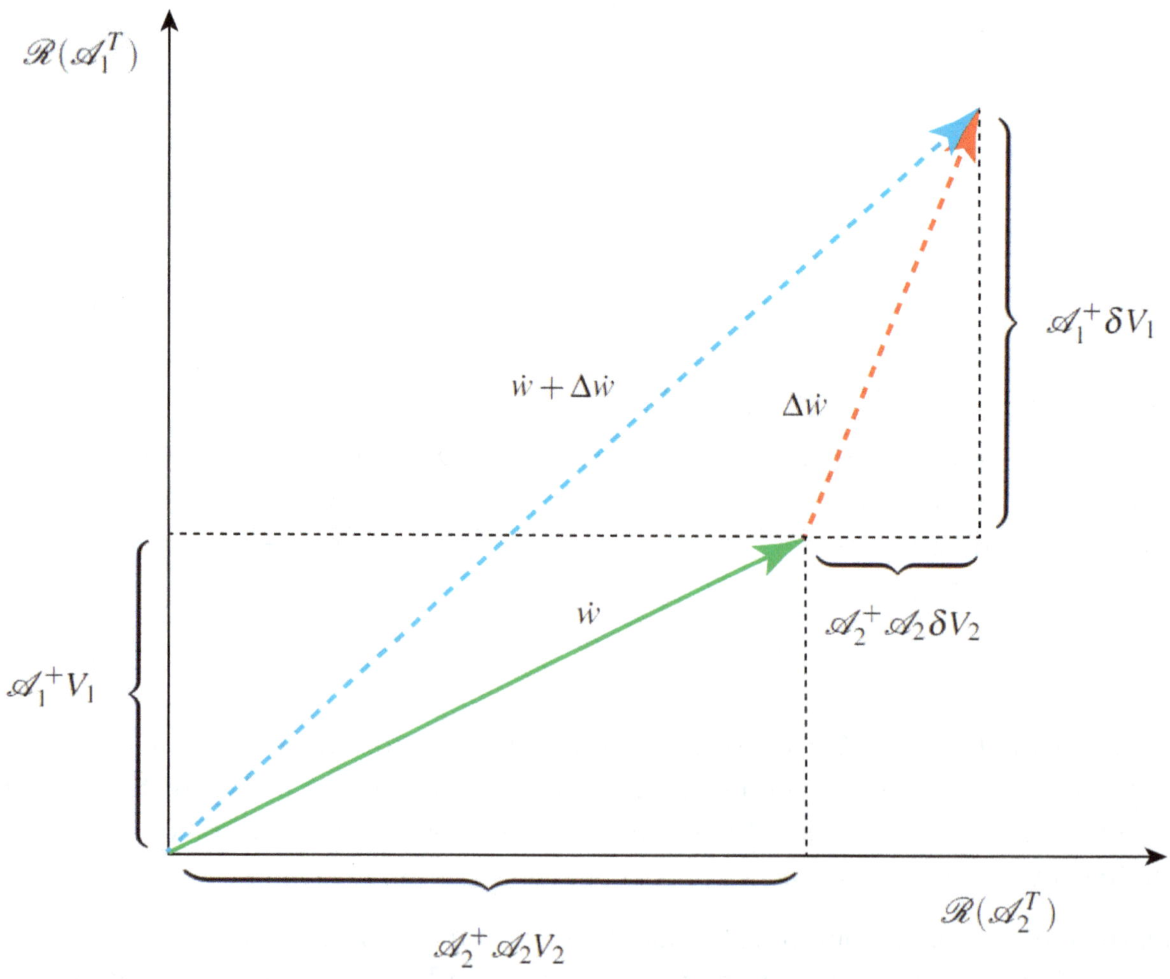

Figure 2. Deviation from the constrained generalized acceleration vector $\dot{w}$.

$$\dot{w} + \Delta\dot{w} = \mathscr{A}_1^+(V_1 + \delta V_1) + \mathscr{A}_2^+\mathscr{A}_2(V_2 + \delta V_2) \tag{48}$$

and is shown in **Figure 2** in dotted blue. On the other hand, the canonical holonomic generalized acceleration vector in terms of the canonical generalized speeds is obtained from (4) and (20) as

$$\dot{w}_h = V_2 \tag{49}$$

where $w_h = Q^{1/2}u$ and u solves (4). Decomposing the expression of $\dot{w}_h$ along $\mathscr{R}(\mathscr{A}_1)$ and $\mathscr{R}(\mathscr{A}_2)$ yields

$$\dot{w}_h = V_2 = \mathscr{P}_1 V_2 + \mathscr{P}_2 V_2 \tag{50}$$

$$= \left(I_{n\times n} - \mathscr{A}_1^+\mathscr{A}_1\right)V_2 + \left(I_{n\times n} - \mathscr{A}_2^+\mathscr{A}_2\right)V_2 \tag{51}$$

$$= \mathscr{A}_2^+\mathscr{A}_2 V_2 + \mathscr{A}_1^+\mathscr{A}_1 V_2. \tag{52}$$

Let us now specify the deviated generalized acceleration vector $\dot{w} + \Delta\dot{w}$ to be $\dot{w}_h$ as shown in **Figure 3**. Equating the two expressions (48) and (52) and solving for δV_1 and δV_2 yield

$$\delta V_1 = \mathscr{A}_1 V_2 - V_1, \tag{53}$$

and

$$\delta V_2 = \mathbf{0}_n. \tag{54}$$

Substituting δV_1 and δV_2 in (47) yields

$$\Delta\dot{w} = \mathscr{A}_1^+(\mathscr{A}_1 V_2 - V_1) \tag{55}$$

which corresponds to the vertical solid red vector in **Fig. (3)**. Notice that $\Delta\dot{w}$ is the shortest among all deviation vectors that end up at $\dot{w}_h$ (two of which are shown in dotted red) by deviating from generalized acceleration vectors that abide by the constraint dynamics given by (23) (two of which are shown in dotted green), i.e.,

$$||\Delta\dot{w}|| = ||\dot{w}_h - \dot{w}|| = \min_i ||\Delta\dot{w}_i|| = \min_i ||\dot{w}_h - \dot{w}_i||, i = 1, 2, \dots \tag{56}$$

where $\dot{w}_i$ satisfies

$$\mathscr{A}_1\dot{w}_i = V_1 \tag{57}$$

and $||x||$ denotes the Euclidean norm of x given by $||x|| = \sqrt{x^T x}$. Moreover, Δw can be expressed in terms of the original set of generalized speeds as

$$\Delta w = Q^{1/2}\Delta u \tag{58}$$

where $\Delta u = u_h - u$ is the difference between holonomic and nonholonomic generalized speeds. Therefore:

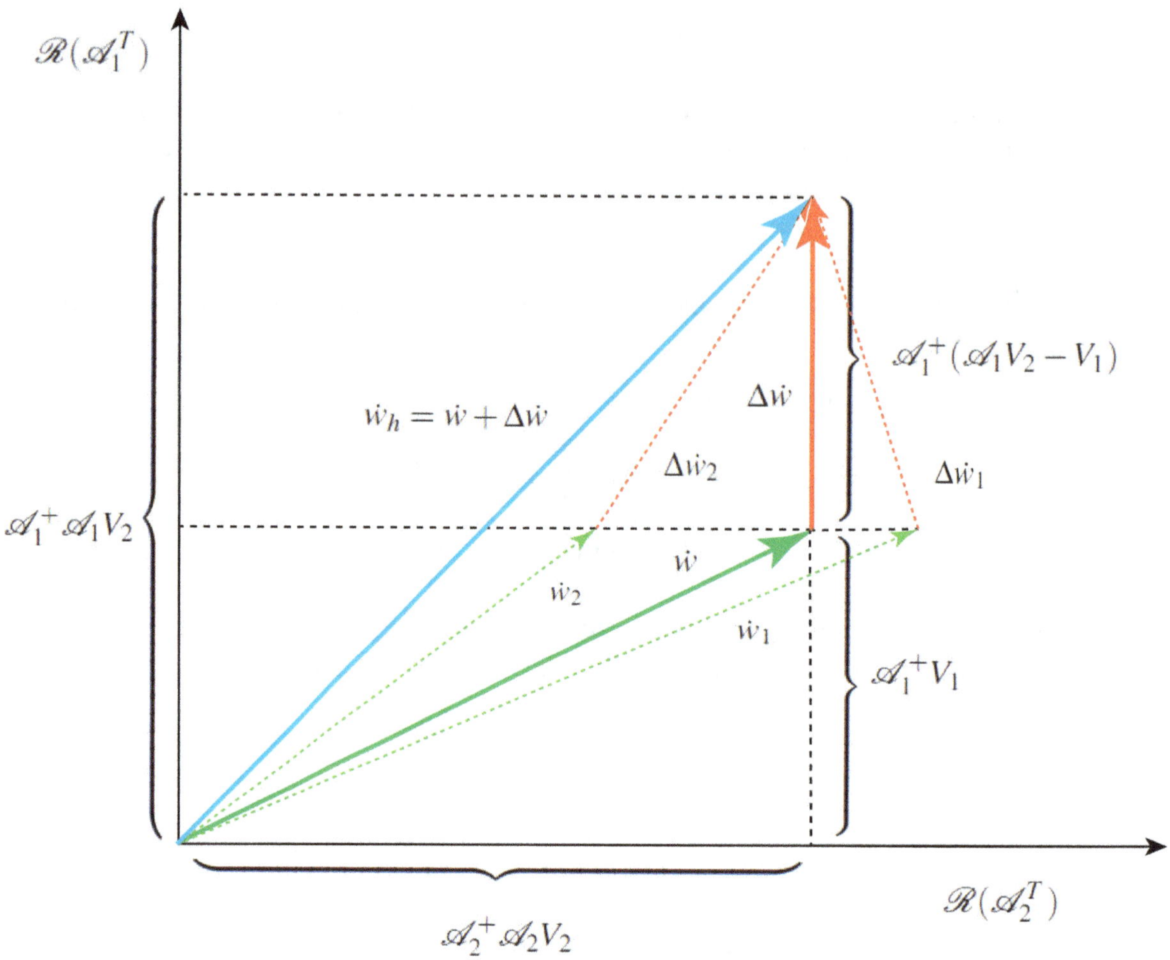

Figure 3. Deviation from the constrained generalized acceleration vector $\dot{w}$.

$$\|\Delta\dot{w}\|=\|Q^{1/2}\Delta\dot{u}+\dot{Q}^{1/2}\Delta u\|=\min_i\|Q^{1/2}\Delta\dot{u}_i+\dot{Q}^{1/2}\Delta u\|, i=1,2,\ldots \tag{59}$$

where $\Delta\dot{u}_i=\dot{u}_h-\dot{u}_i$ and $\dot{u}_i$ satisfies

$$\mathscr{A}_1\dot{w}_i=\mathscr{A}_1\left(Q^{1/2}\dot{u}_i+\dot{Q}^{1/2}u\right)=V_1. \tag{60}$$

Nevertheless, (59) implies that

$$\|Q^{1/2}\Delta\dot{u}\|=\min_i\|Q^{1/2}\Delta\dot{u}_i\|, i=1,2,\ldots., \tag{61}$$

which in terms of the square Euclidean norm implies that

$$\|Q^{1/2}\Delta\dot{u}\|^2=\Delta\dot{u}^TQ^{1/2}Q^{1/2}\Delta\dot{u}=\min_i\left(\Delta\dot{u}_i^TQ^{1/2}Q^{1/2}\Delta\dot{u}_i\right) \tag{62}$$

$$=\Delta\dot{u}^TQ\Delta\dot{u}=\min_i\left(\Delta\dot{u}_i^TQ\Delta\dot{u}_i\right),\ \ i=1,2,\ldots. \tag{63}$$

Eq. (63) is exactly the statement of Gauss' principle of least constraints [8]. The present geometric interpretation of Gauss' principle was first introduced by Udwadia and Kalaba [24].

9. Conclusion

The chapter introduces the canonical generalized inversion dynamical equations of motion for nonholonomic mechanical systems in the framework of Kane's method. The introduced equations of motion use the Greville formula and utilize its geometric structure to produce a full order set of dynamical equations for the nonholonomic system. Moreover, the acceleration form of constraint equations is adopted in a similar manner as in the classical Gibbs-Appell, Udwadia-Kalaba, and Bajodah-Hodges-Chen formulations.

The philosophy on which the present formulation of the dynamical equations of motion is based views the constrained system dynamics of the mechanical system as being composed of a constraint dynamics and a momentum balance dynamics that is unaltered by augmenting the constraints. Inverting both dynamics by means of two Greville formulae and invoking the geometric relations between the resulting two expressions yield the unique natural canonical generalized acceleration vector.

Because the momentum balance dynamics and the acceleration form of constraint dynamics are linear in generalized accelerations, only linear geometric and algebraic mathematical tools are needed to analyze constrained motion of discrete mechanical systems. Also, the present linear analysis is valid in despite of dependencies among the constraint equations and changes in rank that the constraint matrix A may experience because the matrices A_1 and A_2 are always of full row ranks and their m and p rows span two orthogonally complement row spaces. Another advantage of maintaining full row ranks of A_1 and A_2 is that their generalized inverses have explicit and closed form expressions, which alleviate the need for employing numerical methods for computing generalized inverses.

Author details

Abdulrahman H. Bajodah[1*] and Ye-Hwa Chen[2]

*Address all correspondence to: abajodah@kau.edu.sa

1 Aeronautical Engineering Department, King Abdulaziz University, Jeddah, Saudi Arabia

2 School of Mechanical Engineering, Georgia Institute of Technology, Atlanta, Georgia, United States of America

References

[1] d'Alembert J. Traité de Dynamique. Paris; 1743

[2] Lagrange JL. Mechanique Analytique, Mme. Paris: Ve Courcier; 1787

[3] Maggi G. Di alcune nuove forme delle equa- zioni della Dinamica, applicabili ai sistemi anolonomi. Atti Reale Accad Lin Rend Classe Sci Fisiche, Matematiche Naturali. 1901; **5**(10):287-292

[4] Hamel G. Die LagrangeEulersche gleichungen der mechanik. Zeitschrift für angewandte Mathematik und Physik. 1904;**50**:1-57

[5] Gibbs WJ. On the fundamental formulae of dynamics. American Journal of Mathematics. 1879;**2**:49-64

[6] Appell PE. Sur une forme générale des équations de la dynamique. Comptes Rendus de l'Académie des Sciences - Paris. Series I Mathematique. 1899;**129**

[7] Desloge EA. The Gibbs-Appell equations of motion. American Journal of Physics. 1988; **56**(9):1-6

[8] Gauss CF. Über ein Neues Allgemeine Grundgesatz der Mechanik. Journal Für Die Reine Und Angewandte Mathematik. 1829;**4**:232-235

[9] Appell PE. Sur une forme générale des equations de la dynamique et sur le principe de Gauss. Journal für die reine und angewandte Mathematik. 1900;**122**:205-208

[10] Goldstein H, Poole C, Safco J. Classical Mechanics. 3rd ed. Addison Wesley; 2001

[11] Kane TR, Wang CF. On the derivation of equations of motion. Journal of the Society of Industrial and Applied Mathematics. 1965;**13**(2):487-492

[12] Kane TR, Levinson DA. Dynamics: Theory and Applications. McGraw-Hill Series in Mechanical Engineering; 1985

[13] Roithmayr CM, Hodges DH. Dynamics: Theory and Application of Kane's Method. Cambridge University Press; 2016

[14] Lesser M. A geometrical interpretation of Kane's equations. Proceedings of the Royal Society of London. Series A, Mathematical Physical and Engineering Sciences. 1992;**436**:69-87

[15] Udwadia FE, Kalaba RE. Analytical Dynamics: A New Approach. Cambridge University Press; 2007

[16] Udwadia FE, Kalaba RE. A new perspective on constrained motion. Proceedings of the Royal Society A: Mathematical and Physical Sciences. 1992;**439**(1906):407-410

[17] Bajodah AH, Hodges DH, Chen YH. New form of Kane's equations of motion for constrained systems. Journal of Guidance, Control, and Dynamics. 2003;**26**(1):79-88

[18] Greville TNE. The pseudoinverse of a rectangular or singular matrix and its applications to the solutions of systems of linear equations. SIAM Review. 1959;**1**(1):38-43

[19] Ben-Israel A, Greville TNE. Generalized Inverses: Theory and Applications, Springer. 2nd ed. 2003

[20] Bajodah AH, Hodges DH, Chen YH. Inverse dynamics of servo-constraints based on the generalized inverse. Nonlinear Dynamics. 2005;**39**(1–2):179-196

[21] Moore EH. On the reciprocal of the general algebraic matrix. Bulletin of the American Mathematical Society. 1920;**26**:394-395

[22] Penrose R. A generalized inverse for matrices. Proceedings of the Cambridge Philosophical Society. 1955;**51**:406-413

[23] Bernstein DS. Matrix Mathematics: Theory, Facts, and Formulas. 2nd ed. Princeton University Press; 2009

[24] Udwadia FE, Kalaba RE. The geometry of constrained motion. ZAMM-Journal of Applied Mathematics and Mechanics/Zeitschrift für Angewandte Mathematik und Mechanik. 1995;**75**(8):637-640

5

Stability Conditions for a Class of Nonlinear Systems with Delay

Sami Elmadssia and Mohamed Benrejeb

Abstract

This chapter presents an extension and offers a more comprehensive overview of our previous paper entitled "Stability conditions for a class of nonlinear time delay systems" published in "Nonlinear Dynamics and Systems Theory" journal. We first introduce a more complete approach of the nonlinear system stability for the single delay case. Then, we show the application of the obtained results to delayed Lur'e Postnikov systems. A state space representation of the class of system under consideration is used and a new transformation is carried out to represent the system, with delay, by an arrow form matrix. Taking advantage of this representation and applying the Kotelyanski lemma in combination with properties of M-matrices, some new sufficient stability conditions are determined. Finally, illustrative example is provided to show the easiness of using the given stability conditions.

Keywords: nonlinear systems, time delay, arrow matrix, M-matrix, Lur'e Postnikov, stability conditions

1. Introduction

Studying stability of dynamical systems with time delay has received the attention of many researchers from the control community in the past decades, see [1–27] and the references therein. Time-varying delay which varies within an interval with nonzero lower bound is encountered in a variety of engineering applications which spreads from recurrent neural networks to chemical reactors and power systems with loss-less transmission lines. It is therefore more appropriate to study stability analysis and control synthesis of these dynamical systems with time-varying delays as these delays are usually time varying in nature. There are mainly two strategies in obtaining stability conditions. We can obtain delay-independent (i.o.d) results [28, 29] and the references therein, which are applicable to delays of arbitrary size or when there is no information about the delay. In general this lack of information about the delay will result in conservative criteria, especially when the delay is relatively small. Whenever it is possible to include information on the size of the delay, we can get delay-dependent (d.d) conditions which are usually less conservative. Most of the systems described above are nonlinear in practical engineering problems. For this reason, the chapter focuses on determining easy to test sufficient stability conditions for nonlinear systems with time-varying delay [30–33].

New delay dependent stability conditions are derived by employing arrow form state space representation [31–34], Kotelyanski lemma and using tools from M-matrix theory and Lyapunov functional method.

The obtained results are exploited to design a state feedback controller that stabilizes Lur'e systems with time-varying delay and sector-bounded nonlinearity [26, 28, 34]. In fact, Lur'e control systems is considered as one the most important classes of nonlinear control systems and continue to be one of the important problems in control theory that has been studied widely because it has many practical applications [32–36].

The chapter is organized as follows: Section 2 presents the notation used throughout the chapter and some facts on M-matrices that will be needed in proving the obtained results. In sections 3 the main results are given. Application of these results to delayed nonlinear nth order all pole plant and the well-known Lur'e systems, is presented in Section 4. Illustrative example is given in Section 5 and some concluding remarks are provided in Section 6.

2. Notation and facts

Let us fix the notation used. Let $C_n = C([-\tau\ 0], R^n)$ be the Banach space of continuous functions mapping the interval $[-\tau\ 0]$ into R^n with the topology of uniform convergence. Let $x_t \in C_n$ be defined by $x_t(\theta) = x(t+\theta), \theta \in [-\tau\ 0]$ where $x(t) = \left(y(t)\ \ \dot{y}(t)\ \ \ldots\ \ y^{(n-1)}(t)\right)'$. For a given $\varphi \in C_n$, we define $\|\varphi\| = \sup_{-\tau \le \theta \le 0} \|\varphi(\theta)\|, \varphi(\theta) \in R^n$. The functions $a_i(.)$, $b_i(.)$, $i = 1, \ldots, n\text{-}1$ are completely continuous mapping the set $J_a \times C_n^H \times S_\varpi$ into R, where $C_n^H = \{\varphi \in C_n, \|\varphi\| < H\}$, $H > 0, J_a = [a\ \ +\infty), a \in R$ and $S_\varpi = \{\varpi, k_1 \le \varpi \le k_2 / k_1 \le k_2 \in R\}$. In the sequel, we denote $(t, x_t, \varpi) = (.)$.

Now we introduce several useful facts, including some definitions of M-matrices and the Kotelyanski lemma that will be used in subsequent parts of the chapter.

Definition 1. The $n \times n$ matrix $A = \left(a_{i,j}\right)_{1 \le i,j \le n}$ is called an M-matrix if the following conditions are satisfied for $i = 1, 2, \ldots, n$ [34]:

1. $a_{i,i} > 0\ ,\ a_{i,j} \le 0\ \ (i \ne j, j = 1, 2, \ldots, n)$.
2. Successive principal minors of A are positive, i.e.

$$\det\begin{pmatrix} a_{1,1} & \dots & a_{1,i} \\ \vdots & \dots & \vdots \\ a_{i,1} & \dots & a_{i,i} \end{pmatrix} > 0$$

Definition 2. The matrix A is the opposite of an M-matrix if $(-A)$ is an M-matrix. There are many equivalent conditions for characterizing an M matrix. In fact, the following definition is the most appropriate for our purposes [34].

Definition 3. The matrix $A = \left(a_{i,j}\right)_{n\le i,j\le n}$ is called an M-matrix if $a_{i,i} > 0$ $(i = 1, 2, \dots, n)$, $a_{i,j} \le 0$, $i \neq j$, $(i, j = 1, 2, \dots, n)$ and for any vector $\sigma \in R_{+}^{*n}$, the algebraic equation $A'c = \sigma$ has a solution $c = \left(A'\right)^{-1}\sigma \in R_{+}^{*n}$ [34].

Kotelyanski Lemma

The real parts of the eigenvalues of a matrix A, with non-negative off diagonal elements, are less than a real number μ if and only if all those of the matrix M, $M = I_n - \mu A$, are positive, with I_n the $n \times n$ identity matrix [34, 35].

3. Sufficient stability conditions

Our work consists of determining stability conditions for systems described by the following equation:

$$\begin{cases} y^{(n)}(t) + \sum_{i=0}^{n-1} a_i(t, x_t, \varpi) y^{(i)}(t) + \sum_{j=0}^{n} b_j(t, x_t, \varpi) y^{(j)}(t-\tau) = u(t) \\ y^{(i)}(t) = \varphi_i(t), t \in [-\tau\ 0], i = 0, \dots, n-1, \end{cases} \tag{1}$$

where τ is a constant delay and $a_i(.)$, $b_i(.)$, $i = 1, \dots, n-1$ are nonlinear functions.

We start by representing the system (1), under another form. Using the following notation:

$$x_{i+1}(t) = y^{(i)}(t), i = 0, \dots, n-1 \tag{2}$$

we get:

$$\begin{cases} \dot{x}_i(t) = x_{i+1}(t) \quad i = 1, \dots, n-1 \\ \dot{x}_n(t) = -\sum_{i=0}^{n-1} a_i(.) x_i(t) - \sum_{i=0}^{n-1} b_i(.) x_i(t-\tau) \end{cases} \tag{3}$$

or under matrix form:

$$\dot{x}(t) = A(.)x(t) + B(.)x(t-\tau) \tag{4}$$

A(.) and B(.) are $n \times n$ matrices given by:

$$A(.) = \begin{bmatrix} 0 & 1 & \dots & 0 \\ \vdots & \ddots & \ddots & \vdots \\ 0 & 0 & \dots & 1 \\ -a_0(.) & -a_1(.) & \dots & -a_{n-1}(.) \end{bmatrix}, B(.) = \begin{bmatrix} 0 & \dots & 0 \\ \vdots & \dots & \vdots \\ 0 & \dots & 0 \\ -b_0(.) & \dots & -b_{n-1}(.) \end{bmatrix} \quad (5)$$

The regular basis change P transforms the original system to the new one defined by:

$$x(t) = Pz(t), \quad (6)$$

with:

$$P = \begin{bmatrix} 1 & 1 & \dots & 1 & 0 \\ \alpha_1 & \alpha_2 & \dots & \alpha_{n-1} & 0 \\ \vdots & \vdots & \dots & \vdots & \vdots \\ \alpha_1^{n-1} & \alpha_2^{n-1} & \dots & \alpha_{n-1}^{n-1} & 1 \end{bmatrix} \quad (7)$$

The new state space representation is:

$$\dot{z}(t) = F(.)z(t) + D(.)z(t\text{-}\tau) \quad (8)$$

with:

$$F(.) = P^{-1}A(.)P = \begin{bmatrix} \alpha_1 & & & & \beta_1 \\ & \alpha_2 & & & \beta_2 \\ & & \ddots & & \vdots \\ & & & \alpha_{n-1} & \beta_{n-1} \\ \gamma_1(.) & \gamma_2(.) & \dots & \gamma_{n-1}(.) & \gamma_n(.) \end{bmatrix} \quad (9)$$

Elements of the matrix F(.) are defined in [33] by:

$$\begin{cases} \gamma_i(.) = -p_A(\alpha_i, .) \text{ for } i = 1, \dots, n\text{-}1, \\ \gamma_n(.) = -a_{n-1}(.) - \sum_{i=1}^{n-1} \alpha_i \end{cases} \quad (10)$$

where

$$p_A(s, .) = s^n + \sum_{i=0}^{n-1} a_i(.)s^i \quad (11)$$

and

$$\beta_i = \left. \frac{\lambda\text{-}\alpha_i}{Q(\lambda)} \right|_{\lambda = \alpha_i} \text{ for } i = 1, \dots, n\text{-}1 \quad (12)$$

where

$$Q(\lambda) = \prod_{j=1}^{n-1} (\lambda\text{-}\alpha_j) \tag{13}$$

and the matrix D(.) is given by:

$$D(.) = P^{-1}B(.)P = \begin{pmatrix} O_{n-1,n-1} & & & O_{n-1,1} \\ \delta_1(.) & \dots & \delta_{n-1}(.) & \delta_n(.) \end{pmatrix} \tag{14}$$

Elements of the matrix D(.) are defined in [18] by:

$$\begin{cases} \delta_i(.) = \text{-}p_B(\alpha_i, .), i = 1, \dots, n\text{-}1 \\ \delta_n(.) = \text{-}b_{n-1}(.) \end{cases} \tag{15}$$

Based on this transformation and the arbitrary choice of parameters $\alpha_i, i = 1, \dots, n-1$ which play an important role in simplifying the use of aggregate techniques, we give now the main result. Let us start by writing our system in another form. By using the Newton-Leibniz formula

$$x(t\text{-}\tau) = \int_{t-\tau}^{t} \dot{x}(u)du \tag{16}$$

Equation (Eq. 8) becomes

$$\dot{z}(t) = (F(.) + D(.))z(t)\text{-}D(.)\int_{t-\tau}^{t} \dot{x}(\theta)d\theta \tag{17}$$

Let Ω be a domain of R^n, containing a neighborhood of the origin, and $\sup_{J_\tau, \Omega, S_\varpi}$ the suprema calculated for $t \in J_\tau (i.e\ t \geq \tau)$, for functions x with values in Ω, and for ϖ in S_ϖ.

Next, using the special form of system (Eq. (1)) and applying the notation $\sup_{J_\tau, \Omega, S_\varpi} = \sup_{[.]}$, we can announce the following theorem.

Theorem 2.1. The system (Eq. (1)) is asymptotically stable, if there exist distinct parameters $\alpha_i < 0, i = 1, \dots, n\text{-}1,$ such that the matrix $\tilde{F}(.)$ is the opposite of an M-matrix, where $\tilde{F}(.)$ is given by

$$\tilde{F}(.) = \begin{bmatrix} \alpha_1 & & & & |\beta_1| \\ & \alpha_2 & & & |\beta_2| \\ & & \ddots & & \vdots \\ & & & \alpha_{n-1} & |\beta_{n-1}| \\ \tilde{\gamma}_1(.) & \tilde{\gamma}_2(.) & \dots & \tilde{\gamma}_{n-1}(.) & \tilde{\gamma}_n(.) \end{bmatrix} \tag{18}$$

and the elements $\tilde{\gamma}_i(.), i = 1, \dots, n$, are given by

$$\begin{cases} \tilde{\gamma}_i(.) = \dfrac{|\gamma_i(.) + \delta_i(.)| + \tau|\alpha_i|\sup_{[.]} |\delta_i(.)|}{1 - \tau \sup_{[.]} |\delta_n(.)|}, \quad i = 1, \ldots, \text{n-1} \\ \tilde{\gamma}_n(.) = \gamma_n(.) + \delta_n(.) + \dfrac{\tau \sup_{[.]} |\delta_n(.)||\gamma_n(.) + \delta_n(.)|}{1 - \tau \sup_{[.]} |\delta_n(.)|} + \displaystyle\sum_{i=1}^{n} \dfrac{\tau |\beta_i| \sup_{[.]} |\delta_i(.)|}{1 - \tau \sup_{[.]} |\delta_n(.)|} \end{cases} \tag{19}$$

Proof:

We use the following vector norm $p(z) = \left(p_1(z)\, p_2(z)\, p_3(z) \ldots p_n(z) \right)'$, where

$$\begin{cases} p_i(z) = |z_i|, i = 1, \ldots, \text{n-1} \\ p_n(z) = |z_n| + \dfrac{\sum_{i=1}^{n} \sup_{[.]} |\delta_i(.)|}{1 - \tau \sup_{[.]} |\delta_n(.)|} \displaystyle\int_{-\tau}^{0} \int_{t+\theta}^{t} |\dot{z}_i(\vartheta)|\, d\vartheta d\theta \end{cases} \tag{20}$$

with the condition

$$\tau \sup_{[.]} |\delta_n(.)| < 1 \tag{21}$$

Let $V(t)$ be a radially unbounded Lyapunov function given by (Eq. (22)).

$$V(t) = \left\langle (p(z(t)))', w \right\rangle = \sum_{i=1}^{n} w_i p_i(z(t)) \tag{22}$$

where $w \in R_+^n$, $w_i > 0, i = 1, \ldots, n$. First, note that

$$V(t_0) \leq \sum_{i=1}^{n} w_i |z_i(t_0)| + w_n \left(|z_n(t_0)| + \frac{\sup_{[.]} (|\delta_n(.)|)}{1\text{-}\tau \sup_{[.]} (|\delta_n(.)|)} \sup_{[-\tau,0]} |\dot{\varphi}_n| \frac{\tau^2}{2} \right) := r < +\infty$$

and

$$V(t) \geq \sum_{i=1}^{n} w_i |z_i(t)|$$

The right Dini derivative of $V(t)$, along the solution of (Eq. (22)), gives

$$D^+V(t) = \sum_{i=1}^{n} w_i \frac{d^+ p_i(z(t))}{dt^+} \tag{23}$$

For clarification reasons, each element of $\frac{d^+ p_i(z(t))}{dt^+}$, i = 1, …, n is calculated separately. Let us begin with the first (n-1) elements. Because $|z_i| = z_i \text{sign}(z_i)$, we can write, for $i = 1, \ldots, \text{n-1}$,

$$\begin{aligned}\frac{d^+p_i(z(t))}{dt^+}=\frac{d^+|z_i(t)|}{dt^+} &=\frac{d^+z_i(t)}{dt^+}\text{sign}(z_i(t))\\ &=\alpha_i|z_i(t)|+\beta_i z_n(t)\text{sign}(z_i(t))\\ &\le\alpha_i|z_i(t)|+|\beta_i||z_n(t)|\end{aligned}\tag{24}$$

and

$$\frac{d^+p_n(z)}{dt^+}=\frac{d^+|z_n|}{dt^+}+\frac{\sum_{i=1}^{n}\sup_{[.]}|\delta_i(.)|}{1\text{-}\tau\sup_{[.]}|\delta_n(.)|}\frac{d^+}{dt^+}\int_{-\tau}^{0}\int_{t+\theta}^{t}|\dot z_i(v)|\,dv d\theta\tag{25}$$

because

$$\frac{\sum_{i=1}^{n}\sup_{[.]}|\delta_i(.)|}{1-\tau\sup_{[.]}|\delta_n(.)|}\frac{d^+}{dt^+}\int_{-\tau}^{0}\int_{t+\theta}^{t}|\dot z_i(\vartheta)|\,d\vartheta d\theta=\frac{\sum_{i=1}^{n}\sup_{[.]}|\delta_i(.)|}{1\text{-}\tau\sup_{[.]}(|\delta_n(.)|)}\left(\tau|\dot z_i(t)|-\int_{t-\tau}^{t}|\dot z_i(\vartheta)|d\vartheta\right)$$

and

$$\frac{d^+|z_n(t)|}{dt^+}\le(\gamma_n(.)+\delta_n(.))|z_n(t)|+\sum_{i=1}^{n\text{-}1}|\gamma_i(.)+\delta_i(.)||z_i(t)|+\sum_{i=1}^{n}\sup_{[.]}|\delta_i(.)|\int_{t\text{-}\tau}^{t}|\dot z_i(\theta)|d\theta$$

Finally, it is easy to see that equation (Eq. (25)) can be overvalued by the following one

$$\frac{d^+p_n(z)}{dt^+}\le\sum_{i=1}^{n}\tilde\gamma_i(.)|z_i|$$

Then we obtain the following inequality

$$D^+V(t)<\langle\tilde F(.)|z(t)|,w\rangle\tag{26}$$

where $|z(t)|=(\,|z_1(t)|\quad\ldots\quad|z_n(t)|\,)'$, and

$$\tilde F(.)=\begin{bmatrix}\alpha_1 & & & & |\beta_1|\\ & \alpha_2 & & & |\beta_2|\\ & & \ddots & & \vdots\\ & & & \alpha_{n\text{-}1} & |\beta_{n\text{-}1}|\\ \tilde\gamma_1(.) & \tilde\gamma_2(.) & \ldots & \tilde\gamma_{n\text{-}1}(.) & \tilde\gamma_n(.)\end{bmatrix}\tag{27}$$

Because the nonlinear elements of $\tilde F(.)$ are isolated in the last row, the eigenvector $v(t,x_t,\varpi)$ relative to the eigenvalue λ_m is constant [34, 35], where λ_m is such that $\text{Re}(\lambda_m)=\max_i\{\text{Re}(\lambda_i),\lambda_i\in\lambda(\tilde F(.))\}$. Then, in order to have $D^+V(t)<0$, it is sufficient to have $\tilde F(.)$ as the opposite of an M-matrix. Indeed, according to properties of M-matrices, we have $\forall\sigma \in R_{+}^{*n}$, $\exists w \in R_{+}^{*n}$ such that $-(\tilde{F}'(.))^{-1}\sigma = w$. This enables us to write the following equation

$$D^{+}V(t) < \left\langle \left(\tilde{F}(.)|z(t)|\right)', w \right\rangle = \left\langle |z(t)|', \tilde{F}'(.)w \right\rangle = \left\langle |z(t)|', -\sigma \right\rangle = -\sum_{i=1}^{n} \sigma_i |z_i(t)| < 0 \qquad (28)$$

This completes the proof of theorem.

Corollary 2.1. The system (Eq. (1)) is asymptotically stable, if there exist distinct parameters $\alpha_i < 0, i = 1, \ldots, n\text{-}1,$ such that the following condition:

$$\mu(.) + 2\tau\nu(.) - \xi(.) < 0 \qquad (29)$$

is satisfied.

where:

$$\begin{cases} \mu(.) = \gamma_n(.) + \delta_n(.) + \tau \sup_{[.]} |\delta_n(.)| \left(|\gamma_n(.) + \delta_n(.)| - (\gamma_n(.) + \delta_n(.)) \right) \\ \nu(.) = \sum_{i=1}^{n-1} |\beta_i| \sup_{[.]} |\delta_i(.)| \\ \xi(.) = \sum_{i=1}^{n-1} \frac{|\gamma_i(.) + \delta_i(.)| \, |\beta_i|}{\alpha_i} + \end{cases} \qquad (30)$$

Proof:

Basing on definition 1 and definition 2, the choice of $\alpha_k < 0$, $k = 1, \ldots, n\text{-}1$, $\alpha_i \neq \alpha_j$ for $i \neq j$, the condition of signs on the principal minors is as follows

$$\det \begin{pmatrix} -\alpha_1 & & 0 \\ & \ddots & \\ 0 & & -\alpha_i \end{pmatrix} > 0 \,, \quad (i = 1, 2, 3, \ldots, n\text{-}1) \qquad (31)$$

and

$$\det\left(-\tilde{F}(.)\right) = -\left(\tilde{\gamma}_n(.) - \sum_{i=1}^{n-1} \frac{\tilde{\gamma}_i(.)|\beta_i|}{\alpha_i}\right) \prod_{i=1}^{n-1} (-\alpha_i) > 0 \qquad (32)$$

which yields to the following condition

$$\tilde{\gamma}_n(.) - \sum_{i=1}^{n-1} \frac{\tilde{\gamma}_i(.)|\beta_i|}{\alpha_i} < 0 \qquad (33)$$

Replacing each term in (Eq. (33)) of by its expression we get

$$\tilde{\gamma}_n(.)-\sum_{i=1}^{n-1}\frac{\tilde{\gamma}_i(.)|\beta_i|}{\alpha_i}:=\gamma_n(.)+\delta_n(.)+\frac{\tau\sup\limits_{[.]}|\delta_n(.)||\gamma_n(.)+\delta_n(.)|}{1-\tau\sup\limits_{[.]}|\delta_n(.)|}+\frac{\tau\sum\limits_{i=1}^{n-1}|\beta_i|\sup\limits_{[.]}|\delta_i(.)|}{1-\tau\sup\limits_{[.]}|\delta_n(.)|}$$

$$-\sum_{i=1}^{n-1}\frac{\left(|\gamma_i(.)+\delta_i(.)|+\tau|\alpha_i|\sup\limits_{[.]}|\delta_i(.)|\right)|\beta_i|}{\left(1-\tau\sup\limits_{[.]}|\delta_n(.)|\right)\alpha_i}$$

$$=\left(1-2\sup_{[.]}|\delta_n(.)|\right)(\gamma_n(.)+\delta_n(.))+\tau\sum_{i=1}^{n-1}|\beta_i|\sup_{[.]}|\delta_i(.)|$$

$$-\sum_{i=1}^{n-1}\frac{\left(|\gamma_i(.)+\delta_i(.)|-\tau\alpha_i\sup\limits_{[.]}|\delta_i(.)|\right)|\beta_i|}{\alpha_i}$$

which can be re-written as:

$$\begin{aligned}&\mu(.)+\tau\ \nu(.)-\sum_{i=1}^{n-1}\frac{|\gamma_i(.)+\delta_i(.)||\beta_i|}{\alpha_i}-\sum_{i=1}^{n-1}\frac{-\tau\alpha_i\sup\limits_{[.]}|\delta_i(.)||\beta_i|}{\alpha_i}\\&=\mu(.)+\tau\ \nu(.)-\xi(.)+\tau\ \nu(.)\\&=\mu(.)+2\tau\ \nu(.)-\xi(.)\end{aligned}$$

where:

$$\begin{cases}\mu(.)=(1-2\tau\sup\limits_{[.]}(\delta_n(.))(\gamma_n(.)+\delta_n(.))\\ \nu(.)=\sum\limits_{i=1}^{n-1}|\beta_i|\sup\limits_{[.]}|\delta_i(.)|\\ \xi(.)=\sum\limits_{i=1}^{n-1}\frac{|\gamma_i(.)+\delta_i(.)||\beta_i|}{\alpha_i}\end{cases}$$

which completes the proof.

Remark 2.1. If the couple $(p_A(s,.)+p_B(s,.),Q(s))$ forms a positive pair, then there exist distinct negative parameters α_i, $i=1,\ldots,n\text{-}1$, verifying the condition $(\gamma_i(.)+\delta_i(.))\,\beta_i>0$ for $i=1,\ldots,n\text{-}1$.

Using Theorem 2.1 and Remark 2.1, the obtained supremum of time delay is a function of α_i values, $i=1,\ldots,n\text{-}1$. As a result, a sufficient condition for asymptotic stability of our system is when values of the time delay are less than this supremum.

Corollary 2.1. If the couple $(D(s,.)+N(s,.),Q(s))$ forms a positive pair and there exist distinct negative parameters α_i, i = 1, …, n-1, such that:

$$2\tau\left(\left(\gamma_n(.)+\delta_n(.)\right)\sup[.]\,|\delta_n(.)|\text{-}\nu(.)\right)+\frac{D(0,.)+N(0,.)}{Q(0)}>0 \tag{34}$$

then the system (Eq. (1)) is asymptotically stable.

Proof.

According to Remark 2.1, we find that

$$\begin{aligned}\gamma_n(.)+\delta_n(.)\text{-}\sum_{j=1}^{n-1}\frac{\left|\gamma_j(.)+\delta_j(.)\right|\left|\beta_j\right|}{\alpha_j}&=\gamma_n(.)+\delta_n(.)\text{-}\sum_{j=1}^{n-1}\frac{\left(\gamma_j(.)+\delta_j(.)\right)\beta_j}{\alpha_j}\\&=-\frac{D(0,.)+N(0,.)}{Q(0)}\end{aligned}$$

The result of Theorem 2.1 becomes

$$2\tau\left(\left(\gamma_n(.)+\delta_n(.)\right)\sup[.]\,|\delta_n(.)|\text{-}\nu(.)\right)+\frac{D(0,.)+N(0,.)}{Q(0)}>0$$

This completes the proof of corollary.

Remark 2.2

- Theorem 2.1 depends on the new basis change, where parameters α_i of the matrix P are arbitrary chosen such that matrix $T(.)$ is the opposite of an M-matrix. The appropriate choice of the set of free parameters α_i makes the given stability conditions satisfied.

- The theorem takes into account the fact that delayed terms may stabilize our system. Theorem 2.1 can hold even if $p_A(s,.)$ is unstable. This is another advantage as the majority of previously published results assume that $p_A(s,.)$ is linear and stable.

4. Application to delayed nonlinear nth order all pole plant

Consider the complex system S given in **Figure 1**.

$D(s)=p_A(s)$ defined by (Eq. (11)) and $p_B(s)=1$, respectively. In this case $\tilde{f}_i(.)$ are constants and g is a function satisfying the finite sector condition.

Let $\hat{g}$ be a function defined as follows

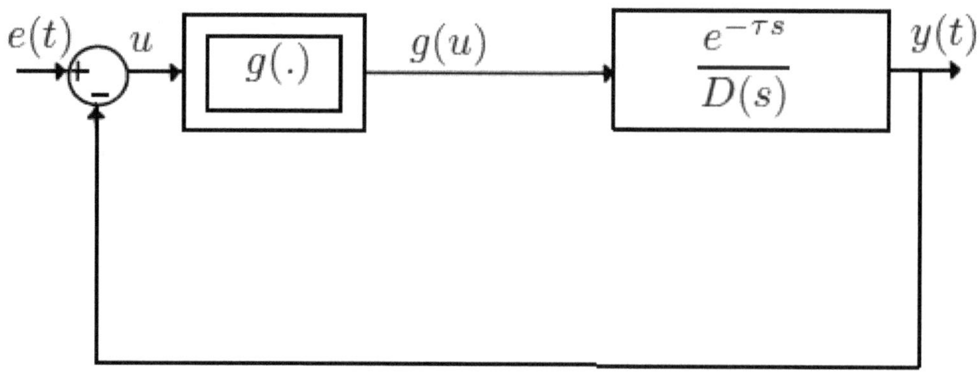

Figure 1. Block diagram of studied system.

$$\hat{g}(e(\theta), y(\theta)) = \frac{g(e(\theta) - y(\theta))}{e(\theta) - y(\theta)}, e(\theta) \neq y(\theta) \ \forall \theta \in [-\tau + \infty[\tag{35}$$

$$\sup_{[.]} |\hat{g}(e(t), y(t))| = \bar{g} \in R_+^*.$$

The presence of delay in the system of **Figure 1** makes stability study difficult. The following steps show how to represent this system in the form of system (Eq. (1)). Then we can write

$$y^{(n)}(t) + \sum_{i=0}^{n-1} a_i \frac{d^i y(t)}{dt^i} = -\hat{g}(e(t-\tau), y(t-\tau))y(t-\tau) + \hat{g}(e(t-\tau), y(t-\tau))e(t-\tau).$$

Using the following notation $\hat{g}(.) = \hat{g}(e(t-\tau), bx(t-\tau))$, therefore

$$y^{(n)}(t) + \sum_{i=0}^{n-1} a_i y^{(i)}(t) + \hat{g}(.)y(t-\tau) = \hat{g}(.)e(t-\tau). \tag{36}$$

It is clear that system (Eq. (36)) is equivalent to system (Eq. (1)) in the special cases $e(\theta) = 0$ and $e(\theta) = -Kx(\theta)$, $x(t) = \left(y(t), \dot{y}(t), \ldots, y^{(n)}(t)\right)'$, $\forall \ \theta \in [-\tau + \infty[$. We will now consider each case separately.

4.1. Case $e(t) = 0$

In case, $e(t) = 0 \ \ \forall t \in [-\tau + \infty[$, the description of the system becomes

$$y^{(n)}(t) + \sum_{i=0}^{n-1} a_i y^{(i)}(t) + \hat{g}(.)y(t-\tau) = 0.$$

This is a special representation of system (Eq. (1)) where $\tilde{f}_i(.) = a_i$, $\tilde{g}_1(.) = \hat{g}(.)$ $\tilde{g}_i(.) = 0$ $\forall$ $i = 2, \ldots, n-1$, $D(s,.) = D(s)$, $N(s,.) = \hat{g}(.)$, $\gamma_n(.) = \gamma_n = -a_{n-1} - \sum_{i=1}^{n-1} \alpha_i$ and $\delta_n(.) = 0$.

A sufficient stability condition for this system is given in the following proposition.

Proposition 4.1. If there exist distinct $\alpha_i < 0\ i = 1, \ldots, n-1$, such that the following conditions

$$\begin{cases} \gamma_n < 0 \\ \mu_1(.) + 2\tau\nu_1(.) - \xi_1(.) < 0 \end{cases} \tag{37}$$

where

$$\begin{cases} \mu_1(.) = \gamma_n \\ \nu_1(.) = \overline{g} \\ \xi_1(.) = \dfrac{|D(\alpha_1) + \widehat{g}(.)||\beta_1|}{\alpha_1} + \displaystyle\sum_{i=2}^{n-1} \dfrac{|D(\alpha_i)||\beta_i|}{\alpha_i} \end{cases} \tag{38}$$

are satisfied. Then the system S is asymptotically stable.

Suppose that $D(s)$ admits n distinct real roots $p_i,\ i = 1, \ldots, n$ among which there are $n-1$ negative ones. By using the fact that $a_{n-1} = -\sum_{i=1}^{n} p_i$, then the choice $\alpha_i = p_i,\ \forall i = 1, .., n-2$ and $\alpha_{n-1} = p_{n-1} + \varepsilon$ permit us to write $\gamma_n = -a_{n-1} - \sum_{i=1}^{n-1} p_i = p_n - \varepsilon$. In this case the last proposition becomes.

Proposition 4.2. If $D(s)$ admits $n-1$ distinct real negative roots such that the following conditions

$$\begin{cases} p_n - \varepsilon < 0 \\ \mu_2(.) + 2\tau\nu_2(.) - \xi_2(.) < 0 \end{cases} \tag{39}$$

are satisfied, where

$$\begin{cases} \mu_2(.) = p_n - \varepsilon \\ \nu_2(.) = \overline{g} \\ \xi_2(.) = \dfrac{|\widehat{g}(.)||\beta_1|}{\alpha_1} + \dfrac{|D(\alpha_{n-1})||\beta_{n-1}|}{\alpha_{n-1}} \end{cases} \tag{40}$$

then the system S is asymptotically stable.

4.2. Case $e(t) = -Kx(t)$

In this case, take $e(t) = -Kx(t)$ with $K = (k_0, k_1, \ldots, k_{n-1})$, then the obtained system has the same form as (Eq. (1)), with $\widehat{g}_1^K(.) = \widehat{g}^K(.)(k_0 + 1)$ and $\widehat{g}_i^K(.) = \widehat{g}^K(.)k_{i-1}$, $i = 2, \ldots, n$.

The stabilizing values of K can be obtained by making the following changes:

$$\gamma_n = -a_{n-1} - \sum_{i=1}^{n-1} \alpha_i, \delta_n^K(.) = -\widehat{g}^K(.)k_{n-1}, \nu_1^K(.) = \overline{g}^K \sum_{i=1}^{n-1} \left|\tilde{N}(\alpha_i)\right| \text{ where } \overline{g}^K = \sup_{[.]} |\widehat{g}^K(.)|$$

$$\text{and } \tilde{N}(\alpha) = (1+k_0) + \sum_{i=1}^{n-1} (b_i + k_i)\alpha^i.$$

Proposition 4.3. If there exist distinct $\alpha_i < 0$, $i = 1, \ldots, n-1$, such that the following conditions

$$\begin{cases} \gamma_n - \widehat{g}K(.)k_{n-1} < 0 \\ \tau < \dfrac{1}{2\overline{g}^K |k_{n-1}|} \\ \mu_1^K(.) + 2\tau\nu_1^K(.) - \xi_1^k(.) < 0 \end{cases} \tag{41}$$

where

$$\begin{cases} \mu_1^K(.) = \left(1 - 2\overline{g}^K \tau |k_{n-1}|\right)\left(\gamma_n + \delta_n^K(.)\right) \\ \nu_1^K(.) = \overline{g}^K \displaystyle\sum_{i=1}^{n-1} |\beta_i||\tilde{N}(\alpha_i)| \\ \xi_1^K(.) = \displaystyle\sum_{i=1}^{n-1} |D(\alpha_i) + \widehat{g}^K(.)\frac{\tilde{N}(\alpha_i)||\beta_i|}{\alpha_i} \end{cases} \tag{42}$$

are satisfied. Then the system S is asymptotically stable.

By a special choice of K the result of proposition 3.3 can be simplified. In fact, if the conditions of this proposition are verified we can choose the vector K such that $D(p_i) = \tilde{N}(p_i)$. In this case we obtain $D(p_i) = \tilde{N}(p_i) = 0$, $\forall\, i = 1, \ldots, n-1$ and $\nu_1(.) = \xi_1(.) = 0$ which yields the following new proposition.

Proposition 4.4. If $D(s)$ admits $n-1$ distinct real negative roots p_i such that the following conditions are satisfied.

$$\begin{cases} \gamma_n - \widehat{g}^K(.)k_{n-1} < 0 \\ \tau < \dfrac{1}{2\overline{g}^K |k_{n-1}|} \\ \mu_1^K(.) < 0 \end{cases} \tag{43}$$

Then the system S is asymptotically stable.

5. Illustrative example

Let us study the same example in [34] defined by **Figure 2** which refer to the dynamics of a time-delayed DC motor speed control system with nonlinear gain, Block diagram of time-delayed DC motor speed control system with nonlinear gain.

where:

- $p_1 = \frac{1}{T_e}$ and $p_2 = \frac{1}{T_m}$ where T_e and T_m are, respectively, electrical constant and mechanical constant.
- τ_f presents the feedback delay between the output and the controller. This delay represents the measurement and communication delays (sensor-to-controller delay).
- τ_c the controller processing and communication delay (controller-to-actuator delay) is placed in the feedforward part between the controller and the DC motor.
- $g(.) : R \to R$ is a function that represents a nonlinear gain.

The process of **Figure 2** can also be modeled by **Figure 1**, where $\tau = \tau_f + \tau_c$.

It is clear that model of **Figure 2** is a particular form of delayed Lurie system in the case where $D(s) = s(s+p_1)(s+p_2) = s^3 + (p_1+p_2)s^2 + p_1p_2s$ and $N(s) = 1$. Thereafter, applying the result of Theorem 2.1, a stability condition of the system is that the matrix $T(.)$ given by:

$$T(.) = \begin{pmatrix} \alpha_1 & 0 & |(\alpha_1 - \alpha_2)^{-1}| \\ 0 & \alpha_2 & |(\alpha_2 - \alpha_1)^{-1}| \\ t_1(.) & t_2(.) & t_3(.) \end{pmatrix}$$

where:

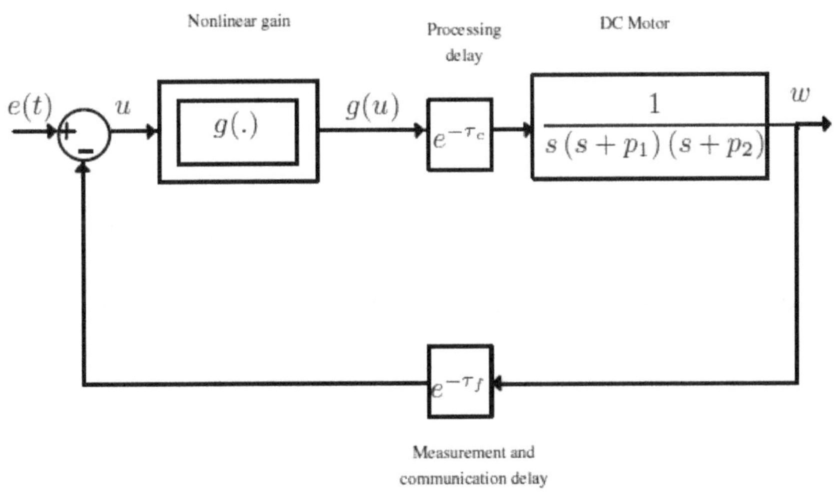

Figure 2. Delayed nonlinear model of DC motor speed control.

$$t_1(.) = |\gamma_1 + \widehat{g}(.)| + \tau|\alpha_1|\overline{g}, \;\; t_2(.) = |\gamma_2|, \;\; t_3(.) = \gamma_3 + \tau|\beta_1|\overline{g}$$

must be the opposite of an M-matrix. By choosing α_i, $i = 1, 2$, negative real and distinct, we get the following stability condition:

$$\gamma_3 + 2\tau|\beta_1|\overline{g} - \frac{|\beta_1||\gamma_1 + \widehat{g}(.)|}{\alpha_1} - \frac{|\beta_2||\gamma_2|}{\alpha_2} < 0$$

For the particular choice of $\alpha_1 = -p_1$ and $\alpha_2 = -p_2 + \varepsilon$, $\varepsilon > 0$.

yields $|\beta_1| = |\beta_2| = |(\varepsilon + p_1 - p_2)^{-1}|$ and we obtain the new stability condition:

$$2\tau\overline{g} + |p_1|^{-1}|\widehat{g}(.)| + |\alpha_2|^{-1}|D(\alpha_2)| < \varepsilon|\varepsilon + p_1 - p_2|$$

Assume that we have this inequality $\overline{g} < |D(\alpha_2)|$, we can find from \ref.{ops} the stabilizing delay given by the following condition:

$$\tau < \frac{1}{2}\left(\frac{\varepsilon|\varepsilon + p_1 - p_2|}{|D(\alpha_2)|} - |p_1|^{-1}| - |\alpha_2|^{-1}\right) \tag{44}$$

By applying the control $e(t) = -Kx(t)$ with $K = (k_0, k_1, k_2)$, we can determine the stabilizing values of K can be obtained by making the following changes:

$$\gamma_3 = -(p_1 + p_2) - \sum_{i=1}^{2} \alpha_i, \delta_1^K(.) = -\widehat{g}^K(.)(k_0 + 1), \delta_i^K(.) = -\widehat{g}^K(.)k_{i-1}, \;\; i = 2, 3$$

$$v_1^K(.) = \overline{g}^K \sum_{i=1}^{2} |\beta_i||\tilde{N}(\alpha_i)| \text{ where } \overline{g}^K = \sup_{[.]} |\widehat{g}^K(.)| \text{ and } \tilde{N}(\alpha) = 1 + k_0 + \sum_{i=1}^{2} k_i\alpha^i$$

If we choose $\alpha_i < 0$, $i = 1, 2$, such that the following conditions

$$D(\alpha_i) = \tilde{N}(\alpha_i) = 0, \forall, i = 1, 2$$

we get

$$\frac{1 + k_0}{k_2} = p_1 + p_2, \;\; \frac{k_1}{k_2} = p_1 p_2$$

and from proposition 3.3 the stabilizing gain values satisfying the following relations:

$$\begin{cases} 0 - \overline{g}^K(.)k_2 < 0 \\ |k_2| < \dfrac{1}{2\tau\overline{g}^K} \end{cases} \tag{45}$$

Finally we find the domain of stabilizing k_0, k_1, k_2 as follows:

$$\begin{cases} 0 < k_2 < \dfrac{1}{2\tau \overline{g}^K} \\ k_1 = p_1 p_2 k_2 \\ \text{and} \\ k_0 = (p_1 + p_2)k_2 - 1 \end{cases} \tag{46}$$

6. Conclusion

In this chapter, a joined and structured procedure for the analysis of delayed nonlinear systems is proven. A complete structured analysis formulation based on the comparison principle and vector norms for the asymptotic stability is presented. Based on the arrow form matrices, and by taking into account for the system parameters, a new stability conditions are synthesized, leading to a practical estimation of the stability domain. In order to highlight the feasibility and the main capabilities of the proposed approach, the case of nonlinear nth order all pole plant and delayed Lur'e Postnikov systems are presented and discussed. In addition, the simplicity of the application of these criteria is demonstrated on model of time-delayed DC motor speed control.

Author details

Sami Elmadssia[1]* and Mohamed Benrejeb[2]

*Address all correspondence to: sami.elmadssia@enit.rnu.tn

1 Higher Institute of Applied Sciences and Technology, Electrical Engineering, Gafsa, Tunisia

2 National Engineering School of Tunis, Tunis, Tunisia

References

[1] Bellman R, Cooke KL. Asymptotic Behavior of Solutions of Differential-Difference Equations. Providence, RI: American Mathematical Society; 1959

[2] Benrejeb M, Abdelkrim MN. On order reduction and stabilization of TSK nonlinear fuzzy models by using arrow form matrix. Systems Analysis Modelling Simulation. 2003;**43**(7): 977-991

[3] Bliman P-A. Extension of Popov absolute stability criterion to non-autonomous systems with delays. International Journal of Control. 2000;**73**(15):1349-1361

[4] Bliman P-A. Lyapunov-Krasovskii functionals and frequency domain: Delay-independent absolute stability criteria for delay systems. International Journal of Robust and Nonlinear Control. 2001;**11**(8):771-788

[5] Chen G, Liu ST. Linearization, stability, and oscillation of the discrete delayed logistic system. IEEE Transactions on Circuits and Systems I: Fundamental Theory and Applications. 2003;**50**(6):822-826

[6] Datta KB. Stability of pulse-width-modulated feedback systems†. International Journal of Control. 1972;**16**(5):977-983

[7] Dugard L, Verriest EI. Stability and Control of Time-Delay Systems. London: Springer; 1998

[8] Gu K, Kharitonov V, Chen J. Stability of Time-Delay Systems. Boston: Birkhäuser; 2003

[9] Hahn W. Stability of Motion. Berlin: Springer; 1967

[10] Hale JK. Theory of Functional Differential Equations. New York: Springer-Verlag; 1977

[11] Han Q-L. Absolute stability of time-delay systems with sector-bounded nonlinearity. Automatica. 2005;**41**(12):2171-2176

[12] Han Q-L. A new delay-dependent absolute stability criterion for a class of nonlinear neutral systems. Automatica. 2008;**44**(1):272-277

[13] He Y, Wu M. Absolute stability for multiple delay general lure control systems with multiple nonlinearities. Journal of Computational and Applied Mathematics. 2003;**159**(2): 241-248

[14] He Y, Wu M, She J-H, Liu G-P. Robust stability for delay lure control systems with multiple nonlinearities. Journal of Computational and Applied Mathematics. 2005;**176**(2): 371-380

[15] Hu H, Wang Z. Dynamics of Controlled Mechanical Systems with Delayed Feedback. 2002

[16] Kamen E. On the relationship between zero criteria for two-variable polynomials and asymptotic stability of delay differential equations. IEEE Transactions on Automatic Control. 1980;**25**(5):983-994

[17] Kharitonov V, Niculescu S-I. On the stability of linear systems with uncertain delay. In: Proceedings of the 2002 American Control Conference (IEEE Cat NoCH37301); 2002

[18] Liu P-L. Delayed decomposition approach to the robust absolute stability of a lure control system with time-varying delay. Applied Mathematical Modelling. 2016;**40**(3):2333-2345

[19] Liu P- L. Robust absolute stability criteria for uncertain Lurie interval time-varying delay systems of neutral type. ISA Transactions. 2016;**60**:2-11

[20] Malek-Zavarei M, Jamshidi M. Sensitivity of linear time-delay systems to parameter variations. Automatica. 1975;**11**(3):315-319

[21] Michiels W, Engelborghs K, Vansevenant P, Roose D. Continuous pole placement for delay equations. Automatica. 2002;**38**(5):747-761

[22] Michiels W, Assche VV, Niculescu S-I. Stabilization of time-delay systems with a controlled time-varying delay and applications. IEEE Transactions on Automatic Control. 2005;**50**(4):493-504

[23] Niculescu S-I, Verriest EI, Dugard L, Dion J-M. Stability and robust stability of time-delay systems: A guided tour. In: Stability and Control of Time-Delay Systems. Lecture Notes in Control and Information Sciences. pp. 1-71

[24] Nuño E, Basañez L, Ortega R. Control of Teleoperators with time-delay: A Lyapunov approach. Topics in Time Delay Systems. Lecture Notes in Control and Information Sciences. 2009:371-781

[25] Richard J-P. Time-delay systems: An overview of some recent advances and open problems. Automatica. 2003;**39**(10):1667-1694

[26] Saadaoui K, Elmadssia S, Benrejeb M. Stabilizing PID controllers for a class of time delay systems. In: PID Controller Design Approaches - Theory, Tuning and Application to Frontier Areas. 2012

[27] Shujaee K, Lehman B. Vibrational feedback control of time delay systems. IEEE Transactions on Automatic Control. 1997;**42**(11):1529-1545

[28] Wang Q, Lam J, Xu S, Gao H. Delay-dependent and delay-independent energy-to-peak model approximation for systems with time-varying delay. International Journal of Systems Science. 2005;**36**(8):445-460

[29] Lehman B, Shujaee K. Delay independent stability conditions and decay estimates for time-varying functional differential equations. IEEE Transactions on Automatic Control. 1994;**39**:1673-1676. DOI: 10.1109/9.310048

[30] Elmadssia S, Saadaoui K, Benrejeb M. New delay-dependent stability conditions for linear systems with delay. In: 2011 International Conference on Communications, Computing and Control Applications (CCCA); 2011

[31] Elmadssia S, Saadaoui K, Benrejeb M. New delay-dependent stability conditions for linear systems with delay. Systems Science & Control Engineering. 2013;**1**(1):2-11

[32] Yang R-M, Wang Y-Z. Stability analysis and estimate of domain of attraction for a class of nonlinear time-varying delay systems. Acta Automatica Sinica. 2012;**38**(5):716-724

[33] Elmadssia S, Saadaoui K, Benrejeb M. New stability conditions for nonlinear time delay systems. In: 2013 International Conference on Control, Decision and Information Technologies (CoDIT); 2013

[34] Elmadssia S, Saadaoui K, Benrejeb M. Stability conditions for a class of nonlinear time delay systems. Nonlinear Dynamics and Systems Theory. 2014;**14**(3):279-291

[35] Elmadssia S, Saadaoui K, Benrejeb M. New stability conditions for nonlinear time varying delay systems. International Journal of Systems Science. 2014;**47**(9):2009-2021

[36] Gil M. On Aizerman-Myshkis problem for systems with delay. Automatica. 2000;**36**(11): 1669-1673

6

Appell-Gibbs Approach in Dynamics of Non-Holonomic Systems

Jiří Náprstek and Cyril Fischer

Abstract

Hamiltonian functional and relevant Lagrange's equations are popular tools in the investigation of dynamic systems. Various generalizations enable to extend the class of problems concerned slightly beyond conventional limits of Hamiltonian system. This strategy is very effective, particularly concerning two-dimensional (2D) and simpler three-dimensional (3D) systems. However, the governing differential systems of most non-holonomic 3D systems suffer from inadequate complexity, when deduced using this way. Any analytical investigation of such a governing system is rather impossible and its physical interpretation can be multivalent. For easier analysis, particularly of systems with non-holonomic constraints, the Appell-Gibbs approach seems to be more effective providing more transparent governing systems. In general, the Appell-Gibbs approach follows from the Gaussian fifth form of the basic principle of dynamics. In this chapter, both Lagrangian and Appell-Gibbs procedures are shortly characterized and later their effectiveness compared on a particular dynamic system of a ball moving inside a spherical cavity under external excitation. Strengths and shortcomings of both procedures are evaluated with respect to applications.

Keywords: Appell-Gibbs function, Lagrangian approach, non-holonomic systems, engineering applications

1. Introduction

The energy contained in a dynamic system is given by a scalar potential $\mathcal{E}(t)$. It is a function of time and system response components (displacement, velocity, and acceleration vectors). Moreover, $\mathcal{E}(t)$ is a function of system parameters, position in a field of forces (potential or not), internal sources of energy and of the system evolution including a residual energy. The total energy of the system increases or decreases accordingly with external excitation and dissipation of energy. The form of energy contained within the system can have a deterministic or stochastic character and similarly also excitation and dissipation.

Considering the mechanical energy only, the total energy increase/decrease of the system with respect to time should be in equilibrium with the energy supplies and energy losses due to dissipation. This relation can be outlined by the following equilibrium:

$$\frac{\mathrm{d}}{\mathrm{d}t}\{\mathcal{E}(t)\} = \mathcal{P}(t) + \mathcal{S}(t), \tag{1}$$

where $\mathcal{P}(t)$ is power supply (excitation energy per unity time) and $\mathcal{S}(t)$ the specific dissipation of energy also per unity time (supposed to be independent on accelerations $\ddot{\mathbf{x}}$). Functions $\mathcal{P}(t), \mathcal{S}(t)$ can dispose in special cases with a superior potential, which, however, cannot be incorporated into the potential part of total energy. Eq. (1) has a scalar character.

The energy is a primary value characterizing the system state and its evolution in time. The function $\mathcal{E}(t)$ and external influences are a background for the derivation of a governing differential system characterizing the system response with respect to initial and boundary conditions. The governing differential system is then deduced from the equivalence of Eq. (1) type using an adequate variational principle. It claims that the form of the system response corresponds with the minimum of energy spent among all admissible shapes of the system reaction. Take a note that many important settings of external forces and dissipation mechanisms do not admit the formulation by means of potentials. In such cases, they should be incorporated separately into the governing differential system using complementary principles and theorems, for example, virtual works, and so on.

We can find in monographs, for example, [1–4, 5] and many others, various formulations of potentials $\mathcal{E}(t)$ and functions $\mathcal{P}(t), \mathcal{S}(t)$ combining the system parameters (physical and geometric) and the system response vectors $\mathbf{x}$-displacements, $\dot{\mathbf{x}}$-velocities, and $\ddot{\mathbf{x}}$-accelerations. They can be selected in individual cases with respect to physical or geometric complexity of the system, components of the response, which are to be found, deterministic or stochastic character of the system and its excitation, and so on.

2. Basic considerations

Approaches commonly applied to construct mathematical models of dynamic systems with multiple degrees of freedoms (MDOF) follow mostly from principles symbolically outlined by Eq. (1). The equation of this type can be deduced using, for instance, a procedure of virtual displacements. They balance the energy flow in every step and subsequently applied minimization steps try to select such response trajectories, which represent a minimum of energy consumption among all admissible shapes. Let us get briefly through Lagrangian and Appell-Gibbs procedures in order to compare their basic properties. Later, we recognize that most of these properties can be regarded as positive or negative in dependence on a particular problem. Therefore, the solution method should be selected in every particular case very sensitively.

Let us remember that the aim of this chapter is a comparison of Lagrangian and Appell-Gibbs approaches effectiveness to process dynamic systems in holonomic and non-holonomic settings and to help estimate which one is more suitable to be employed in a particular case. Despite that the most important features of non-holonomic systems themselves are briefly treated as well, but for thorough evaluation of their properties, special literature should be addressed. Except five monographs cited in introductory section containing a large number of additional relevant references, a vast number of papers have been published concerning the investigation of various properties of non-holonomic systems.

Motion of an MDOF system with n degrees of freedom can be described by a system of n differential equations and l constraints:

$$\begin{aligned} & m_s\ddot{x}_s = X_s + \sum_{r=1}^{l}\lambda_r A_{rs}, \quad s = 1,..,n, \quad \mathbf{x} = [x_s],\ \mathbf{X} = |X_s|,\quad \mathbf{x},\mathbf{X}\in\mathbb{R}^n, \quad (a) \\ & \sum_{s=1}^{n} A_{rs}\dot{x}_s + B_r = 0, \quad r = 1,..,l, \quad \boldsymbol{\lambda} = [\lambda_r],\ \mathbf{B} = [B_r],\quad \boldsymbol{\lambda},\mathbf{B}\in\mathbb{R}^l, \quad (b) \\ & \mathbf{A} = [A_{rs}],\quad \mathbf{A}\in\mathbb{R}^{l\times n}. \end{aligned} \tag{2}$$

Vector $\mathbf{X}$ represents external forces, while $\boldsymbol{\lambda}$ are unknown multipliers. The summation in Eq. (2a) characterizes influence of constraints (holonomic and non-holonomic) related with constraints (Eq. (2b)). These constraints reduce the number of the original degrees of freedom from n to $k = n - l$. The system (Eq. (2)) includes $n + l$ differential equations for $\mathbf{x}$ and $\boldsymbol{\lambda}$ unknown functions t, which can be determined, provided $\mathbf{x}, \dot{\mathbf{x}}$ are given in an initial point t_0. If the system (Eq. (2b)) is fully integrable, it provides l functions $f_r = f_r(\mathbf{x}, t), r = 1,..,l$ and constraints can be formulated as $f_r = f_r(\mathbf{x}, t) = c_r$. They are exclusively of a geometric character and the system is holonomic. Corresponding constraints are formulated in displacements only. In principle, l components of $\mathbf{x}$ can be eliminated and then remains to analyze the system with $n - l$ unknowns. Then, it can be considered $\boldsymbol{\lambda} \equiv 0$, and the second part on the right side of Eq. (2a) vanishes. The system with holonomic constraints takes the form:

$$m_s\ddot{x}_s = X_s, \quad f_r = f_r(\mathbf{x}, t) = c_r, \quad s = 1,..,k, \quad r = 1,..,l, \quad k = n - l. \tag{3}$$

However, frankly speaking, such an operation is possible rather exceptionally. In general, the full form of Eq. (2) should be treated, despite the system is holonomic. If some (or all) of constraints (Eq. (2b)) are not integrable, then the system is non-holonomic. In practice, we encounter these cases when the formulation of constraints includes velocities (more often velocities only).

We should remember that the non-holonomic constraints introduced in Eq. (2b) represent the most simple version of such constraints, as they are linear and in velocity. Many applications, for example, robotics, wind engineering, automotive systems, plasma physics, and so on, present more complicated types of non-holonomic constraints. Notifications to nonlinear non-holonomic constraints in velocity are given in elderly monographs [2, 3]. Later, many papers have appeared presenting results of systematic research at this field originating from particular physical or engineering problems, for example, [6–8], where higher derivatives of velocities in non-holonomic constraints are discussed. These attributes have been reflecting also in pure mathematical studies with respect to control theory and systems with a delayed feedback, see, for example, series [9–11] dealing with generalized Lagrange-d'Alembert-Poincaré equations or other studies devoted to non-holonomic reduction and related problems, see, for example, [12] and many others.

Let us realize now that the virtual work of every constraint force should vanish in the meaning as follows:

$$\lambda_r \sum_{s=1}^{n} A_{rs}\delta x_s = 0, \quad r = 1, .., l. \tag{4}$$

Therefore, we have with respect to Eq. (2a):

$$\sum_{s=1}^{n} (m_s\ddot{x}_s - X_r)\delta x_s = 0. \tag{5}$$

This equation holds for any arbitrary virtual displacements and represents a generalization of the principle of virtual works in statics and of the d'Alembert principle. The important issue is that it does not include any reactions of constraints. It has been well investigated in the study. For many details, see monographs, for example, [1, 3] and many others.

Let us consider that velocities in constraints Eq. (2b) are increased by virtual increments $\delta\dot{\mathbf{x}}$, so that they read

$$\sum_{s=1}^{n} A_{rs}(\dot{x}_s + \delta\dot{x}_s) + B_r = 0, \quad r = 1, 2, .., l, \tag{6}$$

Deducting from Eq. (6), the initial state (Eq. (2b)) holds

$$\sum_{s=1}^{n} A_{rs}\delta\dot{x}_s = 0, \quad r = 1, 2, .., l. \tag{7}$$

Virtual increments of velocities $\delta\dot{\mathbf{x}}$ fit into constraints requested for constraints (Eq. (4)) and, consequently, in Eq. (5) the δx_s can be replaced by $\delta\dot{x}_s$:

$$\sum_{s=1}^{n} (m_s\ddot{x}_s - X_s)\delta\dot{x}_s = 0. \tag{8}$$

We revisit Eq. (2b) and perform differentiation with respect to t:

$$\sum_{s=1}^{n} \left(A_{rs}\ddot{x}_s + \frac{\mathrm{d}A_{rs}}{\mathrm{d}t}\dot{x}_s \right) + \frac{\mathrm{d}B_r}{\mathrm{d}t} = 0, \quad r = 1, .., l, \tag{9}$$

where $\mathrm{d}/\mathrm{d}t$ represents the operator $\partial/\partial t + \sum_{i=1}^{n} \dot{x}_i \partial/\partial x_i$. Considering two possible movements of the system in identical initial state and velocities in time t, but with different accelerations $\ddot{\mathbf{x}}$ and $\ddot{\mathbf{x}} + \delta\ddot{\mathbf{x}}$, then we obtain with respect to Eq. (9):

$$\sum_{s=1}^{n} A_{rs}\delta\ddot{x}_r = 0, \quad r = 1, .., l. \tag{10}$$

It means that virtual accelerations $\delta\ddot{\mathbf{x}}$ satisfy constraints requested similarly like virtual displacements or velocities following Eqs. (4) or (7). Therefore, we can write:

$$\sum_{s=1}^{n} (m_r\ddot{x}_s - X_s)\delta\ddot{x}_s = 0. \tag{11}$$

Some authors call relations (Eqs. (5), (8), and (11)) as the first, second, and third form of the system equation, see, for example, [3] and others.

Let us multiply each equation in the system (Eq. (2a)) by velocities $\dot{x}_s$. Summing them together, one obtains

$$\sum_{s=1}^{n} m_s\ddot{x}_s\dot{x}_s = \sum_{s=1}^{n} X_s\dot{x}_s + \sum_{r=1}^{l}\sum_{s=1}^{n} \lambda_r A_{rs}\dot{x}_s, \tag{12}$$

which can be rewritten in the form

$$\frac{\mathrm{d}\mathcal{T}}{\mathrm{d}t} + \frac{\mathrm{d}\mathcal{V}}{\mathrm{d}t} = \sum_{s=1}^{n} \tilde{X}_s\dot{x}_s, \tag{13}$$

where $\mathcal{T}, \mathcal{V}$ are kinetic and potential energy, respectively, and $\tilde{X}_s$ are forces, which cannot be included into the potential energy $\mathcal{V}$. In other words, relation Eq. (13) indicates that the change of the full energy (kinetic and potential) is equivalent to power (work on velocities) of all forces $\tilde{X}_s$, which do not contribute to the potential energy $\mathcal{V}$. Relation Eq. (13) corresponds to equilibrium condition Eq. (1), where functions of excitation and dissipation $\mathcal{P}(t), \mathcal{S}(t)$ correspond with the influence of non-potential forces $\tilde{X}_s$.

3. Lagrange's equations

The original coordinates $\mathbf{x}$ should be replaced with respect to Lagrangian coordinates $\mathbf{q}$. The reason is that they represent the most inherent coordinates respecting the real movement of the system and configuration of external forces. Let us write basic coordinates as functions of Lagrangian ones:

$$x_r = x_r(q_1, .., q_n, t), \quad r = 1, .., n. \tag{14}$$

It can be easily shown

$$\frac{\partial \dot{x}_r}{\partial \dot{q}_s} = \frac{\partial x_r}{\partial q_s}, \quad \frac{\partial \dot{x}_r}{\partial q_s} = \frac{\mathrm{d}}{\mathrm{d}t}\left(\frac{\partial x_r}{\partial q_s}\right). \tag{15}$$

We reconsider Eq. (5) where the virtual displacement $\delta \mathbf{x}$ is replaced by

$$\delta x_r = \sum_{s=1}^{n} \frac{\partial x_r}{\partial q_s} \delta q_s, \quad r = 1, .., n \quad \Rightarrow \quad \sum_{s=1}^{n} \left\{ \sum_{r=1}^{n} (m_r \ddot{x}_r - X_r) \frac{\partial x_r}{\partial q_s} \right\} \delta q_s = 0. \tag{16}$$

The last Eq. (16) can be modified using Eqs. (15), which implies

$$\sum_{s=1}^{n} \left\{ \sum_{r=1}^{n} m_r \left[\frac{\mathrm{d}}{\mathrm{d}t} \left(\dot{x}_r \frac{\partial \dot{x}_r}{\partial \dot{q}_s} \right) - \dot{x}_r \frac{\partial \dot{x}_r}{\partial q_s} \right] - \sum_{r=1}^{n} X_r \frac{\partial x_r}{\partial q_s} \right\} \delta q_s = 0, \tag{17}$$

This equation can be rewritten now in the form:

$$\sum_{s=1}^{n} \left\{ \frac{\mathrm{d}}{\mathrm{d}t} \left(\frac{\partial \mathcal{L}}{\partial \dot{q}_s} \right) - \frac{\partial \mathcal{L}}{\partial q_s} - Q_s \right\} \delta q_s = 0, \tag{18}$$

where it has been denoted:

$$\mathcal{L} = \sum_{r=1}^{n} \dot{x}_r \frac{\partial \dot{x}_r}{\partial q_s}, \quad Q_s = \sum_{r=1}^{n} X_r \frac{\partial x_r}{\partial q_s}. \tag{19}$$

Inspecting the polynomial $\mathcal{L}$, we recognize that it consists of the polynomial of the second and first degrees of components $\dot{\mathbf{q}}$ (coefficients are still functions of displacements $\mathbf{q}$ and time t) and the absolute part without any velocity components $\dot{\mathbf{q}}$. We can now assign the first part to the kinetic energy $\mathcal{T}$, while the part without velocities to the potential energy $\mathcal{V}$. So that $\mathcal{L}$ can be understood as the Lagrange function as usually defined

$$\mathcal{L} = \mathcal{T} - \mathcal{V}, \tag{20}$$

provided the dynamic system studied is holonomic and no constraints are applied. In such a case, all variations δq_s are independent and Eq. (18) can be fulfilled only if every coefficient in curly brackets vanishes individually. Consequently, we obtain Lagrange's equations in the form:

$$\frac{\mathrm{d}}{\mathrm{d}t} \left(\frac{\partial \mathcal{T}}{\partial \dot{q}_s} \right) - \frac{\partial \mathcal{T}}{\partial q_s} + \frac{\partial \mathcal{V}}{\partial q_s} = Q_s, \quad s_1, .., n, \tag{21}$$

where Q_s are generalized external forces as functions of $\mathbf{q}$ and t. These forces are basically linear transforms of original forces X_r, see Eq. (19). If holonomic constraints are inserted, then the number of remaining degrees of freedom is lower ($k = n - l$). Nevertheless, if there is possibility to define the system after elimination of inactive DOFs, then we can consider formally $k = n$ again and Eq. (21) remains in force.

Let us suppose now that the system includes l non-holonomic constraints and those holonomic, which cannot be eliminated. Whatever is the reason for that, it still holds $k + l = n$. These constraints are described by constraints in Lagrange's coordinates (analogous with Eq. (2b)) as follows:

$$\sum_{s=1}^{n} C_{rs}\dot{q}_s + D_r = 0, \quad r = 1, .., l, \quad \mathbf{D} = [D_r], \mathbf{D} \in \mathbb{R}^l \ , \quad \mathbf{C} = [C_{rs}], \mathbf{C} \in \mathbb{R}^{l \times n}, \tag{22}$$

This time, the variations δq_s are not fully independent and only those components, which satisfy conditions:

$$\sum_{s=1}^{n} C_{rs}\delta q_s = 0 \quad r = 1, 2, .., l, \tag{23}$$

can be regarded as independent.

In such a case, the right side of Eq. (21) should be completed:

$$\frac{\mathrm{d}}{\mathrm{d}t}\left(\frac{\partial \mathcal{T}}{\partial \dot{q}_s}\right) - \frac{\partial \mathcal{T}}{\partial q_s} + \frac{\partial \mathcal{V}}{\partial q_s} = Q_s + \sum_{r=1}^{l} \lambda_r C_{rs}, \quad s = 1, .., n. \tag{24}$$

To the system, Eq. (24) should be attached l constraints Eq. (22). So that, finally we have the system of $n + l$ equations with unknowns $\mathbf{q}$ and $\boldsymbol{\lambda}$. Multipliers $\boldsymbol{\lambda}$ are linearly related with forces in constraints. In particular cases, multipliers $\boldsymbol{\lambda}$ can be physically interpreted, for instance, they can have a meaning of reactions of a body moving along a given trajectory. Very knowledgeable explanation about manipulation and interpretation of Lagrange's multipliers from the viewpoint of a general theory as well as of employment in particular cases can be found in the monograph concerning non-holonomic systems, see [2]. For additional information and a large overview of additional literature resources, see [5].

The real dynamic system is always influenced by energy dissipation. Some simple models can be introduced using Rayleigh function $\mathcal{R}$, see, for example, [3]. This way is typically applicable, if linear viscous damping is considered and the Rayleigh function has a quadratic form in velocities $\dot{q}_s$. We can include this factor symbolically into Eq. (24), which reads now:

$$\frac{\mathrm{d}}{\mathrm{d}t}\left(\frac{\partial \mathcal{T}}{\partial \dot{q}_s}\right) - \frac{\partial \mathcal{T}}{\partial q_s} + \frac{\partial \mathcal{V}}{\partial q_s} = Q_s + \frac{\partial \mathcal{R}}{\partial \dot{q}_s} + \sum_{r=1}^{l} \lambda_r C_{rs}, \quad s_1, .., n. \tag{25}$$

Hence, the completed system Eqs. (22) and (25) with $n + l$ unknowns can be considered. However, we should be aware that this supplement is rather intuitive and does not follow from any rigorous derivation, although in practice it is widely and successfully used. Nevertheless, comparison of this system with the general relation Eq. (1) introduced in Section 1 is obvious. Take a note that more sophisticated versions of Lagrange's equations have been developed inspired by physical problems; see, for instance, generalized Lagrange-d'Alembert-Poincaré equations discussed in [11].

Let us add that many details (internal mechanisms and inclusion into governing system) concerning more sophisticated models of damping can be found in monographs of the rational dynamics, for example, [1, 4]. See also papers oriented to practical aspects of the damping either of natural, for example, rheological, aeroelastic origin, or intentionally included in order to achieve the highest damping effectiveness, for instance [13].

4. Appell-Gibbs function and equation system

Although the Appell-Gibbs approach is not referred so often in the study as the Lagrangian procedure, there are some monographs treating the analytical dynamics, for example, [1, 3], where detailed features of this method are explained. Moreover, journal papers can be found where special aspects of the Appell-Gibbs approach are discussed. A close relation of the fifth Gaussian form and the Gibbs equations from the viewpoint of Dynamics is studied, for example, [14, 15], important remarks for application are concerned in [16, 17], as well as possibilities of extension for systems with time-dependent masses [18] are indicated.

Let us briefly outline principal steps leading to the Appell-Gibbs differential system with respect to essentials ascertained and introduced in Section 2. We should be aware that generalized external forces Q_s, introduced in Eq. (19), follow in principle only k degrees of freedom, which remained free; thereafter, l constraints have been applied and the original number n of DOFs has been reduced to $k = n - l, 0 < l \leq n$. However, due to complicated relations inside the dynamic system, this fact is rather impossible to be employed in basic coordinates $x_s, s = 1, .., n$ and Lagrange's coordinates $q_s, s = 1, .., n$ should be addressed, as we have also seen in previous Section 3. Nevertheless, it is worthy to involve only such coordinates q_s, which correspond to k remained DOFs. It can be easily expressed in Lagrange's coordinates, unlike basic coordinates x_s. So that, as the first step, we reformulate some expressions of Section 2 concerning the transform from basic to Lagrange's coordinates.

Velocities $\dot{x}_r, r = 1, .., n$ should be evaluated with respect to the fact that coordinates x_r are functions of all Lagrange's coordinates $q_s, s = 1, .., k$ and time t, see Eq. (14):

$$\dot{x}_r = \sum_{s=1}^{k} \alpha_{rs}\dot{q}_s + \alpha_r, \quad r = 1, .., n, \quad \text{where}: \quad \alpha_{rs} = \frac{\partial x_r}{\partial q_s}, \quad \alpha_r = \frac{\partial x_r}{\partial t}, \tag{26}$$

which also implies

$$\delta\dot{x}_r = \sum_{s=1}^{k} \alpha_{rs}\delta\dot{q}_s, \quad r = 1, .., n, \tag{27}$$

Differentiation of Eq. (26) with respect to time gives

$$\ddot{x}_r = \sum_{s=1}^{k} \alpha_{rs}\ddot{q}_s + \sum_{s=1}^{k} \frac{d\alpha_{rs}}{dt}\dot{q}_s + \frac{d\alpha_r}{dt}, \quad \frac{d}{dt} = \frac{\partial}{\partial t} + \sum_{m=1}^{k} \dot{q}_m \frac{\partial}{dt}, \quad r = 1, .., n. \tag{28}$$

The incremented acceleration vector, when keeping velocities and displacements, can be formulated as follows:

$$\ddot{x}_r + \delta\ddot{x}_r = \sum_{s=1}^{k} \alpha_{rs}(\ddot{q}_s + \delta\ddot{q}_s) + \sum_{s=1}^{k} \frac{d\alpha_{rs}}{dt}\dot{q}_s + \frac{d\alpha_r}{dt}, \quad r = 1, .., n. \tag{29}$$

Deducting Eq. (28) from Eq. (29), one obtains

$$\delta\ddot{x}_r = \sum_{s=1}^{k} \alpha_{rs}\delta\ddot{q}_s, \quad r = 1, .., n. \tag{30}$$

Hence, it can be written

$$\sum_{r=1}^{n} X_r\delta\ddot{x}_r = \sum_{s=1}^{k} \left(\sum_{r=1}^{n} X_r\alpha_{rs} \right) \delta\ddot{q}_s = \sum_{s=1}^{k} Q_s\delta\ddot{q}_s, \tag{31}$$

where Q_s are identical generalized forces, as they have been defined in Eq. (19). With reference to Eq. (11), we can reformulate this equation as follows:

$$\sum_{r=1}^{n} m_r\ddot{x}_r\delta\ddot{x}_r \quad - \quad \sum_{s=1}^{k} Q_s\delta\ddot{q}_s \quad = \quad 0. \tag{32}$$

This relation will be used later, see Eq. (36).

As a principal step of this section, we define now the Gibbs function $\mathcal{G}$ concentrating "acceleration energy" included in all n DOFs as follows:

$$\mathcal{G} \quad = \quad \frac{1}{2}\sum_{r=1}^{n} m_r\ddot{x}_r^2, \tag{33}$$

When we pass from basic to Lagrange's coordinates, only k active coordinates remain in force and so the expression Eq. (33) can be rewritten:

$$\mathcal{G} \quad = \quad \frac{1}{2}\sum_{r=1}^{k} m_r\ddot{q}_r^2, \tag{34}$$

Expressions Eqs. (33) and (34) differ only in terms independent from accelerations.

Let us introduce the function $\mathcal{H}$:

$$\mathcal{H} = \mathcal{G} - \sum_{s=1}^{k} Q_s \ddot{q}_s, \tag{35}$$

and evaluate its virtual increment:

$$\begin{aligned}\delta\mathcal{H} = \delta\left(\mathcal{G} - \sum_{s=1}^{k} Q_s \ddot{q}_s\right) &= \frac{1}{2}\sum_{r=1}^{n} m_r(\ddot{x}_r + \delta\ddot{x}_r)^2 - \frac{1}{2}\sum_{r=1}^{n} m_r \ddot{x}_r^2 - \sum_{s=1}^{k} Q_s \delta\ddot{q}_s \\ &= \frac{1}{2}\sum_{r=1}^{n} m_r(\delta\ddot{x}_r)^2 + \left(\sum_{r=1}^{n} m_r \ddot{x}_r \delta\ddot{x}_r - \sum_{s=1}^{k} Q_s \delta\ddot{q}_s\right).\end{aligned} \tag{36}$$

The last parenthesis vanishes due to the relation Eq. (32). Therefore, if $\delta\ddot{\mathbf{x}} \neq 0$, then the function $\delta\mathcal{H}$ is always positive:

$$\delta\mathcal{H} = \delta\left(\mathcal{G} - \sum_{s=1}^{k} Q_s \ddot{q}_s\right) > 0, \tag{37}$$

which implies that accelerations $\ddot{q}_s, s = 1, .., k$ should lead to a minimum of the function $\mathcal{H}$, which means:

$$\frac{\partial \mathcal{G}}{\partial \ddot{q}_r} = Q_r, \quad r = 1, .., k. \tag{38}$$

The energy dissipation terms R_x, R_y, R_z should be added to the right side of Eq. (38). At this moment, the conformity of Eq. (38) with the equivalence Eq. (1) is well pronounced, similar like in the previous section. The system Eq. (38) should be completed by geometric constraints:

$$\dot{q}_r = \sum_{s=1}^{k} \beta_{rs} \dot{q}_s + \beta_r \quad r = k+1, .., n. \tag{39}$$

Equations (38) and (39) are the Gibbs-Appell differential system including n equations, which can be written in the normal form and hence it is suitable to be immediately investigated using common methods.

The differential system (Eqs. (38) and (39)) represents the simplest and in the same time the most general form of equations of the dynamic system movement. The form of this system is very simple, and it can be used with the same effectiveness to the investigation of holonomic as well as non-holonomic systems, as the constraints can represent non-holonomic but also holonomic type of constraints. Unlike the Lagrangian approach, the non-holonomic or non-eliminable constraints do not augment the number of differential equations.

Procedure of the Appell-Gibbs equations employment in particular cases is obvious, looking back at this section. In the first step, the "so called kinetic energy of accelerations" $\frac{1}{2}\sum_{r=1}^{N} m_r \ddot{x}_r^2$

is composed using n acceleration components of the vector $\ddot{\mathbf{x}}$. It represents the Appell-Gibbs function $\mathcal{G}$. In a general case, this function includes also all coordinates $\mathbf{x}$ and velocities $\dot{\mathbf{x}}$. Nevertheless, it is important that $\mathcal{G}$ in Lagrange's coordinates contains only k selected components of accelerations $\ddot{\mathbf{q}}$. Anyway, all n components of $\dot{\mathbf{q}}$ and $\mathbf{q}$ are still included as a result of a transformation from basic to Lagrange's coordinates.

It is worthy to remind that the differentiation outlined in Eq. (38) is very easy in a particular case. Indeed, let us realize that $\mathcal{G}$ can be symbolically expressed as a sum of quadratic function of accelerations $\ddot{q}_s, s = 1, .., k \rightarrow \mathcal{G}_2$, linear function of these components $\mathcal{G}_1$ and function without accelerations $\mathcal{G}_0$. Differentiating $\mathcal{G}_2$, one obtains the relevant acceleration component in a linear form, which will be moved onto the left side together with a coefficient, which can be a function of all velocities and displacements $\dot{q}_s, q_s, s = 1, .., n$. Differentiation of $\mathcal{G}_1$ leads to acceleration-free coefficients and $\mathcal{G}_0$ can be omitted leading to zeroes. Sometimes, the so-called reduced Appell-Gibbs function $\mathcal{G}^*$ is defined where $\mathcal{G}_0$ is a priori omitted.

In the second step, the work of k given forces $\mathbf{Q}$ on k virtual displacements $\mathbf{q}$ is carried out. It has the form $\sum_{s=1}^{k} Q_s \delta q_s$. We substitute now back into Eqs. (38) and add $l = n - k$ geometric constraints following Eqs. (39). So we obtain $k + l = n$ differential equations for n components of the vector $\ddot{\mathbf{q}}(t)$. Take a note that no unknown multipliers $\boldsymbol{\lambda}$ emerge here, which on the other hand increases the number of unknowns in a Lagrangian approach.

The procedure working with accelerations instead with velocities provides much simpler governing differential system. Unlike velocities, the acceleration components in the Appell-Gibbs function are included only in a few parts of energy expression. Therefore, all parts including only velocity and displacement components disappear during the differentiation of the Appell-Gibbs function with respect to $\ddot{q}_r, r = 1, .., k$, and therefore they can be considered beforehand as unimportant.

Investigating problems with rotations, we work with Lagrange's coordinates ω, which represent in fact velocities. So that by solving the abovementioned differential system, the displacements and velocities ω emerge as results. Rotations themselves remain unattended. May be, it is a forfeit for a relative simplicity of the governing system in comparison with the Lagrangian approach. However, this shortcoming is mostly apparent only. The main part of the result represents usually displacement components, which are obtained without restrictions. Together with velocities ω, they represent a full set of information needed to get through the shape of trajectories of the system response including rotation (illustrative example will be presented later in Section 6). If detailed rotations (not only velocities) are still needed, a subsequent integration can be performed independently using differential relations between rotation velocity vector ω and (for instance) Euler angles, see monographs [1, 3, 4] and others. They provide a detailed description of time history of a body orientation as a function of time t. This step can be useful, for instance, when a detailed animation is needed for presentation purposes.

5. Planar movement of a ball in a spherical cavity

5.1. Engineering motivation

Passive vibration absorbers of various types are very widely used in civil engineering. TV towers, masts, and other slender structures exposed to wind excitation are usually equipped by such devices. Conventional passive absorbers are of the pendulum type. Although they are very effective and reliable, they have several disadvantages limiting their application.

These shortcomings can be avoided using the absorber of the ball type. The basic principle comes out of a rolling movement of a metallic ball of a radius r inside of a rubber-coated cavity of a radius $R > r$. This system is closed in an airtight case, see, for instance, **Figure 1**. First papers dealing with the theory and practical aspects of ball absorbers have been published during the last decade, see [13, 19].

5.2. Planar layout of the system, Lagrangian procedure

The version, when the ball is forced to move solely in a vertical plane, has been thoroughly studied using Lagrangian approach in [20, 21] and other detailed papers dealing not only with theoretical aspects but also with experimental verification in the laboratory and in situ examining absorbers installed on real structures.

The cavity is fixed to a vibrating structure. Their dynamic character is represented by a linear single degree of freedom (SDOF) system represented by a mass M. Inside of the cavity, the ball m in a vertical plane is moving, that is two degrees of freedom (TDOF) system should be investigated, as it is outlined in **Figure 2**. It follows from geometric relations:

$$R \cdot \varphi = r(\psi + \varphi) \Rightarrow r\psi = \varrho_r \varphi, \tag{40}$$

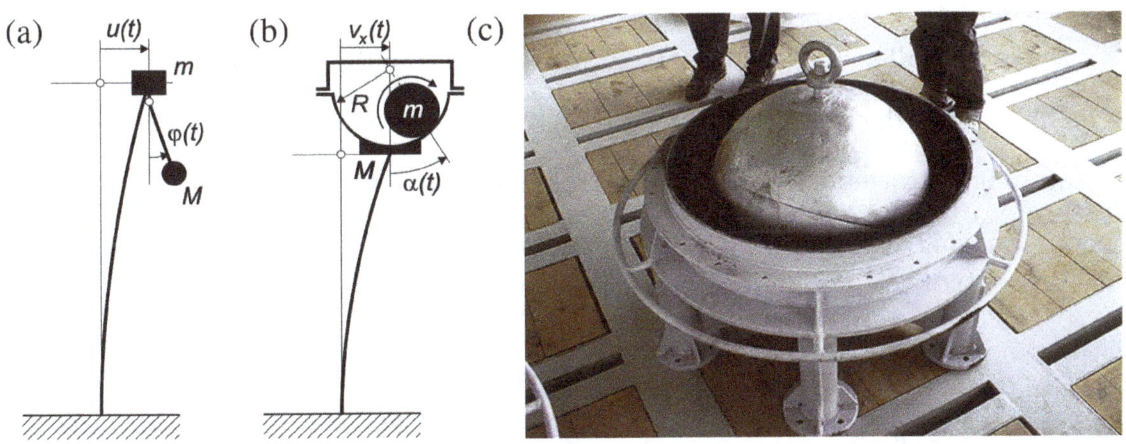

Figure 1. Dynamic scheme of (a) spherical pendulum absorber, (b) ball absorber, and (c) ball absorber during testing in a dynamic laboratory, see [19].

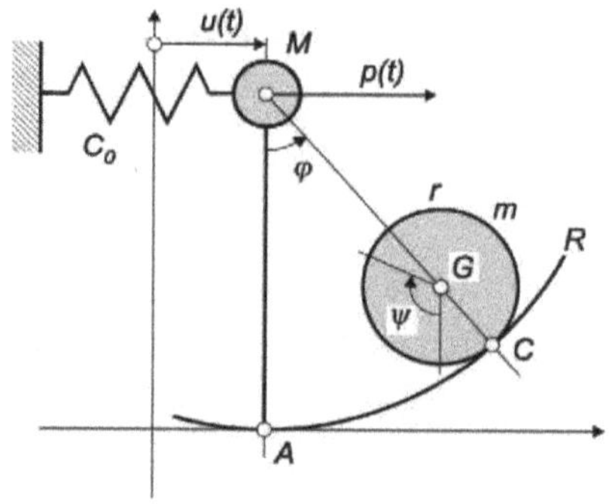

Figure 2. Basic scheme of a system.

where $\rho_r = R - r$. It holds for vertical or horizontal components of a displacement and velocity of the internal ball center:

$$\left.\begin{array}{llcl} \text{horiz.}: & u + \varrho_r \cdot \sin\varphi & \Rightarrow & \dot{u} + \varrho_r \dot{\varphi} \cos\varphi, \\ \text{vert.}: & \varrho_r \cdot \cos\varphi & \Rightarrow & -\varrho_r \dot{\varphi} \sin\varphi. \end{array}\right\} \tag{41}$$

Kinetic energy of a moving system of the ball m and the cavity M can be written in a form:

$$\mathcal{T} = \frac{1}{2} m \left[(\dot{u} + \varrho_r \dot{\varphi} \cos\varphi)^2 + \varrho_r^2 \dot{\varphi}^2 \sin^2\varphi \right] + \frac{1}{2} J \dot{\psi}^2 + \frac{1}{2} M \dot{u}^2 = \frac{1}{2} (m + M) \dot{u}^2 + m \varrho_r \dot{u} \dot{\varphi} \cos\varphi + \frac{m}{2\kappa} \varrho_r^2 \dot{\varphi}^2, \tag{42}$$

where $m/\kappa = m + J/r^2 \ \Rightarrow \ \kappa = 5/7$, while the potential energy is given by an expression:

$$\mathcal{V} = mg\varrho_r (1 - \cos\varphi) + \frac{1}{2} C u^2. \tag{43}$$

The damping should be introduced in a form of a simple Rayleigh function:

$$\mathcal{R} = \frac{1}{2} \left(M b_u \dot{u}^2 + m b_\varphi \varrho_r^2 \dot{\varphi}^2 \right). \tag{44}$$

m, M – mass of the ball m, mass of the cavity, M representing the protected structure;

J – inertia moment of the ball m;

b_u, b_φ – damping coefficients (logarithmic decrements, linear viscous damping);

Expressions Eqs. (42), (43), and (44) should be put into Lagrange's equations of the second type, see Eqs. (24) or (25) and monographs, for example, [1, 3, 4] and others:

$$\begin{array}{c} \sum_{r=1}^{n} \left\{ \frac{d}{dt} \left(\frac{\partial \mathcal{T}}{\partial \dot{q}_r} \right) - \frac{\partial \mathcal{T}}{\partial q_r} + \frac{\partial \mathcal{V}}{\partial q_r} + \frac{\partial \mathcal{R}}{\partial \dot{q}_r} \right\} \delta q_r = P_r(t), \\ q_1 = u = \zeta \cdot \varrho_r \ , \quad q_2 = \varphi \ , \quad P_u(t) = p(t) \cdot M\varrho_r \ , \quad P_\varphi(t) = 0, \end{array} \tag{45}$$

which give the governing equations of the system:

$$\begin{aligned} \ddot{\varphi} + \kappa b_{\varphi}\dot{\varphi} + \kappa\omega_m^2 \sin\varphi + \kappa\ddot{\zeta} \cdot \cos\varphi &= 0, & (a) \\ \mu\ddot{\varphi}\cos\varphi - \mu\dot{\varphi}^2 \sin\varphi + (1+\mu)\ddot{\zeta} + b_u\dot{\zeta} + \omega_M^2\zeta &= p(t), & (b) \\ \mu = m/M, \quad \omega_M^2 = C/M, \quad \omega_m^2 = g/\varrho_r. & & (c) \end{aligned} \tag{46}$$

Equation (46) describes 2D movement of a ball absorber under excitation by the force $P(t)$ at any arbitrary deviation amplitudes including incidental transition through a limit cycle toward an open regime.

5.3. Illustration of some planar system features

Analysis of the governing system (Eqs. (46)) has been done in a couple of papers, for example, [20, 21]. Investigation has been carried out using the harmonic or multi-harmonic balance method, see, for example, [22, 23], respectively.

The system is auto-parametric, see, for example, [24] and other resources. Very rich overview of a theoretical basis of auto-parametric systems can be found in [25]. Expecting a single mode response, the Harmonic balance-based methods are applicable. Following approximate expressions for excitation and response can be written (cf., e.g., [22]):

$$\begin{aligned} p(t) &= p_0 \sin(\omega t), \\ \varphi(t) &= \alpha \sin(\omega t) + \beta \cos(\omega t), \\ \zeta(t) &= \gamma \sin(\omega t) + \delta \cos(\omega t). \end{aligned} \tag{47}$$

Having four new variables $\alpha = \alpha(t), \beta = \beta(t), \gamma = \gamma(t), \delta = \delta(t)$ instead of two original unknowns $\varphi(t), \zeta(t)$, two additional conditions can be freely chosen:

$$\dot{\alpha} \sin(\omega t) + \dot{\beta} \cos(\omega t) = 0, \qquad \dot{\gamma} \sin(\omega t) + \dot{\delta} \cos(\omega t) = 0. \tag{48}$$

After substituting Eqs. (47) and (48) into Eqs. (46) and substituting the $\sin\varpi$ and $\cos\varphi$ functions by two terms of Taylor expansion, the harmonic balance procedure gives the differential system for unknown amplitudes $\mathbf{Z} = (\alpha, \beta, \gamma, \delta)^T$, see, for example, [21, 23]:

$$\mathbf{M}(\mathbf{Z})\dot{\mathbf{Z}} = \mathbf{F}(\mathbf{Z}). \tag{49}$$

System (49) for amplitudes $\mathbf{Z}(t)$ is meaningful if they are functions of a "slow time," in other words, if their changes within one period $2\pi/\omega$ are small or vanishing and individual steps of the harmonic balance operation are acceptable. The matrix $\mathbf{M}$ and the right-hand side vector $\mathbf{F}$ have the following form:

$$\mathbf{M} = \begin{pmatrix} 0 & -\omega & -\frac{1}{4}\alpha\beta\kappa\omega & \frac{1}{8}\kappa\omega A_{\alpha} \\ \omega & 0 & -\frac{1}{8}\kappa\omega A_{\beta} & \frac{1}{4}\alpha\beta\kappa\omega \\ -\frac{1}{8}\mu\omega A_{\beta} & \frac{1}{4}\alpha\beta\mu\omega & (\mu+1)\omega & 0 \\ -\frac{1}{4}\alpha\beta\mu\omega & \frac{1}{8}\mu\omega A_{\alpha} & 0 & -(\mu+1)\omega \end{pmatrix}, \tag{50}$$

$$\mathbf{F} = \frac{1}{48}\begin{pmatrix} 6A_0\kappa(3\gamma\omega^2 - \alpha\omega_m^2) + 12\omega^2(\kappa(\alpha\beta\delta + (8-\beta^2)\gamma) - 4\alpha) - 48\beta\kappa\omega b_\varphi \\ 6A_0\kappa(\delta\omega^2 - \beta\omega_m^2) + 12\omega^2(\alpha\gamma\kappa + \beta\delta\kappa - 4)\beta + 48\alpha\kappa\omega b_\varphi \\ \omega^2(A_0(A_0+22)\beta\mu - 16(3\delta(\mu+1) - 4\beta\mu)) + 48(\gamma\omega b_u + \delta\omega_M^2) \\ \omega^2(A_0(A_0+22)\alpha\mu - 16(3\gamma(\mu+1) - 4\alpha\mu)) - 48(\delta\omega b_u - \gamma\omega_M^2 + p_0) \end{pmatrix}, \tag{51}$$

where $A_0 = \alpha^2 + \beta^2 - 8, \quad A_\alpha = 3\alpha^2 + \beta^2 - 8, \quad A_\beta = \alpha^2 + 3\beta^2 - 8.$

Let us consider stationary response of the system. In this case, the derivatives $\mathrm{d}\mathbf{Z}/dt$ vanish and the right-hand side has to vanish too. Eq. (49) degenerates to the form of

$$\mathbf{F}(\mathbf{Z}) = 0 \tag{52}$$

Thus, to identify the stationary solutions, the zero solution points of **F**, depending on the excitation frequency and amplitude, should be traced. In the same time, the signum and the zero points of the Jacobian det(**JF**) have to be checked. The negative value of the Jacobian for a particular point indicates that the corresponding solution is stable, whereas when Jacobian vanishes a bifurcation could occur.

The curve $\mathbf{F}(\alpha, \beta, \gamma, \delta, \omega) = 0$, projected into the planes (ω, R) or (ω, S) (for $S^2 = \gamma^2 + \delta^2$), forms the resonance curves known from the analysis of linear oscillators. However, correspondence of this curve to the original Eq. (46) is limited to the case of stationary response. It is necessary to remind that limits of stationarity of the response cannot be determined from properties of Eq. (52) itself. The complete Eq. (49) has to be taken into account for this purpose.

With respect to actual experiences regarding passive vibration absorbers and some interesting properties of system (46), the following reference input data have been introduced:

$$M = 10.0; \;\; m = 2.0; \;\; \varrho_r = 0.71; \;\; b_\varphi = 0.1; \;\; b_u = 0.2; \;\; C = 140; \;\; p_o = 0.5 \div 2.5. \tag{53}$$

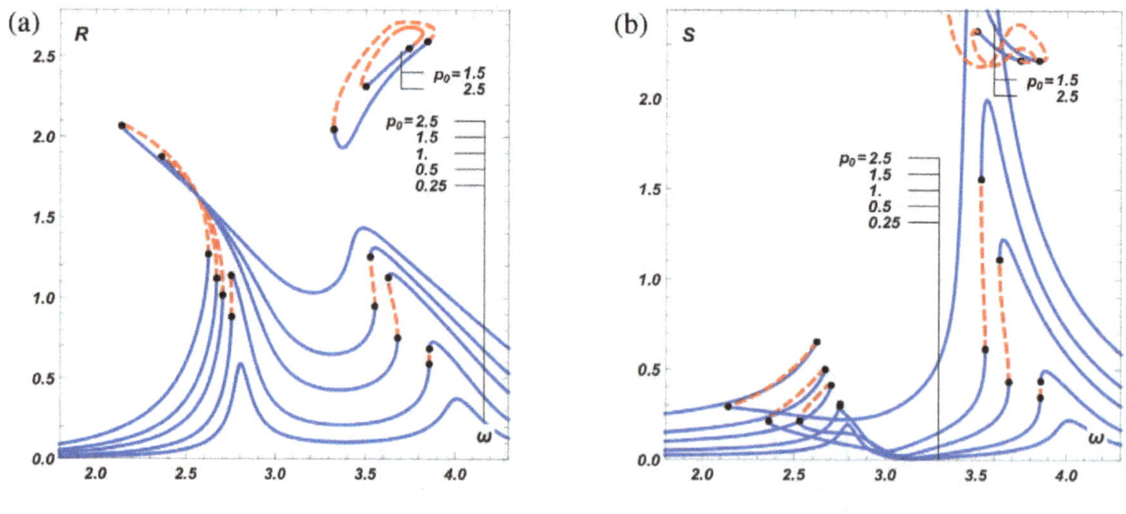

Figure 3. Nonlinear resonance curves describing the stationary response of the system for excitation amplitudes $p_0 = 0.25, 0.5, 1, 1.5, 2.5$. Stable branches are shown as solid blue curves, unstable parts are indicated as the red dashed curves. Amplitudes, see Eq. (47), $R = \sqrt{\alpha^2 + \beta^2}$ are shown in the left part of the figure, amplitudes $S = \sqrt{\gamma^2 + \delta^2}$ are on the right.

Utilizing Eqs. (52) and (51), the nonlinear resonance curves describing the stationary response of system (46) can be obtained. A set of such curves for excitation amplitudes $p_0 = 0.25, 0.5, 1, 1.5, 2.5$ is shown in **Figure 3**. It is obvious for the first view the nonlinear character manifesting oneself by a dependence of a position of extreme points on an amplitude of excitation force. This effect is visible predominantly in a neighborhood of a conventional "linear" natural frequency of the absorber, although also the second natural frequency corresponding to the original natural frequency of the structure is affected. The resonance curves are typical for a system with "softening" nonlinearities.

6. Spatial version of the system, Appell-Gibbs procedure

6.1. Gibbs function

The spatial version of the ball absorber on the basis of rational dynamics has been widely investigated by authors of this chapter, see, for example, [26, 27]. Lagrangian approach and Appell-Gibbs procedures have been discussed in these papers combining analytical and numer-ical methods. Some important issues will be roughly outlined and for details see cited papers.

Unlike the planar version discussed in the previous section, the Appell-Gibbs approach is used to formulate the governing nonlinear differential system. The authors tried to formulate the spatial version using the Lagrangian procedure as well, see [28]. Although the governing system of the respective holonomic system has been successfully assembled, the further analysis appeared very cumbersome, and therefore, it has been given up to follow this way. Thus, the Appell-Gibbs approach is used to formulate the governing system. Its structure is much more transparent and represents a wider option of analytical-numerical investigation of detailed properties of the ball trajectories within the cavity.

With respect to Sections 2 and 4, the first step represents to construct the Appell-Gibbs function (often referred to as an energy acceleration function) defined as follows:

$$\mathcal{G} = \frac{1}{2}M\left(\ddot{u}_{Gx}^2 + \ddot{u}_{Gy}^2 + \ddot{u}_{Gz}^2\right) + \frac{1}{2}J\left(\dot{\omega}_x^2 + \dot{\omega}_y^2 + \dot{\omega}_z^2\right), \tag{54}$$

where M is the mass of the ball, J is central inertia moment of the ball with respect to point G, $\boldsymbol{\omega}$ the angular velocity vector of the ball with respect to its center G, $\mathbf{u}_G$ the displacement of the ball center with respect to absolute origin O, C contact point of the ball and cavity, A moving origin related with the cavity in its bottom point, see **Figure 4**. Coordinates $\mathbf{x} = [x, y, z]$ are Cartesian coordinates with origin in the point O. Hence, it holds:

$$\begin{aligned} \mathbf{u}_G &= \mathbf{u}_A + \mathbf{u}_C + \mathbf{u}_n, \qquad \mathbf{u}_n = r \cdot \mathbf{n} \\ \dot{\mathbf{u}}_G &= \dot{\mathbf{u}}_A + \underbrace{\dot{\mathbf{u}}_C + r \cdot \dot{\mathbf{u}}_n,}_{\rho\dot{\mathbf{u}}_C,} \qquad \rho = 1 - r/R, \quad (\text{cf. Eq. } (40): \quad \varrho_r = R - r), \end{aligned} \tag{55}$$

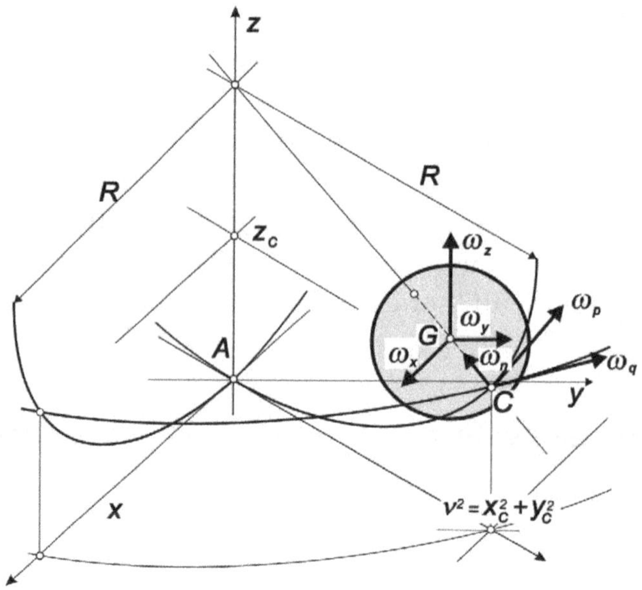

Figure 4. Ball rotation vector in moving coordinates.

where $\mathbf{u}_A$ is the displacement of the moving origin A with respect to absolute origin O, $\mathbf{u}_C$ the displacement of the contact point C with respect to moving origin A, $\mathbf{u}_n$ the displacement of the ball center G with respect to contact point C, $\mathbf{n}$ the cavity normal unit vector in point C.

Geometry of the cavity (radius R) with respect to moving origin A is given by equation:

$$x_A^2 + y_A^2 + (z_A - R)^2 = R^2, \tag{56}$$

where $\mathbf{x}_A = \left[x_A, y_A, z_A\right]$ are Cartesian coordinates with origin in the moving origin A.

Using Pfaff theorem and adopting a conjecture of non-sliding contact between the ball and the cavity, the respective non-holonomic constraints of "perfect" rolling can be deduced after a longer manipulation:

$$\begin{aligned} \dot{u}_{Gx} &= \dot{u}_{Ax} + \rho\left(\omega_y(u_{Cz} - R) - \omega_z u_{Cy}\right), \\ \dot{u}_{Gy} &= \dot{u}_{Ay} + \rho(\omega_z u_{Cx} - \omega_x(u_{Cz} - R)), \\ \dot{u}_{Gz} &= +\rho\left(\omega_x u_{Cy} - \omega_y u_{Cx}\right), \end{aligned} \tag{57}$$

where $\rho = 1 - r/R$.

In order to substitute for accelerations $\mathbf{u}_o$ into the Appell function (Eq. (54)), let us differentiate constraints Eqs. (57).

Several manipulations provide expressions for components of the ball center acceleration $\ddot{\mathbf{u}}_G$, which consist of acceleration in the moving origin $A : \ddot{\mathbf{u}}_A$ representing the given external kinematic excitation and acceleration related to the point A being given by an expression: $\rho\ddot{\mathbf{u}}_C$:

$$\begin{aligned}\ddot{u}_{Gx} &= \ddot{u}_{Ax} + \rho\left(\dot{\omega}_y(u_{Cz} - R) - \dot{\omega}_z u_{Cy}\right) + \rho\left(\omega_y \dot{u}_{Cz} - \omega_z \dot{u}_{Cy}\right),\\ \ddot{u}_{Gy} &= \ddot{u}_{Ay} + \rho(\dot{\omega}_z u_{Cx} - \dot{\omega}_x(u_{Cz} - R)) + \rho\left(\omega_y \dot{u}_{Cz} - \omega_z \dot{u}_{Cy}\right),\\ \ddot{u}_{Gz} &= \ddot{u}_{Az} + \rho\left(\dot{\omega}_x u_{Cy} - \dot{\omega}_y u_{Cx}\right) + \rho\left(\omega_x \dot{u}_{Cy} - \omega_y \dot{u}_{Cx}\right),\end{aligned} \tag{58}$$

Because the kinematic excitation is supposed to be horizontal, $\ddot{u}_{Az} = 0$ into Eqs. (58) should be substituted.

Expressions Eqs. (58) are to be substituted into Eq. (54). Thereby, we obtain the Appell-Gibbs function $\mathcal{G}$ for the system investigated. The function $\mathcal{G}$ can be significantly simplified keeping only terms including second-time derivatives $\ddot{\mathbf{u}}_G$ and $\dot{\boldsymbol{\omega}}$, which represent second-time derivatives of respective rotations. This step provides the reduced Appell-Gibbs function $\mathcal{G}^r$. Using $\mathcal{G}^r$, one can write the Appell-Gibbs differential system:

$$\partial\mathcal{G}^r/\partial\dot{\omega}_x = F_{Gx},\ \ \partial\mathcal{G}^r/\partial\dot{\omega}_y = F_{Gy},\ \ \partial\mathcal{G}^r/\partial\dot{\omega}_z = F_{Gz}, \tag{59}$$

where $\mathbf{F}_G$ is the external force vector acting in ball center G. Vector $\mathbf{F}_G$ is determined subsequently using the virtual displacements principle. Let us introduce the quasi-coordinates $\varphi_x,\ \varphi_y,\ \varphi_z$ where $\omega_x = \dot{\varphi}_x,\ \ \omega_y = \dot{\varphi}_y,\ \ \omega_z = \dot{\varphi}_z$. The only external force acting in the ball center is the gravity. Therefore, the elementary work performed can be expressed as

$$\delta F_G = -mg \cdot \delta u_{Gz}. \tag{60}$$

Virtual displacement δu_{Gz} can be determined using the third non-holonomic constraint in Eqs. (57). It holds

$$\delta u_{Gz} = \rho\left(u_{Cy}\delta\varphi_x - u_{Cx}\delta\varphi_y\right), \tag{61}$$

and therefore

$$\delta F_G = -mg\rho\left(u_{Cy}\delta\varphi_x - u_{Cx}\delta\varphi_y\right). \tag{62}$$

At the same time, the elementary work can be expressed in terms of quasi-coordinates:

$$\delta F_G = F_{Gx}\delta\varphi_x + F_{Gy}\delta\varphi_y + F_{Gz}\delta\varphi_z. \tag{63}$$

Comparing coefficients at respective virtual components $\delta\varphi_x, \delta\varphi_y, \delta\varphi_z$, in Eqs. (61) and (63), one obtains

$$F_{Gx} = -\rho mg \cdot u_{Cy}, F_{Gy} = \rho mg \cdot u_{Cx}, F_{Gz} = 0. \tag{64}$$

The damping will be introduced later in Section 6.3 in order to separate energy conservative approach and enable to discuss various stationary regimes with respect to parameter and excitation settings.

6.2. Governing system

Carrying out the differentiation outlined in Eqs. (59), and respecting Eqs. (64), it can be written after some adaptations:

$$\begin{aligned}
J_s\dot{\omega}_x - u_{Cx}\dot{\Omega}_s &= \big((\ddot{u}_{Ay} + \rho(\omega_z\dot{u}_{Cx} - \omega_x\dot{u}_{Cz}))(u_{Cz} - R) \\
&\qquad - u_{Cy}(g + \rho(\omega_x\dot{u}_{Cy} - \omega_y\dot{u}_{Cx}))\big), \\
J_s\dot{\omega}_y - u_{Cy}\dot{\Omega}_s &= \big(-(\ddot{u}_{Ax} + \rho(\omega_y\dot{u}_{Cz} - \omega_z\dot{u}_{Cy}))(u_{Cz} - R) \\
&\qquad + u_{Cx}(g + \rho(\omega_x\dot{u}_{Cy} - \omega_y\dot{u}_{Cx}))\big), \\
J_s\dot{\omega}_z - (u_{Cz} - R)\dot{\Omega}_s &= \big((\ddot{u}_{Ax} + \rho(\omega_y\dot{u}_{Cz} - \omega_z\dot{u}_{Cy}))u_{Cy} \\
&\qquad - (\ddot{u}_{Ay} + \rho(\omega_z\dot{u}_{Cx} - \omega_x\dot{u}_{Cz}))u_{Cx}\big),
\end{aligned} \tag{65}$$

where

$$\begin{aligned}
\dot{\Omega}_s &= u_{\dot{C}x}\dot{\omega}_x + u_{Cy}\dot{\omega}_y + (u_{Cz} - R)\dot{\omega}_z, \\
J_s &= \underbrace{(J + m\rho^2 R^2)}_{\substack{\text{mass interia moment of the ball} \\ \text{with respect to center of the cavity}}}/m\rho^2.
\end{aligned} \tag{66}$$

It can be shown that $\dot{\Omega}_s = 0$ and therefore the second column on the left side of the system Eqs. (65) should be omitted. External excitations are specified by movement or acceleration in the point A. Hence, kinematic excitation in A is given as follows:

$$E_{Ax}(t) = \ddot{u}_{Ax}/\rho, \;\; E_{Ay}(t) = \ddot{u}_{Ay}/\rho, E_{Az}(t) = 0, \tag{67}$$

as it can be seen in Eqs. (65). Provided we need to investigate the response processes in a vertical plane, only one component remains non-zero and the second vanishes as well.

In order to obtain the system Eqs. (65) in the form with first-time derivatives concentrated on the left side, the first derivatives $\dot{\mathbf{u}}_C$ in its right sides should be expressed in displacements $\mathbf{u}_C$ using non-holonomic constraints Eqs. (57):

$$\begin{aligned}
\dot{u}_{Cx} &= \omega_y(u_{Cz} - R) - \omega_z u_{Cy}, \\
\dot{u}_{Cy} &= \omega_z u_{Cx} - \omega_x(u_{Cz} - R), \\
\dot{u}_{Cz} &= \omega_x u_{Cy} - \omega_y u_{Cx}.
\end{aligned} \tag{68}$$

Therefore, we obtained the system of six non-linear ODEs (Eqs. (65) and (68)) in a normal form with six unknown functions of time: $u_{Cx}, u_{Cy}, u_{Cz}, \omega_x, \omega_y, \omega_z$. Vector $\mathbf{u}_C$ depicts displacements of the contact point and can be used to study the movement of the ball from a global point of view. Detailed behavior of the ball as a rotating body is given by angular velocities $\boldsymbol{\omega}$. If the time history of rotation should be traced, then a subsequent run is necessary to obtain rotations by means of Euler angles as solution of the system of three ODEs with an input of angular velocities $\boldsymbol{\omega}$.

6.3. Influence of the damping

Influence of the damping will be taken into account. Basically, two sources of the energy dissipation are ruling in the system: (1) dissipation due to air dynamic resistance and (2) energy loss in contact of the cavity and rolling ball. The former one can be neglected with respect to obvious geometric configuration of the device and relative velocity ball/cavity. Concerning the latter one, complicated energy dissipating processes are ruling in contact of the ball with cavity. Nevertheless, supposing that no slipping arises in the contact, the dissipation process can be approximated as proportional to relevant components of the angular velocity vector $\boldsymbol{\omega}$ and the quality of the cavity/ball contact. Considering the obvious setting, the respective material coefficients characterizing the rolling movement of the ball can be considered constant regardless of the direction in the tangential plane to the cavity in the point C, see **Figure 4**. The coefficient determining the rotation resistance around the normal vector $\mathbf{n}$ in the contact point C is different as a rule. Therefore, the resistance moment vector $\mathbf{D}$ can be expressed in moving coordinates p, q, n, see **Figure 4**, as follows:

$$\mathbf{D} = \left[D_p, D_q, D_n\right]^T. \tag{69}$$

Components of the above vector can be written in a form as follows:

$$D_p = \kappa_r \cdot \omega_p, \quad D_q = \kappa_r \cdot \omega_q, \quad D_n = \kappa_s \cdot \omega_n, \tag{70}$$

where κ_r, κ_s are coefficients of "viscous resistance" of rolling and spinning. Their meaning is: the moment for a unity rotation per second, that is (Nms/rad).

Turning of the vector $\mathbf{D}_G = \left[D_{Gx}, D_{Gy}, D_{Gz}\right]^T$ expressed in (xyz) coordinates into the vector $\mathbf{D}$ can be written as

$$\mathbf{D} = \mathbf{T}_C \cdot \mathbf{D}_G, \tag{71}$$

The transformation matrix T_C reads

$$\mathbf{T}_C = \begin{bmatrix} \frac{x_C(-z_C+R)}{Rv}, & \frac{y_C(-z_C+R)}{Rv}, & \frac{v}{R} \\ \frac{-y_C}{v}, & \frac{x_C}{v}, & 0 \\ \frac{-x_C}{R}, & \frac{-y_C}{R}, & \frac{-z_C+R}{R} \end{bmatrix} \tag{72}$$

where $v^2 = x_C^2 + y_C^2$. The matrix $\mathbf{T}_C$ is orthogonal and, therefore, the inverse transformation goes using matrix $\mathbf{T}_C^{-1} = \mathbf{T}_C^T$, in particular:

$$\mathbf{D}_G = \mathbf{T}_C^T \cdot \mathbf{D}, \tag{73}$$

Components of the vector $\mathbf{D}_G$ should be incorporated onto the right side of Eqs. (59), where right sides should be completed. It means that the elementary work δF_G following Eq. (60) must be completed by a negative dissipating work due to $\mathbf{D}_G$.

$$\delta F_G = -mg \cdot \delta u_{Gz} - \mathbf{D}_G \cdot \delta \boldsymbol{\varphi}, \tag{74}$$

Repeating the further derivation like in Section 6.2, one can revisit the system Eqs. (65) and (68), where the right sides are completed and instead (Eqs. (65)) they read:

$$\begin{aligned} J_s\dot{\omega}_x &= ((\ddot{u}_{Ay} + \rho(\omega_z\dot{u}_{Cx} - \omega_x\dot{u}_{Cz}))(u_{Cz} - R) - u_{Cy}(g + \rho(\omega_x\dot{u}_{Cy} - \omega_y\dot{u}_{Cx}))) - D_{Gx}/m, \\ J_s\dot{\omega}_y &= (-(\ddot{u}_{Ax} + \rho(\omega_y\dot{u}_{Cz} - \omega_z\dot{u}_{Cy}))(u_{Cz} - R) + u_{Cx}(g + \rho(\omega_x\dot{u}_{Cy} - \omega_y\dot{u}_{Cx}))) - D_{Gy}/m, \\ J_s\dot{\omega}_z &= ((\ddot{u}_{Ax} + \rho(\omega_y\dot{u}_{Cz} - \omega_z\dot{u}_{Cy}))u_{Cy} - (\ddot{u}_{Ay} + \rho(\omega_z\dot{u}_{Cx} - \omega_x\dot{u}_{Cz}))u_{Cx}) - D_{Gz}/m. \end{aligned} \tag{75}$$

Terms $D_{Gx}/m, D_{Gy}/m, D_{Gz}/m$ which are linear functions of $\omega_x, \omega_y, \omega_z$ determine the viscous type of the damping, although intensity in individual coordinates is variable depending on the position of the ball within the cavity.

6.4. Ball trajectories within the fixed cavity due to initial conditions

A large program of a ball trajectory investigation within a spherical cavity has been performed using the differential system (Eqs. (68) and (75)). Basically, it consists of two groups which are briefly illustrated in this and the next subsections. The first group concerns the fixed cavity (no excitation is applied). The only source of energy introduced is given by the initial deflection of the ball from equilibrium position in the point A ("southern pole"), or in other words by non-homogeneous initial conditions.

Differential system (Eqs. (68) and (75)) admits a number of singular solutions which can serve as separating limits of zones within which regular solutions exhibit certain character of trajectory shape. Some of them can be found analytically from the differential system taking into account their special properties concerning individual response component along the trajectory as a whole or in certain points of these curves. For details, special papers should be referred. Take a note that most of them emerge when no damping is considered. The reason is that the trajectory should be quasi-periodic (or cyclic-stationary), which is impossible when damping is respected and no external energy supply is considered. Trajectories start in a certain point on a meridian into which the ball is elevated. Then, it is thrown horizontally along the cavity parallel circle. Let us mention a few of the most important:

1. *circular trajectory in horizontal plane.* No initial spin is considered $(\omega_{n0} = 0)$. The impulse applied corresponds with the initial velocity $\boldsymbol{\omega} = [\omega_{ps}, 0, 0]$, where it holds for ω_{ps}:

$$\omega_{ps} = \frac{gM\rho u_{Cz0}(2R - u_{Cz0})}{(J + M\rho^2 R^2)(R - u_{Cz0})}. \tag{76}$$

This case is the most important and can be called separating circle (SC).

2. *circular trajectory in inclined plane*, see **Figure 5**, $\boldsymbol{\omega}_0 = [100.0, 0.0, 0.0]$. This state is exactly valid for $\omega_{p0} \to \infty$. The space spiral type trajectory changes from SC upwards successively into the upper hemisphere. Before the limit state for $\omega_{p0} \to \infty$ is reached, the osculation plane of the trajectory can be recognized. It rotates around the vertical axis with a descending angular velocity as far as it vanishes and osculation and operating planes coincide.

3. *trajectory of "* *"* see **Figure 5**, $\boldsymbol{\omega}_0 = [5.817, 0.0, 5.0537]$. Cases, when the $\omega_{n0} = 5.0537$ takes a shape visible in the picture. The apexes of this curve correspond to $\boldsymbol{\omega} = [0.0, 0.0, \omega_{n0}]$ and $\mathbf{u} = [0.0, 0.0, 0.0]$, which is a clue to find forms and parameters of this special case. This trajectory is reached from SC, increasing the initial spin velocity until the limit value. If it is lower, the trajectory has the spiral form. For a higher value, it became a curly form, see **Figure 5**, $\boldsymbol{\omega}_0 = [5.817, 0.0, 10.0]$. The limit state for infinite initial spin represents the ball apparently fixed in the initial point and not moving neither horizontally nor vertically.

Let us have a look at the bottom two pictures in **Figure 5**. They respect the influence of the damping. Coefficients κ_r, κ_s are different as it corresponds to conditions in the real system. The left demonstrates trajectory for positive initial spin and the right for negative initial spin. The transition through limit cases mentioned earlier is visible. The trajectory obviously finishes in the bottom "southern pole" of the cavity.

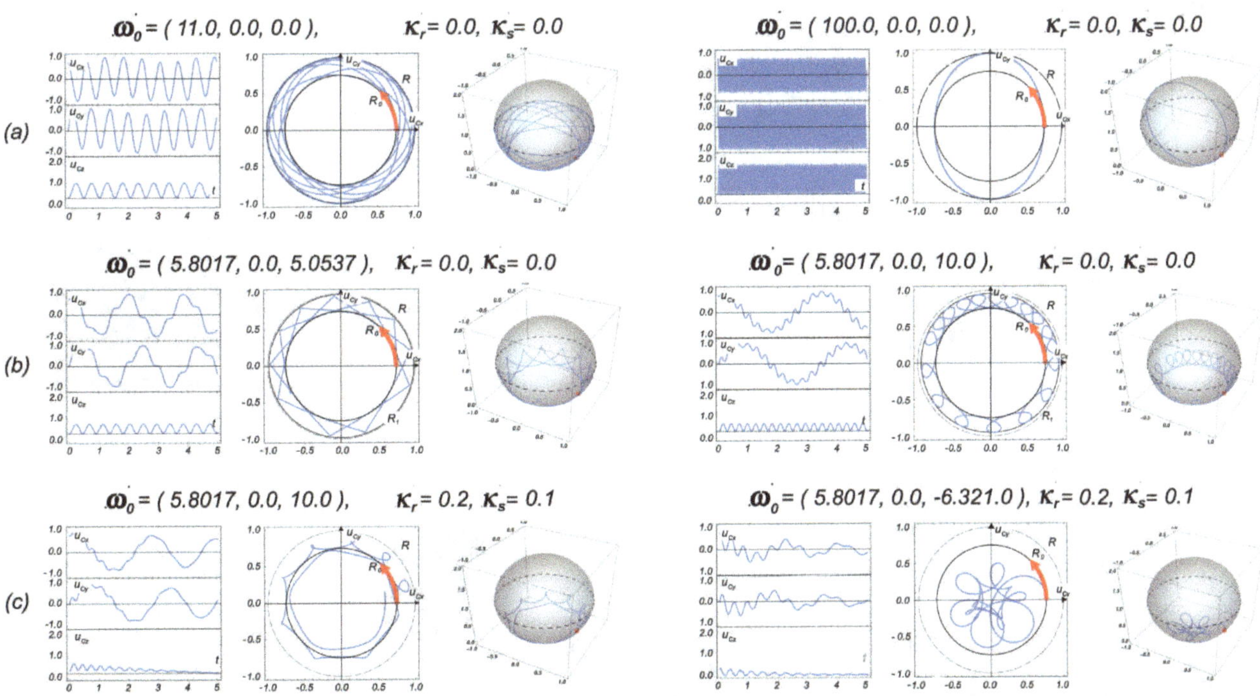

Figure 5. Illustration of the ball trajectories; cavity is not excited; energy only supply is due to non-homogeneous initial condition; in every triplet: movement time history of the contact point C: u_{Cx}, u_{Cy}, u_{Cz}; vertical view of trajectories u_{Cx}, u_{Cy} components; axonometric view of trajectories; parameters above triplets: initial values of $\boldsymbol{\omega}_0 = [\omega_{p0}, \omega_{q0}, \omega_{n0}]$ and damping parameters: κ_r, κ_s; line (a): no spin, no damping, line (b): spin considered, line (c): spin and damping considered.

6.5. Ball trajectories within kinematically excited cavity

The second group of tests deals with the cavity which is undergone to kinematic excitation in a horizontal plane (only one-direction excitation is reported here).

Two extensive series of tests demonstrate the auto-parametric character of the system. In the first series, the response has been evaluated separately for every excitation frequency ω starting from homogeneous initial conditions. **Figure 6** shows some selected results of numerical simulations which follow from the differential system (Eqs. (68) and (75)). We briefly point out a couple of features visible in **Figure 6**. In the picture (a), we can see the maximal horizontal amplitude of the ball trajectory, when the cavity is kinematically excited in the horizontal plane in x direction. The solid curve represents $\max|u_{Cx}|$ and the dashed curve is $\max|u_{Cy}|$ as functions of the exciting frequency ω. We can see that in the interval $\omega \in (0, 2.84)$, the semi-trivial solution is stable and so $u_y = 0$. The point $\omega = 2.84$ is a beginning of the resonance zone, which spans in $\omega \in (2.84, 2.99)$, where auto-parametric resonance occurs and amplitudes of both response components are commeasurable. For $\omega > 2.99$, the semi-trivial solution is regained. Samples of the trajectory shape are plotted in picture (b) for four frequencies $\omega = 2.84, 2.88, 2.92, 2.96$. Their vertical views demonstrate the character of the semi-trivial and the auto-parametric resonance states. Take a note that the trajectory since $\omega = 2.94$ is a simple ellipse-like curve, which does not exhibit any symptom of a chaotic process. Compare this finding with analysis concerning the sweeping up and down excitation frequency for ω around and above B_2 bifurcation point (BP) (see **Figure 7** and explanation later in this subsection).

The second series has been controlled by sweeping the excitation frequency up and down in a large interval and in several detailed regimes in the area of the auto-parametric resonance

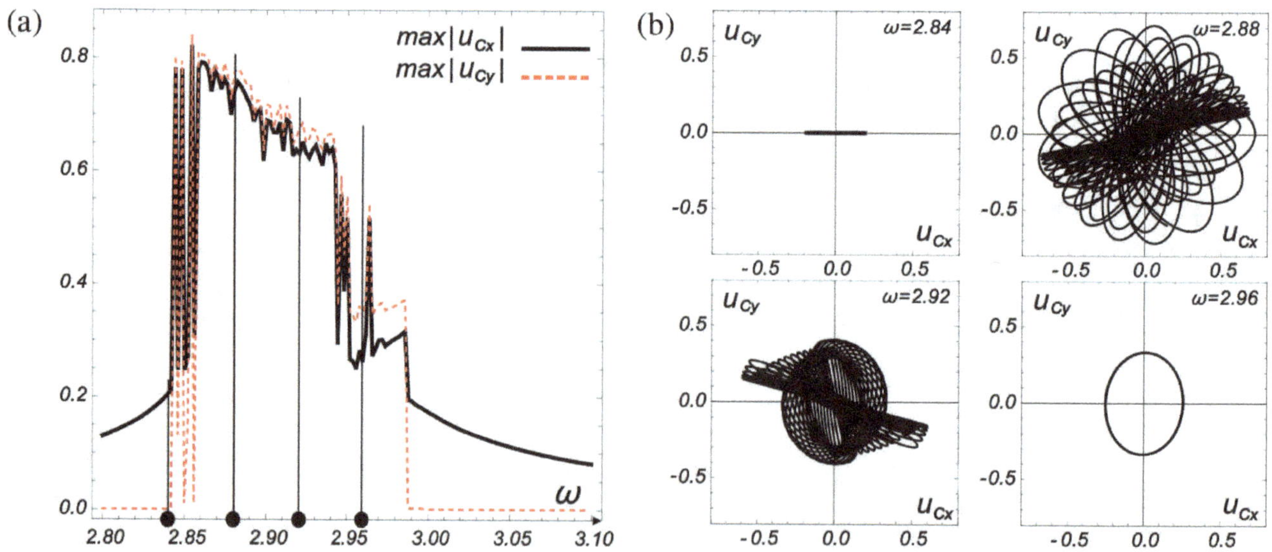

Figure 6. Response of the ball in the resonance and adjacent zones due to harmonic horizontal excitation of the cavity: (a) amplitude of the displacement as a function of the excitation frequency; (b) vertical views of the ball trajectory for frequencies $\omega = 2.84, 2.88, 2.92, 2.96$.

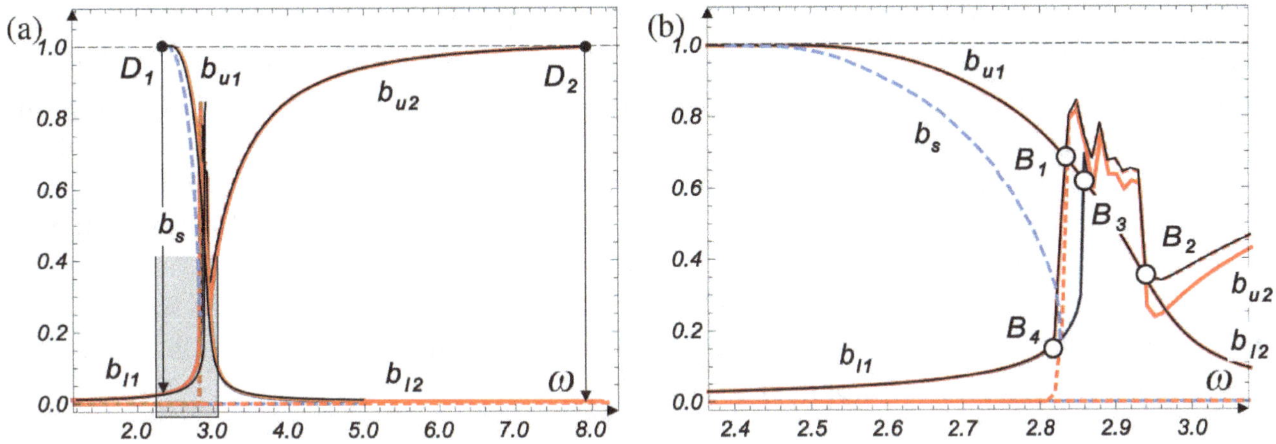

Figure 7. Amplitudes of the ball displacement under cavity harmonic excitation, when the frequency is swept up and down: (a) amplitudes overview in the interval $\omega \in (1.0, 8.0)$, (b) zooming in the interval $\omega \in (2.4, 3.1)$; curves: solid red—$\max|u_{Cx}|$, dashed red—$\max|u_{Cy}|$, solid black—absolute displacement amplitude, blue dashed—attraction boundary between b_{u1} and b_{l1}.

zone. A few of the results are visible in **Figure 7**. Picture (a) demonstrates amplitudes $\max|u_{Cx}|$ (solid curves) and $\max|u_{Cy}|$ (dashed curve) and the total amplitude u_{Cr} in the interval $\omega \in (1.0, 8.5)$. Picture (b) is the magnified detail of picture (a) within the interval $\omega \in (2.80, 3.05)$ in order to make visible the resonance zone.

Let us pay attention to bifurcation points (BPs). There are obviously concentrated in the resonance zone. In principle, they can be classified into two categories. The most important reveal B_1 and B_2. In the latter one, two branches start. The lower one b_{l2} approaches zero for $\omega \to \infty$ which indicates the non-moving ball in the vertical view. This branch takes place in the vertical plane and basically has a form of semi-trivial solution. Its stability increases with rising $\omega > \omega_{B2}$ as it follows from decreasing negative values of the Lyapunov exponent and of inspection of the relevant stability basins. The upper branch b_{u2} is spatial. It follows from the resonance zone where the spatial response type has a chaotic character. The relevant attractor reveals as an annular concentric area with diminishing width with increasing ω. The trajectory very quickly approaches a circular form in the horizontal plane. Its level with respect to the vertical axis rises and approaches "equatorial" position. However, the stability of this trajectory decreases, and we can see in **Figure 7** that around $\omega = 8.0$ even numerical perturbations of the integration process can overcome the stability limit (despite very small integration step) and the response trajectory falls down to the lower branch in the point D_2. Its position is not fixed. If hypothetically zero perturbation occurs, it could shift to infinity and approach together with the branch b_{u2} the asymptote at the level $R = 1$. Observing black $\max|u_{Cx}|$ and red $\max|u_{Cy}|$ parts of b_{u2}, we can see that they are getting coincide with increasing ω. It means that trajectory approaches the circle with radius $R = 1$.

Let us briefly discuss the shape of the response amplitudes for ω below BP B_1 and B_4. The BP B_1 is reached sweeping up along the branch b_{l1}, when it loses planar character passing through B_4. In such a case, the spatial response type emerges, exhibiting a chaotic response since B_1. This fact is obvious also looking at the dashed red curve representing u_{Cy}, which is trivial as far as B_4 and can bring the system from the semi-trivial solution into the auto-parametric resonance starting B_4. Take a note that passing BP B_4, planar response can remain in force, if any perturbation is avoided. It meets in B_3 the branch b_{l2} following also a planar path for swept up ω. Branch b_{u1} starts in B_1. Its stability rapidly decreases with descending ω. The point D_1 illustrates its limited extent in sub-resonance zone. This feature is visible observing the curve b_s, which represents a limit separating the area of attraction to b_{u1} and to b_{l1}. Take a note that b_s starts in B_4 and approaches D_1, despite hypothetically it goes together with D_1 as far as the vertical axis in the point $R = 1$. The b_s can be earned from stability basins for ω in the adequate interval for initial value $\omega_p = 0$. It corresponds to amplitude of u_{Cx} as a testing value for decision about affiliation to b_{u1} or b_{l1} attractiveness.

The interval between B_1 and B_2 includes the spatial response, see non-trivial amplitude $\max|u_{Cy}|$. The spatial response has a chaotic character, as it has been already outlined in the previous paragraph, when commenting the branch b_{u2}.

7. Conclusion

The common physical origin of Lagrangian and Appell-Gibbs approaches has been shown. It originates from the equilibrium of energy-level evolution in time on one side and power supply together with energy dissipation on the other side. Various formulations of this principle lead finally to different variational principles, although they follow from the same minimization of the energy spent to system response portrait. Comparing individual sections of the chapter, we can see that each one of commonly used procedures based on particular energy formulations is preferable for a certain type of problems. It can be concluded that there does not exist a single universal approach which should be recommended.

Some detailed properties of both approaches have been demonstrated in Sections 5 and 6. Both of them discuss non-holonomic problem of the ball movement within the spherical cavity under external excitation. The former one deals with a simple planar problem and shows that the Lagrangian approach is easily applicable to obtain reasonable results as far as a wide parametric discussion, which enable to earn a detailed insight into the system dynamic properties. The latter alternative represents the full space problem with six DOFs and three non-holonomic constraints. Some earlier studies tried to formulate this problem also in Lagrangian style using Lagrange's multipliers. Finally, it proved that the relevant governing differential system is too complex and does not enable appropriate detailed analysis of dynamic properties of the system. Therefore, the space problem outlined in Section 6 has been formulated using Appell-Gibbs approach. Transparent results have been obtained as needed for practical purposes in a device design and in further study of multi-body system dynamics. Take a note regarding the classification of singular solutions and their applicability for detailed analysis, stability of various regimes of the system under kinematic excitation, transitions among semi-trivial, auto-parametric, chaotic, and other states typical for nonlinear system. Let us add that both 2D and 3D problems have been investigated respecting the full nonlinearity without any simplifications of transcendent functions and thus enabling to study all effects without any limitations in amplitudes.

A certain shortcomings which apparently follow from the knowledge of rotation velocities only (no rotations themselves are calculated) can be disregarded, when displacements have been obtained. The rotation velocities represent mostly satisfactory information. Nevertheless, if rotations are still needed, there exist several variants of a simple differential system (following rotation vector definition) relating velocity and rotation vector components. This system can be subsequently easily solved, when necessary. A hidden complexity of the Lagrangian approach follows from an implicit connection of both parts, which are independent when Appell-Gibbs procedure is applied.

Acknowledgements

The kind support of the Czech Science Foundation project No. 17-26353 J and of the RVO 68378297 institutional support are gratefully acknowledged.

Author details

Jiří Náprstek* and Cyril Fischer

*Address all correspondence to: naprstek@itam.cas.cz

Institute of Theoretical and Applied Mechanics CAS, Prague, Czech Republic

References

[1] Lurie A. Analytical Mechanics (in Russian). Moscow: GIFML; 1961

[2] Nejmark JI, Fufaev NA. Dynamics of nonholonomic systems (in Russian). Moscow: Nauka; 1967. English translation—A.M.S. Translations of Mathematical Monographs (Book) Vol. 33. Providence: AMS; 1972

[3] Pars LA. A treatise on analytical dynamics. 2nd ed. Woodbridge: Ox Bow Press; 1972

[4] Hamel G. Theoretische Mechanik. Berlin: Springer; 1978. DOI: 10.1007/978-3-642-88463-4

[5] Bloch AM. Nonholonomic Mechanics and Control. New York: Springer; 2003. DOI: 10.1007/b97376

[6] Rubio Hervas J, Reyhanoglu M. Controllability and stabilizability of a class of systems with higher-order nonholonomic constraints. IFAC Journal Automatica. 2015;**54**:229-234. DOI: 10.1016/j.automatica.2015.02.006

[7] Bloch AM, Reyhanoglu M, McClamroch NH. Control and stabilization of nonholonomic dynamic systems. IEEE Transactions on Automatic Control. 1992;**37**(11):1746-1757. DOI: 10.1109/9.173144

[8] Reyhanoglu M, van der Schaft AJ, McClamroch NH, Kolmanovsky I. Dynamics and control of a class of underactuated mechanical systems. IEEE Transactions on Automatic Control. 1999;**44**(9):1663-1671. DOI: 10.1109/9.788533

[9] Cendra H, Ferraro S. A nonholonomic approach to isoparallel problems and some applications. Dynamic Systems. 2006;**21**(4):409-437. DOI: 10.1080/14689360600734112

[10] Cendra H, Grillo S. Generalized nonholonomic mechanics, servomechanisms and related brackets. AIP, Journal of Mathematical Physics. 2006;**47**:2209-2238. DOI: 10.1063/1.2165797

[11] Cendra H, Ferraro S, Grillo S. Lagrangian reduction of generalized nonholonomic systems. Journal of Geometry and Physics. 2008;**58**:1271-1290. DOI: 10.1016/j.geomphys.2008.05.002

[12] Bates L, Sniatycki J. Nonholonomic reduction. Reports on Mathematical Physics. 1993;**32**: 99-115. DOI: 10.1016/0034-4877(93)90073-N

[13] Pirner M. Dissipation of kinetic energy of large-span bridges. Acta Technica CSAV. 1994; **39**:407-418

[14] Lewis AD. The geometry of the Gibbs-Appell equations and gauss' principle of least constraints. Reports on Mathematical Physics. 1996;**38**:11-28. DOI: 10.1016/0034-4877(96)87675-0

[15] Udwadia F, Kalaba R. The explicit Gibbs-Appell equation and generalized inverse forms. Quarterly of Applied Mathematics. 1998;**56**:277-288

[16] Desloge E. The Gibbs-Appell equation of motion. American Journal of Physics. 1988;**56**: 841-846. DOI: 10.1119/1.15463

[17] Emami M, Zohoor H, Sohrabpour S. Solving high order nonholonomic systems using Gibbs-Appell method. In: Balan V, editor. Proceedings of the International Conference Differential Geometry and Dynamical Systems. Mangalia: Geometry Balkan Press; 2009. pp. 70-79

[18] Yong-Fen Q. Gibbs-Appells equations of variable mass nonlinear nonholonomic mechanical systems. Applied Mathematics and Mechanics. 1990;**11**:973-983. DOI: 10.1007/BF02115681

[19] Pirner M, Fischer O. The development of a ball vibration absorber for the use on towers. Journal of the International Association for Shell and Spatial Structures. 2000;**41**(2):91-99

[20] Náprstek J, Pirner M. Non-linear behaviour and dynamic stability of a vibration spherical absorber. In: Smyth A et al., editors. Proceedings of the 15th ASCE Engineering Mechanics Division Conference. New York: Columbia University; 2002. CD #150. 10 p

[21] Náprstek J, Fischer C, Pirner M, Fischer O. Non-linear dynamic behaviour of a ball vibration absorber. In: Papadrakakis M, Fragiadakis M, Plevris V, editors. Proceedings of the 3rd International Conference on Computational Methods in Structural Dynamics and Earthquake Engineering. Corfu: COMPDYN; 2011. ID #180. 14 p

[22] Xu Z, Cheung YK. Averaging method using generalized harmonic functions for strongly non-linear oscillators. Journal of Sound and Vibration. 1994;**174**(4):563-576. DOI: 10.1006/jsvi.1994.1294

[23] Ren Y, Beards CF. A new receptance-based perturbative multi-harmonic balance method for the calculation of the steady state response of non-linear systems. Journal of Sound and Vibration. 1994;**172**(5):593-604. DOI: 10.1006/jsvi.1994.1201

[24] Haxton RS, Barr A. The autoparametric vibration absorber. Journal of Engineering for Industry. 1972;**94**:119-125. DOI: 10.1115/1.3428100

[25] Tondl A. Quenching of self-excited vibrations. Prague: Academia; 1991

[26] Náprstek J, Fischer C. Dynamic behavior and stability of a ball rolling inside a spherical surface under external excitation. In: Zingoni A, editor. Proceedings of the 6th International Conference on Structural Engineering, Mechanics and Computation, SEMC. London:Taylor & Francis; 2016. pp. 214-219

[27] Náprstek J, Fischer C. Non-holonomic dynamics of a ball moving inside a spherical cavity. In: Vestroni F et al., editors. Proceedings of the EURODYN 2017—10th International Conference on Structural Dynamics. Procedia Engineering. 2017;**199**:613-618. DOI: 10.1016/j.proeng.2017.09.105

[28] Náprstek J, Fischer C. Dynamic response of a heavy ball rolling inside a spherical dish under external excitation. In: Zolotarev I, editor. Proceedings of the Engineering Mechanics. Svratka. Prague: IT ASCR; 2013. CD #46. pp. 96-106

7

Non-Linear Behaviours in the Dynamics of Some Biostructures

Emil Anton, Anna Gavrilut, Maricel Agop and
Daniel Timofte

Abstract

Various differentiable models are frequently used to describe the dynamics of complex systems (see the kinetic models, fluid models, etc.). Given the complexity of all the physical phenomena involved in the dynamics of such systems, it is required to introduce the dynamic variable dependencies both on the space-time coordinates and on the scale resolutions. Therefore, in this case an adequate theoretical approach may be the use of non-linear physical models either in the form of the Scale Relativity Theory or of the Extended Scale Relativity Theory, i.e., the Scale Relativity Theory with an arbitrary constant fractal dimension. In the framework of the Extended Scale Relativity Theory, fractal velocity field is described both by topological solitons of kink type and by non-topological soliton varieties of breather type. Applications for the blood flow are proposed. The results revealed the directional flow toward the walls, which can explain the thickening effect which is one of the source of arteriosclerosis.

Keywords: complex system, fractal, non-linear, bio-structure, non-differentiable

1. Introduction

Bio-structure is a complex non-Newtonian fluid made of: plasma and formed cells, cholesterol vesicles and other suspended elements [1]; thus, the laws of fractal physics are completely applicable to sanguine circulation. For conformity, the perfect Newtonian fluid is a fluid in which viscosity is independent of the shear stress, thus having no relation to the sanguine fluid.

However, not only the complex structure of bio-structure justifies the using of fractality, but also the complex structure of the arterial system, with its multiple ramifications, which generate turbulence areas and interruptions of the linear flowing that make classical physics not applicable in this context. We can thus discuss about multi-fractality: a morphological one due to complex structure of the arterial tree as well as a functional one due to bio-structure flow "regimes" [2]. Also, the stress of a visco-elastic fluid, unlike the Newtonian fluid, depends not only on the actually stress applied but on the one applied during previous deformation of the fluid [3, 4].

Standard theoretical models usually used in complex fluid dynamics and particularly that of flow through bio-structure vessels are ambiguous [5]. The assessment that the entities contained by bio-structure move along continuous and differentiable courses proves to be false, as it cannot comprise the entire variety of dynamics that are induced by the flowing of bio-structure through the vessel system (from the separation of bio-structure components through turbulence regimes to bio-structure-bio-structure vessel interactions).

In this context, we propose a new hypothesis according to which bio-structure structural unit move on continuous but non-differentiable curves (and particularly on fractal curves). We cannot predict the entirety of the bio-structure-vessel system, bio-structure-organic tissues, etc., or at elementary level, bio-structure entity-bio-structure structural unit (i.e., lymphocyte – granulocyte, lymphocyte – platelet and others) interactions. This is why accepting the above stated hypothesis comes up as a simple, elegant and efficient solution, the impossibility of predicting all these interactions that take place being substituted by the use of fractality [6].

We are led to the dynamics of a special type of fluid, free of interactions, in which the stream lines are continuous and non-differentiable curves.

Multiple physical models have been developed in the attempt to explain the dynamics of bio-structure flow and its physiological and pathological changes on the course of the entire arterial trunk, starting from the big elastic arteries and continuing with the small arteries and arterioles – resistance vessels -, following with the bio-structure capillaries (with arterial and venous components) and, backwards, with the post capillary veins, then with the middle and large veins – capacitance vessels – ensuring the anti-gravitational mobilization of bio-structure.

Thus, the hypothesis of the geometric risk factor in the development of the circulatory system's suffering has been proposed. This initially promising theory [7] proved its drawbacks that derive from a non-differentiable, Euclidean approach to the dynamics of bio-structure flow and its effects on the vessels' wall. The proposed counterbalance is represented by fractal physics laws [4, 8] that offer a great amount of freedom due to accepting the relativity of the complex fluids' behavior (bio-structure belongs to this category).

Correspondingly, the theoretical models that describe the complex fluids' dynamics are sophisticated [4, 8]. However, these models can be simplified since the complexity of the interaction process imposes various temporal resolution scales. Also, one should take into account the fact that the pattern evolution imposes different freedom degrees [3, 28].

One could develop new theoretical models assuming the fact that the complex fluids displaying chaotic behavior are recognized to acquire self-similarity in association with strong fluctuations at all possible space-time scales [4, 9]. One can replace the deterministic trajectories by collections of potential trajectories for temporal scales which are large with respect to the inverse of the highest Lyapunov exponent (see, e.g., [10, 11]). Also, the concept of definite positions can be replaced by that of probability density. An interesting example is represented by that of collisions processes in complex system, where the dynamics of the particles can be described by fractal (non-differentiable) curves.

Since fractality (non-differentiability) is a universal property of complex fluids, one should build a fractal (non-differentiable) physics. In this context, replacing the complexity of the interactions processes by non-differentiability, it is not necessary anymore to use the classical "arsenal" of quantities from the differentiable physics. This was developed in the Scale Relativity Theory (SRT) [12, 13] and in the non-standard Scale Relativity Theory (NSRT), i.e., Scale Relativity Theory with arbitrary constant fractal dimension [14]. In the framework of SRT or NSRT let us suppose that the complex system structural unit motions take place on continuous but non-differentiable curves (called fractal curves). In this way, all physical phenomena that are involved in the dynamics depend on the space-time coordinates and also on the space-time scales resolution. In this context, the physical quantities describing the complex systems dynamics could be considered to be fractal functions [13, 14]. Additionally, the complex system entities could be reduced to and identified with their own trajectories. In this way, the complex system behaves as a special interaction-less "fluid" by means of its geodesics in a non-differentiable (fractal) space.

In the present chapter, various bio-structure flow dynamics are analyzed aiming to propose new mechanisms for the genesis and evolution of different bio-structure-related pathologies (arterial occlusion, cholesterol deposition, etc.).

2. Material and method

Assuming that the motions of bio-structure's structural units take place on continuous but non-differentiable curves (fractal curves), the following consequences emerge [13, 14]:

(i) Any continuous but non-differentiable curve of bio-structure's structural units (bio-structure fractal curve) is scale resolution δt dependent. This means that when δt tends to zero, its length tends to infinity.

Let us recall that a curve is non-differentiable if it satisfies the Lebesgue theorem [9]. This means that when the scale resolution goes to zero, its length becomes infinite. In consequence, in this limit, a curve has a zigzag form and consequently it has the property of self-similarity in every one of its points. Since every part reflects the whole, this can be translated into a holography property [9];

(ii) The physics of bio-structure phenomena are related to the behavior of a set of functions during the zoom operation of the resolution scale δt. Through the substitution principle, δt

can be identified with dt, that is, $\delta t \equiv dt$. Consequently, it will play the role of an independent variable. We shall use the notation dt for the usual time as in the Hamiltonian biostructure dynamics;

(iii) The dynamics of bio-structure's structural units are described by means of fractal variables. Since the differential time reflection invariance of any dynamical variable is broken, these functions depend on the space-time coordinates and also on the resolution scale. In consequence, one can define two derivatives of the variable field $Q(t, dt)$ at any point of the bio-structure fractal curve:

$$\begin{aligned} \frac{d_+Q(t,dt)}{dt} &= \lim_{\Delta t \to 0_+} \frac{Q(t+\Delta t, \Delta t) - Q(t, \Delta t)}{\Delta t} \\ \frac{d_-Q(t,dt)}{dt} &= \lim_{\Delta t \to 0_-} \frac{Q(t, \Delta t) - Q(t-\Delta t, \Delta t)}{\Delta t} \end{aligned} \tag{1}$$

The "+" sign corresponds to forward processes of bio-structure's structural units, while the "−" sign correspond to the backwards ones;

(iv) The differential of the spatial coordinate field $dX^i(t, dt)$ by means of which we can describe the bio-structure dynamics, is expressed as the sum of the two differentials, one of them being scale resolution independent (differential part $d_\pm x^i(t)$ and the other one being scale resolution dependent (fractal part $d_+\zeta^i(t)$) i.e.,

$$d_\pm X^i(t) = d_\pm x^i(t) + d_\pm \zeta^i(t) \tag{2}$$

(v) The non-differentiable part of the spatial coordinate field, by means of which we can describe the bio-structure dynamics, satisfies the fractal equation [14]:

$$d_\pm \zeta^i(t, dt) = \lambda_\pm^i (dt)^{1/D_F} \tag{3}$$

where $\lambda_\pm^i$ are constant coefficients that help to specify the fractalization type which describes the bio-structure dynamics. Also, D_F defines the fractal dimension of the bio-structure non-differentiable curve.

In this way, the bio-structure processes imply dynamics on geodesics having different fractal dimensions. This variety of fractal dimensions of the bio-structure geodesics is a result of the bio-structure's structure. For $D_F = 2$, quantum type processes are generated in bio-structure dynamics [15]. For $D_F < 2$, correlative type processes are induced and for $D_F > 2$ non-correlative type processes are generated [6, 12, 13].

(vi) The differential time reflection invariance of any dynamical variable of the bio-structure is recovered by combining the derivatives d_+/dt and d_-/dt in the non-differentiable operator

$$\frac{\hat{d}}{dt} = \frac{1}{2}\left(\frac{d_+ + d_-}{dt}\right) - \frac{i}{2}\left(\frac{d_+ - d_-}{dt}\right) \tag{4}$$

This is a natural result of the complex prolongation procedure applied to bio-structure dynamics [14, 16]. Applying now the non-differentiable operator to the spatial coordinate field, by means of which we can describe the bio-structure dynamics, yields the complex velocity field of the bio-structure.

$$\widehat{V}^i = \frac{\widehat{d}X^i}{dt} = V^i - iU^i, i = \sqrt{-1} \tag{5}$$

with

$$V^i = \frac{1}{2}\frac{d_+X^i + d_-X^i}{dt}, U^i = \frac{1}{2}\frac{d_+X^i - d_-X^i}{dt} \tag{6}$$

The real part V^i of the bio-structure complex velocity field is differentiable and scale resolution independent (differentiable velocity field). The imaginary part U^i is non-differentiable and scale resolution dependent (fractal velocity field).

(vii) If we have no external constraint, one can find an infinite number of fractal curves (geodesics) relating any pair of points. This happens on all scales of bio-structure dynamics. Then, in the fractal space of the bio-structure, all the structural units are substituted with the geodesics themselves so that any external constraint can be interpreted as a selection of bio-structure geodesics. The infinity of geodesics in the bundle, their non-differentiability and the two values of the derivative imply a generalized statistical fluid-like description of the bio-structure dynamics. Then, the average values of the bio-structure variables must be considered in the previously mentioned sense, so the average of $d_\pm X^i$ is

$$\langle d_\pm X^i \rangle \equiv d_\pm x^i \tag{7}$$

with

$$\langle d_\pm \zeta^i \rangle = 0 \tag{8}$$

The previous relation (8) implies that the average of the fractal fluctuations is null.

(viii) One can describe the bio-structure dynamics by means of a scale covariant derivative. Its explicit form can be obtained as follows. We assume that the bio-structure fractal curves are immersed in a 3-dimensional space. We also suppose that X^i is the spatial coordinate field of a point on this fractal curve. Let us also consider a variable field $Q(X^i, t)$ and the following Taylor expansion up to the second order

$$d_\pm Q(X^i, t) = \partial_t Q dt + \partial_i Q d_\pm X^i + \frac{1}{2}\partial_l \partial_k Q d_\pm X^l d_\pm X^k \tag{9}$$

These relations are valid at any point and more for the points X^i on the bio-structure fractal curve which we have selected in Eq. (9). From here, forward and backward average values for bio-structure variables from Eq. (9) become

$$\langle d_\pm Q \rangle = \langle \partial_t Q dt \rangle + \langle \partial_i Q d_\pm X^i \rangle + \frac{1}{2} < \partial_l \partial_k Q d_\pm X^l d_\pm X^k > \tag{10}$$

We suppose that the average values of the all variable field Q and its derivatives coincide with themselves and the differentials $d_{\pm}X^i$ and dt are independent. Therefore, the average of their products coincides with the product of averages. Consequently, Eq. (10) becomes

$$d_{\pm}Q = \partial_t Q dt + \partial_i Q < d_{\pm}X^i > + \frac{1}{2}\partial_l\partial_k Q \langle d_{\pm}X^l d_{\pm}X^k \rangle \tag{11}$$

Even the average value of $d_{\pm}\zeta^i$ is null, for the higher order of $d_{\pm}\zeta^i$ the situation can still be different. Let us focus on the averages $\langle d_{\pm}\zeta^l d_{\pm}\zeta^k \rangle$. Using Eq. (3) we can write

$$\langle d_{\pm}\zeta^l d_{\pm}\zeta^k \rangle = \pm\lambda_{\pm}^l \lambda_{\pm}^k (dt)^{\left(\frac{2}{D_F}\right)-1} dt \tag{12}$$

where the sign + corresponds to $dt > 0$ and the sign – corresponds to $dt < 0$.

Then, Eq. (11) takes the form:

$$d_{\pm}Q = \partial_t Q dt + \partial_i Q < d_{\pm}X^i > + \frac{1}{2}\partial_l\partial_k Q d_{\pm}x^l d_{\pm}x^k \pm \frac{1}{2}\partial_l\partial_k Q \left[\lambda_{\pm}^l \lambda_{\pm}^k (dt)^{\left(\frac{2}{D_F}\right)-1} dt\right] \tag{13}$$

If we divide by dt and neglect the terms that contain differential factors (for details, see the method from [13, 14]) we obtain:

$$\frac{d_{\pm}Q}{dt} = \partial_t Q + v_{\pm}^i \partial_i Q \pm \frac{1}{2}\lambda_{\pm}^l \lambda_{\pm}^k (dt)^{\left(\frac{2}{D_F}\right)-1} \partial_l\partial_k Q \tag{14}$$

where $v_{+}^i = \frac{d_{+}x^i}{dt}$, $v_{-}^i = \frac{d_{-}x^i}{dt}$.

These relations also allow us to define the operators

$$\frac{d_{\pm}}{dt} = \partial_t + v_{\pm}^i \partial_i \pm \frac{1}{2}\lambda_{\pm}^l \lambda_{\pm}^k (dt)^{\left(\frac{2}{D_F}\right)-1} \partial_l\partial_k \tag{15}$$

Using Eqs. (4), (5), and (15), let us calculate the differentiable operator

$$\frac{\hat{d}Q}{dt} = \partial_t Q + \hat{V}^i \partial_i Q + \frac{1}{4}(dt)^{\left(\frac{2}{D_F}\right)-1} D^{lk} \partial_l\partial_k Q \tag{16}$$

where

$$\begin{gathered} D^{lk} = d^{lk} - i\bar{d}^{lk} \\ d^{lk} = \lambda_{+}^l \lambda_{+}^k - \lambda_{-}^l \lambda_{-}^k, \ \bar{d}^{lk} = \lambda_{+}^l \lambda_{+}^k + \lambda_{-}^l \lambda_{-}^k \end{gathered} \tag{17}$$

Eq. (16) also allows us to define the covariant derivative in the bio-structure dynamics

$$\frac{\widehat{d}}{dt} = \partial_t + \widehat{V}^i \partial_i + \frac{1}{4}(dt)^{\left(\frac{2}{D_F}\right)-1} D^{lk} \partial_l \partial_k \quad (18)$$

Let us now consider the principle of scale covariance (the physics laws – bio-structure dynamics specific – are invariant with respect to scale transformations) and postulate that the passage from the classical (differentiable) physics to the fractal (non-differentiable) physics can be implemented by replacing the standard time derivative d/dt by the non-differentiable operator $\widehat{d}/dt$. In this way, this operator has the role of a scale covariant derivative. More precisely, it is used to write the bio-structure dynamics fundamental equations in the same form as in the classic (differentiable) case. In these conditions, applying the operator (18) to the complex velocity field (5), with no external constraint, the bio-structure geodesics take the form:

$$\frac{\widehat{d}\widehat{V}^i}{dt} = \partial_t \widehat{V}^i + \widehat{V}^l \partial_l \widehat{V}^i + \frac{1}{4}(dt)^{\left(\frac{2}{D_F}\right)-1} D^{lk} \partial_l \partial_k \widehat{V}^i = 0 \quad (19)$$

This means that the local acceleration $\partial_t \widehat{V}^i$, the convection $\widehat{V}^l \partial_l \widehat{V}^i$ and the dissipation $D^{lk} \partial_l \partial_k \widehat{V}^i$, make their balance at any point of the bio-structure fractal curve. Moreover, the presence of the complex coefficient of viscosity-type $4^{-1}(dt)^{\left(\frac{2}{D_F}\right)-1} D^{lk}$ in the bio-structures dynamics specifies that it is a rheological medium. So, it has memory, as a datum, by its own structure.

If the fractalization is achieved by Markov type stochastic processes, which involve Lévy type movements [9, 13, 17] of the bio-structure structural units, then:

$$\lambda_+^i \lambda_+^l = \lambda_-^i \lambda_-^l = 2\lambda \delta^{il} \quad (20)$$

where δ^{il} is the Kronecker's pseudo-tensor.

Under these conditions, the equation of bio-structure geodesics takes the simple form

$$\frac{\widehat{d}\widehat{V}^i}{dt} = \partial_t \widehat{V}^i + \widehat{V}^l \partial_l \widehat{V}^i - i\lambda(dt)^{\left(\frac{2}{D_F}\right)-1} D^{lk} \partial_l \partial_k \widehat{V}^i = 0 \quad (21)$$

or more, by separating the motions on differential and fractal scale resolutions,

$$\begin{aligned} \frac{\widehat{d}V_D^i}{dt} &= \partial_t V_D^i + V_D^l \partial_l V_D^i - \left[V_F^l + \lambda(dt)^{\left(\frac{2}{D_F}\right)-1} \partial^l \right] \partial_l V_F^i = 0 \\ \frac{\widehat{d}V_F^i}{dt} &= \partial_t V_F^i + V_D^l \partial_l V_F^i + \left[V_F^l + \lambda(dt)^{\left(\frac{2}{D_F}\right)-1} \partial^l \right] \partial_l V_D^i = 0 \end{aligned} \quad (22)$$

Using the standard procedure from [18], the bio-structure dynamics at fractal scale resolution can be described by means of the following equations:

$$\partial_t V_F^i + V_F^l \partial_l V_F^i = \lambda (dt)^{\left(\frac{2}{D_F}\right)-1} \partial^l \partial_l V_F^i \tag{23}$$

$$\partial_i V_F^i = 0 \tag{24}$$

Eq. (23) corresponds to the specific impulse conservation law at fractal scale resolution, while Eq. (24) corresponds to the states density conservation law at fractal scale resolution (we consider that the density of the bio-structure at fractal scale resolution is constant – incompressible bio-structure).

Since this equation system is non-linear, one could find relatively difficult finding the solutions for these equations [19, 20]. However, in the particular case of a stationary flow in a plane symmetry (x, y), there is an analytical solution of this system. Then, for $V_F = (V_x, V_y, 0)$, Eqs. (23) and (24) take the form:

$$V_x \frac{\partial V_x}{\partial x} + V_y \frac{\partial V_x}{\partial x} = \lambda (dt)^{\left(\frac{2}{D_F}\right)-1} \frac{\partial^2 V_x}{\partial y^2} \tag{25}$$

$$\frac{\partial V_x}{\partial x} + \frac{\partial V_y}{\partial y} = 0 \tag{26}$$

The boundary conditions of the flow are:

$$\lim_{y \to 0} V_y(x, y) = 0, \ \lim_{y \to 0} \frac{\partial V_x}{\partial y} = 0, \ \lim_{y \to \infty} V_x(x, y) = 0 \tag{27}$$

and the flux momentum per length unit is constant

$$\Theta = \rho \int_{-\infty}^{+\infty} V_x^2 dy = const. \tag{28}$$

Using the method from [18–20] for solving Eqs. (25) and (26), with the conditions (27) and (28), the following solutions result:

$$V_x = \frac{\left[1.5\left(\frac{\Theta}{6\varrho}\right)^{\frac{2}{3}}\right]}{\left[\lambda (dt)^{\left(\frac{2}{D_F}\right)-1} x\right]^{1/3}} \cdot \text{sech}^2 \cdot \frac{\left[(0.5y)\left(\frac{\Theta}{6\varrho}\right)^{\frac{1}{3}}\right]}{\left[\lambda (dt)^{\left(\frac{2}{D_F}\right)-1} x\right]^{2/3}} \tag{29}$$

$$V_y = \frac{\left[4.5\left(\frac{\Theta}{6\varrho}\right)^{\frac{2}{3}}\right]}{\left[3\lambda (dt)^{\left(\frac{2}{D_F}\right)-1} x\right]^{1/3}} \cdot \left[\frac{\left[y\left(\frac{\Theta}{6\varrho}\right)^{\frac{1}{3}}\right]}{\left[\lambda (dt)^{\left(\frac{2}{D_F}\right)-1} x\right]^{\frac{2}{3}}} \text{sech}^2 \cdot \frac{\left[(0.5y)\left(\frac{\Theta}{6\varrho}\right)^{\frac{1}{3}}\right]}{\left[\lambda (dt)^{\left(\frac{2}{D_F}\right)-1} x\right]^{\frac{2}{3}}} - \tanh \frac{\left[(0.5y)\left(\frac{\Theta}{6\varrho}\right)^{\frac{1}{3}}\right]}{\left[\lambda (dt)^{\left(\frac{2}{D_F}\right)-1} x\right]^{\frac{2}{3}}}\right. \tag{30}$$

Relations (29) and (30) suggest that the bio-structure velocity field is highly non-linear through topological solitons of kink type (tanh), various non-topological solitons of breather type ($sech^2$), and through topological – non-topological soliton mixtures of kink-breather type ($sech^2$-tanh). Given the structural complexity of the bio-structure (which is given by its various structural units, that retains their own velocity field) an accurate way of writing relations (29) and (30) will be the one in which we assign indexes for each component.

For $y = 0$, we obtain in relation (29) the flow critical velocity of the bio-structure in the form

$$V_x(x, y=0) = V_c = \frac{\left[1.5\left(\frac{\Theta}{6\varrho}\right)^{\frac{2}{3}}\right]}{\left[\lambda(dt)^{\left(\frac{2}{D_F}\right)-1}x\right]^{1/3}} \tag{31}$$

while taking into account (31), relation (28) becomes

$$\Theta = \rho \int_{+\infty}^{+\infty} V_x^2(x,y)dy = \int_{-d_c}^{+d_c} V_c^2(x,0)dy, \tag{32}$$

so that the critical cross section of the strains lines tube of the bio-structure is given by:

$$d_c(x, y=0) = \frac{\Theta}{2\rho V_c^2} = 2.42\left[\lambda(dt)^{\left(\frac{2}{D_F}\right)-1}x\right]^{\frac{2}{3}}\left(\frac{\rho}{\Theta}\right)^{1/3} \tag{33}$$

Relations (29) and (30) can be strongly simplified if we introduce the normalized quantities

$$\zeta = \frac{x}{x_0}, \eta = \frac{y}{y_0}, u = \frac{V_x}{w_0}, v = \frac{V_y}{w_0}, \Omega = \frac{\left(\frac{\Theta}{6\varrho}\right)^{\frac{2}{3}}}{w_0\left[\lambda(dt)^{\left(\frac{2}{D_F}\right)-1}x_0\right]^{1/3}}, \omega = \frac{\left(\frac{\Theta}{6\varrho}\right)^{\frac{2}{3}}y_0}{\left[\lambda(dt)^{\left(\frac{2}{D_F}\right)-1}x_0\right]^{2/3}} \tag{34}$$

where x_0, y_0, w_0 are specific lengths and the specific velocity, respectively, of the laminar flow of the bio-structure. It results that

$$u(\zeta,\eta) = \frac{1.5\Omega}{\zeta^{\frac{1}{3}}}\operatorname{sech}^2\left(\frac{0.5\Omega\omega\eta}{\zeta^{\frac{2}{3}}}\right), \tag{35}$$

$$v(\zeta,\eta) = \frac{4.5^{2/3}}{3^{\frac{1}{3}}}\cdot\frac{\Omega}{\zeta^{\frac{1}{3}}}\left[\frac{\omega\eta}{\zeta^{\frac{2}{3}}}\operatorname{sech}^2\left(\frac{0.5\Omega\omega\eta}{\zeta^{\frac{2}{3}}}\right) - \tanh\left(\frac{0.5\Omega\omega\eta}{\zeta^{\frac{2}{3}}}\right)\right] \tag{36}$$

We present in **Figures 1a, b** and **2a, b** the dependence of the normalized velocity field u on the normalized spatial coordinates ξ, η for various nonlinearity degrees (ω = 0.3; 6). The results

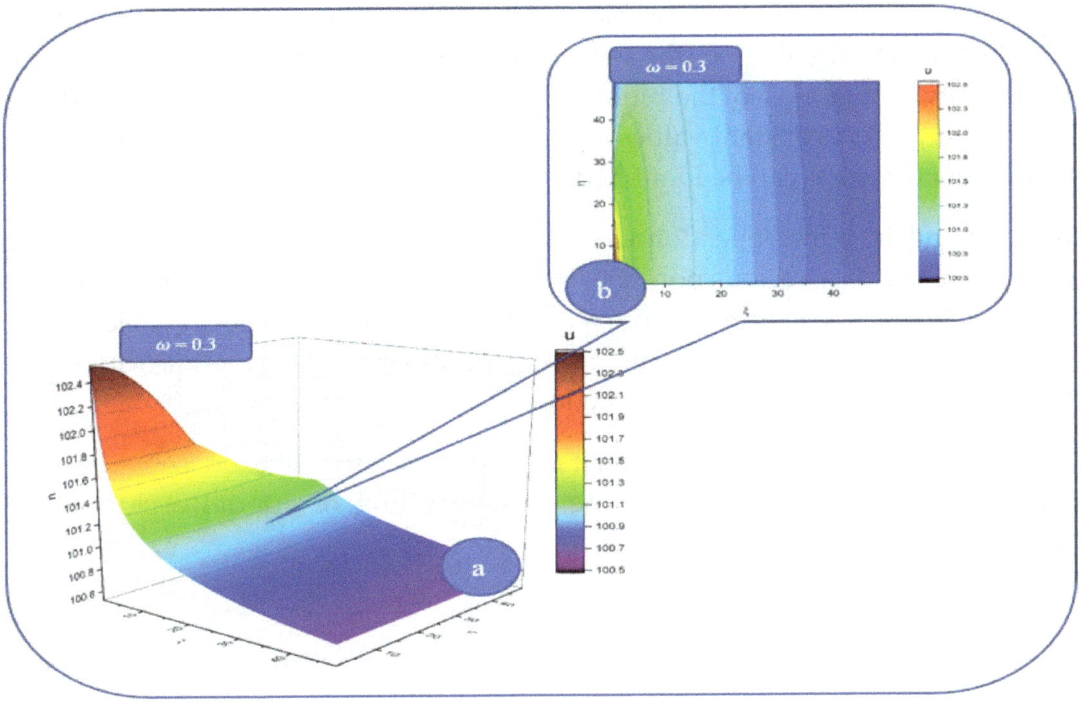

Figure 1. The dependence of the normalized velocity field u on the normalized spatial coordinates ξ, η for the non-linearity degree $\omega = 0.3$: (a) 3D representation; contour plot (b).

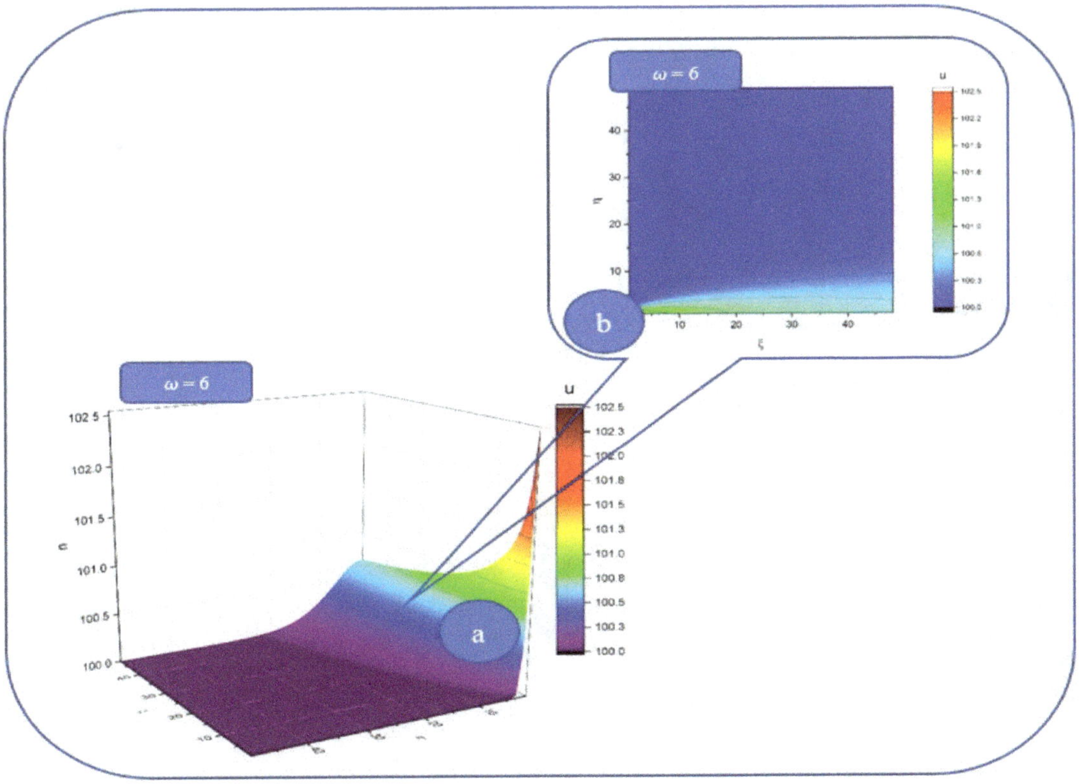

Figure 2. The dependence of the normalized velocity field u on the normalized spatial coordinates ξ, η for the non-linearity degree $\omega = 6$: (a) 3D representation; contour plot (b).

showcase that the velocity field on the bio-structure flow direction (ξ) is affected in a weak manner by the nonlinearity degree (the velocity always decreases on the flow axes regardless of the nonlinearity degree). Also, the bio-structure flow direction (η) is strongly affected. Bio-structure flow starts from constant values on the η axes and with the increase of ω, preferential bio-structure flow direction can be identified.

In **Figures 3a, b** and **4a, b** the dependences of the normalized velocity field v on the normalized spatial coordinates ξ, η for various non-linearity degrees ($\omega = 0.3$; 6) are represented. For small non-linearity degrees, the variations (increase/decrease) of the velocity field have similar behaviors on both directions (ξ, η), while for higher values of the non-linearity degree these variations are only focused on a single direction (ξ).

Taking the above into account, the force that the bio-structure will exercise to the walls of the flow vessels is of great importance for the understanding of arterial occlusion and other circulatory system diseases.

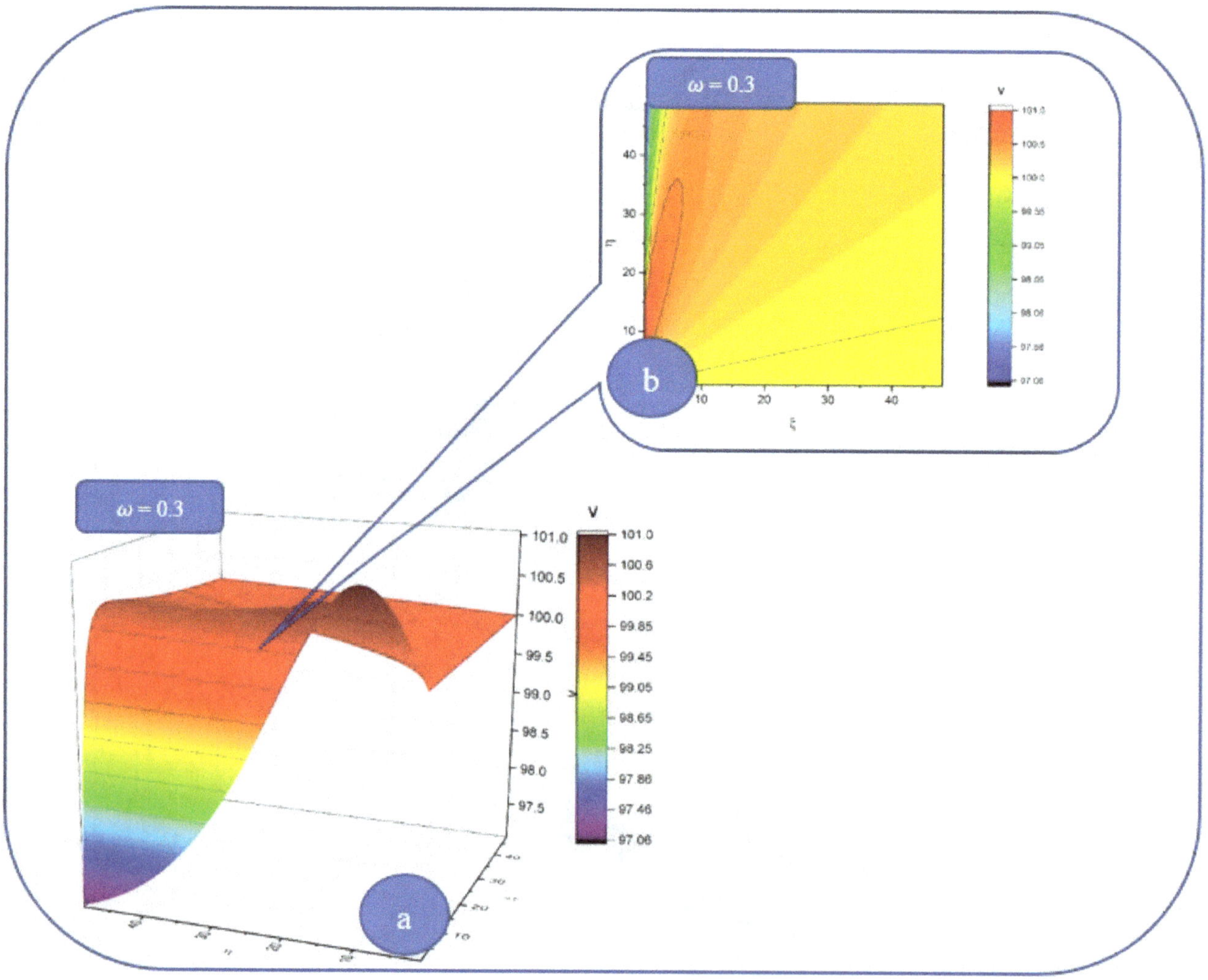

Figure 3. The 3D representation (a) and the contour plot (b) of the normalized velocity field v on the normalized spatial coordinates ξ, η for the nonlinearity degree $\omega = 0.3$.

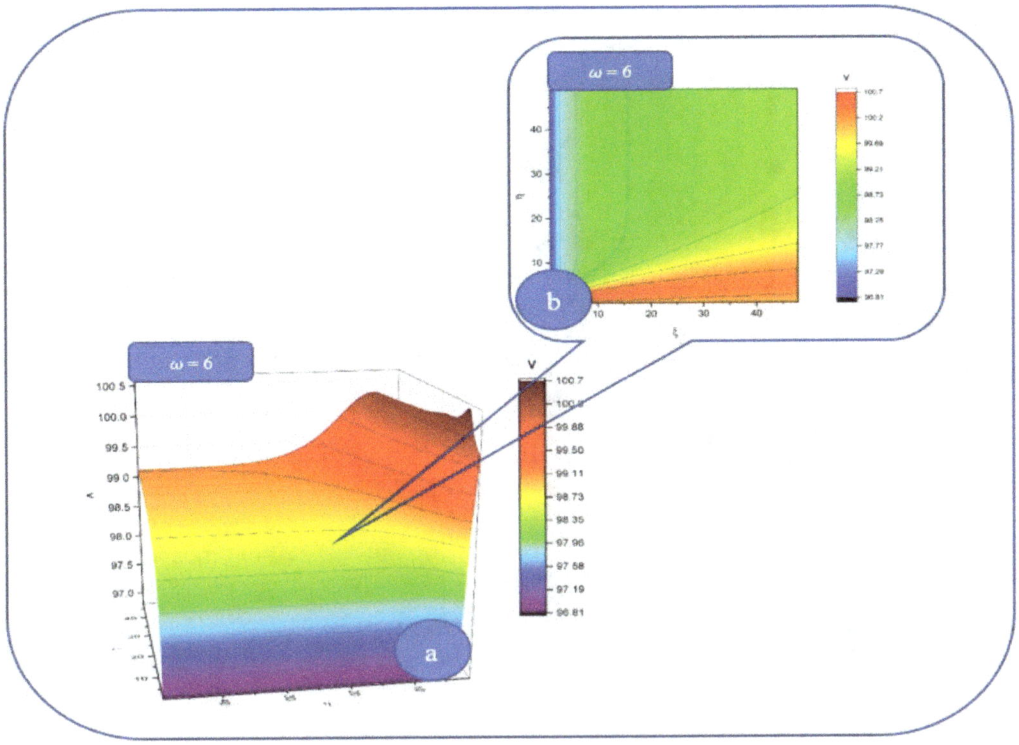

Figure 4. The 3D representation (a) and the contour plot (b) of the normalized velocity field v on the normalized spatial coordinates ξ, η for the non-linearity degree $\omega = 6$.

In our case the normalized force is given by the relation:

$$
\begin{aligned}
f(\zeta,\eta) &= \partial_\eta u - \partial_\zeta v \\
&= -1.5\frac{\Omega \mathrm{sech}^{2}\left(\frac{0.5\omega\eta}{\zeta^{\frac{2}{3}}}\right)\tanh\left(\frac{0.5\omega\eta}{\zeta^{\frac{2}{3}}}\right)\omega}{\zeta} - \frac{1}{\zeta^{\frac{1}{3}}} \\
&\cdot\left(0.9\Omega\left(-\frac{2}{3}\frac{\omega\eta\mathrm{sech}^2\left(\frac{0.5\omega\eta}{\zeta^{\frac{2}{3}}}\right)}{\zeta^{\frac{5}{3}}} + \frac{0.66\omega^2\eta^2\mathrm{sech}^2\left(\frac{0.5\omega\eta}{\zeta^{\frac{2}{3}}}\right)\tanh\left(\frac{0.5\omega\eta}{\zeta^{\frac{2}{3}}}\right)\omega}{\zeta^{\frac{7}{3}}}\right.\right. \\
&\left.\left. + \frac{0.33\left[\left(1-\tanh^2\left(\frac{0.5\omega\eta}{\zeta^{\frac{2}{3}}}\right)\right)\omega\eta\right]}{\zeta^{\frac{5}{3}}}\right)\right) \\
&+ \frac{0.3\Omega\left(\frac{\omega\eta\mathrm{sech}^2\left(\frac{0.5\omega\eta}{\zeta^{\frac{2}{3}}}\right)}{\zeta^{\frac{2}{3}}} - \tanh\left(\frac{0.5\omega\eta}{\zeta^{\frac{2}{3}}}\right)\right)3^{3/2}}{\zeta^{\frac{4}{3}}}
\end{aligned}
\tag{37}
$$

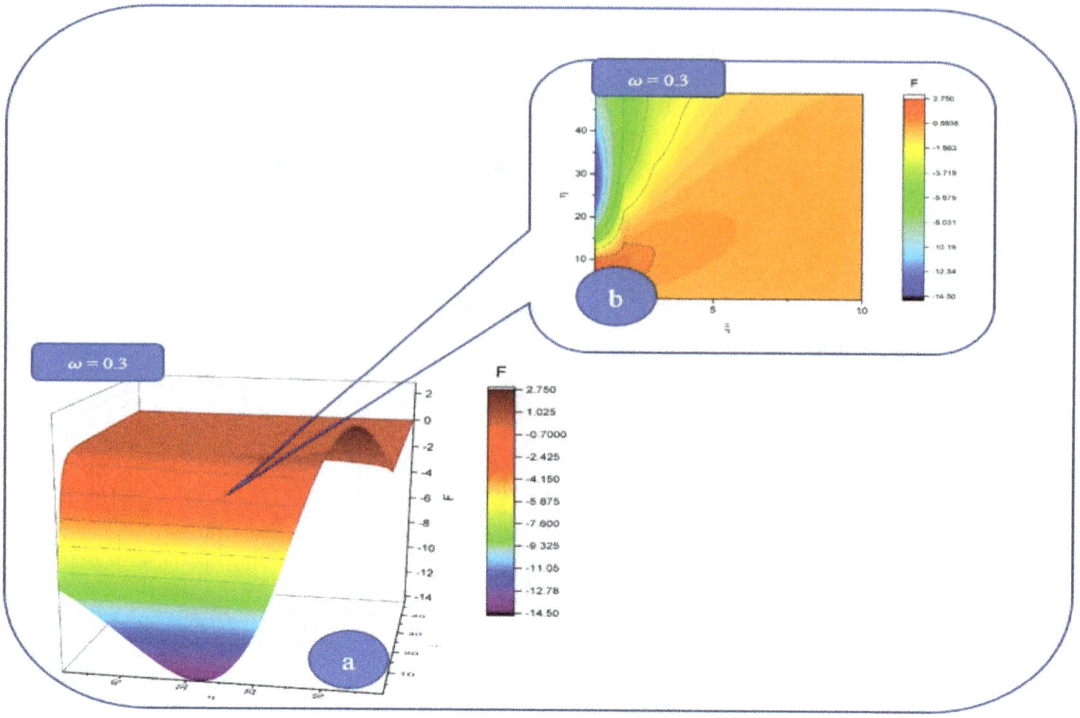

Figure 5. The dependence of the normalized force field F of a bio-structure flow on the vessels, on the normalized spatial coordinates ξ, η for two resolution scales: 3D representation (a); contour plot (b) for the non-linearity degree $\omega = 0.3$.

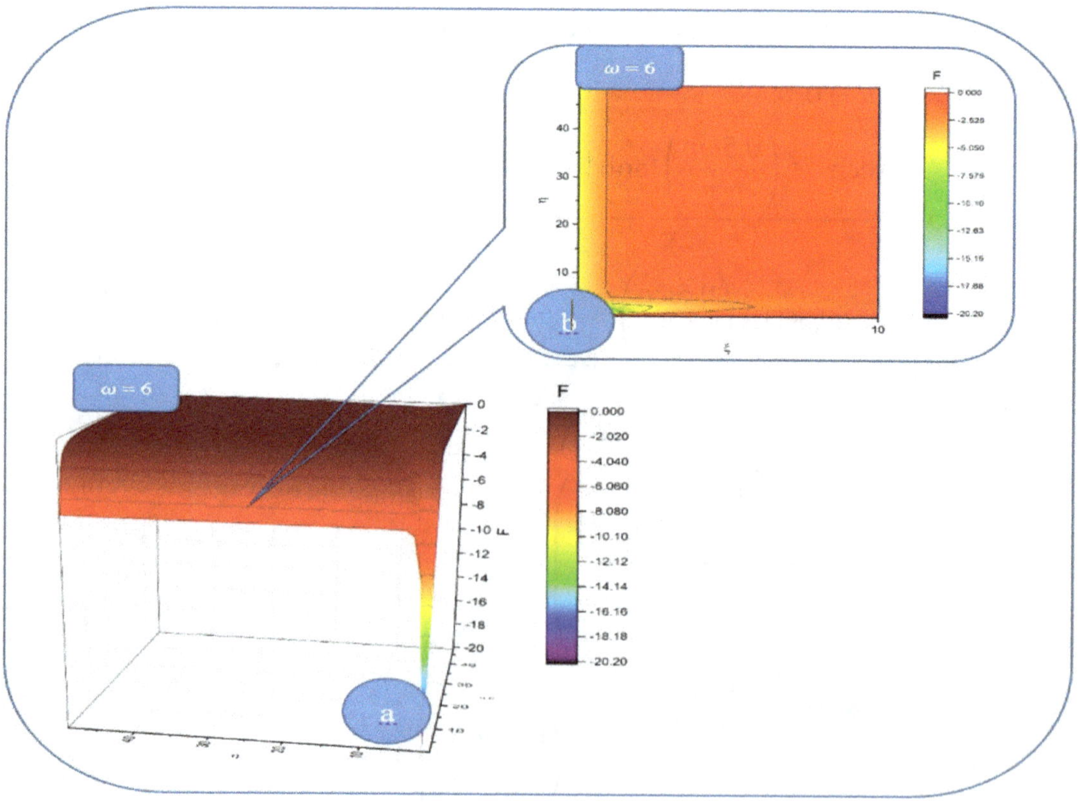

Figure 6. The dependence of the normalized force field F of a bio-structure flow on the vessels, on the normalized spatial coordinates ξ, η for two resolution scales: 3D representation (a) and contour plot (b), for the non-linearity degree $\omega = 6$.

In **Figures 5a**, **b** and **6a**, **b** the normalized force field evolution on the two-flow direction (ξ, η) for various non-linear degrees is represented. It results that with the increase of the non-linearity of the bio-structure the force toward the walls increases.

3. Discussions

The theory proposed in this chapter explains from a fractal viewpoint the atherogenesis process [21], basically "molding" to the classical anatomical and histopathological descriptions and completely respecting the process they postulate. In consequence, the fractal physics model sustains already accumulated morpho-pathological information and research. There are plenty of electronic and optical microscopy images that describe the spatial-temporal hologram of the phenomenon; we can thus discuss about the non-fractal – fractal and micro-scopic – macroscopic translation through holographically reproducible auto-similarity [21]. In this way, we affirm that fractality is the mathematical and semantic quintessence for defining atherogenesis, a process that can be physically characterized by fractal physics. This physics becomes in this situation more of a component rather than an explanation for the complex biological system represented by the atheroma plaque [22, 23].

In what concerns the recovery of such biological diseases, there are a huge number of techniques. We recall that external electrical stimulation can cause changes in the bio-structure vessels. Although atherosclerosis cause vasodilatation in the affected area and bio-structure flow remains unchanged for an extended period of time, the vascular wall stiffness will increase the pulse pressure. The purpose of the study developed in [24, 25] was to measure the effects of electrical stimulation (ES) on bio-structure flow and bio-structure pressure. All subjects received electrical stimulation at intensity sufficient to produce torque equal to 15% of the predetermined maximal voluntary contraction of their right quadriceps femor is muscle. The conclusions were that the increase in bio-structure flow occurred within 5 min after the onset of ES and dropped to resting levels within 1 min after a 10-min period of ES [25]. Kinesiotherapy or Kinesitherapy or kinesiatrics, is the therapeutic treatment of disease by passive and active muscular movements (as by massage) and of exercise [26]. From the physiotherapeutic viewpoint, an efficient treat-ment is directed toward improving bio-structure flow and also decreasing the disparity between the demand for bio-structure and its supply [22]. An effective vascular rehabilitation training program for improving walking efficiency and vascular remodeling in patients with diabetic atherosclerosis suffering from intermittent claudication could be a supervised treadmill walking exercise combined with Allen-Burger exercises [23].

4. Conclusions

The present chapter proposes a fractal model for the dynamics analysis of bio-structure flows. The fractal hydrodynamic equations were obtained and applied for the laminar flow of bio-structure.

A second application was proposed for bio-structure flow, and of cholesterol deposition on the vessel walls. The results revealed the directional flow toward the walls. This could explain in our opinion the thickening effect which is one of the sources of arteriosclerosis. Moreover, our model imposes redefinition of "good" and "bad" cholesterol (which are traditionally associated with HDL and LDL respectively); instead they should be replaced by the following notions: specific cholesterol entities, associated with a certain non-differentiable curve, that have a major endothelial impact and specific cholesterol entities which have no or low endothelial impact.

There is currently a great number of works describing matter organization and behavior in all of its variations, from which we mention [27]. We consider that our bio-structure flow model could also be used to further development in the study of other complex systems dynamics (such as pulmonary and metabolic diseases or environmental systems). Moreover, possible therapeutic treatments can be developed, e.g., new drug release mechanisms.

Conflict of interest

The authors declare no conflict of interest.

Author details

Emil Anton[1], Anna Gavrilut[2]*, Maricel Agop[3] and Daniel Timofte[4]

*Address all correspondence to: gavrilut@uaic.ro

1 Department Obstetrics Gynecology, Gr. T. Popa University of Medicine and Pharmacy, Iasi, Romania

2 Faculty of Mathematics, Alexandru Ioan Cuza University from Iasi, Romania

3 Department of Physics, Gheorghe Asachi Technical University of Iasi, Romania

4 Department of Surgery, Gr. T. Popa University of Medicine and Pharmacy, Iasi, Romania

References

[1] Popescu DM. Hematologie Clinică (in Romanian). Bucharest: Medical Publishing House; 1999

[2] Tesloianu ND. PAD and COPD in Smokers. Iași: Ars Longa; 2014

[3] Badii R, Politi A. Complexity: Hierarchical Structure and Scaling in Physics. Cambridge: Cambridge University Press; 1997

[4] Michel OD, Thomas BG. Mathematical Modelling for Complex Fluids and Flows. NewYork: Springer; 2012

[5] Schwartz SM, Murry CE. Proliferation and the monoclonal origins of atherosclerotic lesions. Annual Review of Medicine. 1998;**49**:437-460

[6] Nottale L. Fractals and the quantum theory of space time. International Journal of Modern Physics A. 1989;**4**:5047-5117

[7] Friedman MH, Deters OJ, Mark FF, et al. Arterial geometry affects hemodinamics. A potential risk factor for atherosclerosis. Atherosclerosis. 1983;**46**:225-231

[8] Thomas YH. Multi-Scale Phenomena in Complex Fluids: Modelling, Analysis and Numerical Simulations. Singapore: World Scientific Publishing Company; 2009

[9] Mandelbrot BB. The Fractal Geometry of Nature (Updated and Augm. Ed.). New York: W.H. Freeman; 1983

[10] Cristescu CP. Non-linear Dynamics and Chaos. Theoretical Fundaments and Applications (in Romanian). Bucharest: Romanian Academy Publishing House; 2008

[11] Federer J, Aharoner A. Fractals in Physics. Amsterdam: North Holland; 1990

[12] Nottale L. Fractal Space-Time and Microphysics: Towards a Theory of Scale Relativity. Singapore: World Scientific; 1995

[13] Nottale L. Fractal Space-Time. A New Approach to Unifying Relativity and Quantum Mechanics. London: Imperial College Press; 2011

[14] Merches I, Agop M. Differentiability and Fractality in Dynamics of Physical Systems. Singapore: World Scientific; 2016

[15] El Naschie MS, Rossler OE, Prigogine I. Quantum Mechanics, Diffusion and Chaotic Fractals. Oxford: Elsevier; 1995

[16] Cresson J. Non-differentiable deformations of $\mathbb{R}^n$. International Journal of Geometric Methods in Modern Physics. 2006;**3**:13-95

[17] Gouyet JF. Physique et Structures Fractales. Paris: Masson; 1992

[18] Solovastru L-G, Ghizdovat V, Nedeff V, Lazar G, Eva L, Ochiuz L, Agop M, Popa RF. Non-linear effects at differentiable-non-differentiable scale transition in complex fluids. Journal of Computational and Theoretical Nanoscience. 2016;**13**:1-6

[19] Batchelor GK. An Introduction to Fluid Dynamics. Cambridge: Cambridge University Press; 2000

[20] Landau LD, Lifshitz EM. Fluid Mechanics. 2nd ed. Oxford: Butterworth Heinemann; 1987

[21] Tesloianu ND, Ghizdovat V, Agop M. Flow Dynamics Via Non-differentiability and Cardiovascular Diseases. Saarbrücken: Scholar's Press; 2015

[22] Adelaide EC. Atherosclerosis Obliterans. The Australian Journal of Physiotherapy. 1958; **4**:19-22

[23] Haas TL, Lloyd PG, Yang H-T, Terjung RL. Exercise training and peripheral arterial disease. Comprehensive Physiology. 2012;**2**:2933-3017

[24] Singh RB, Mengi SA, Xu Y-J, Arneja AS, Dhalla NS. Pathogenesis of atherosclerosis: A multifactorial process. Experimental and Clinical Cardiology. 2002;**7**:40-53

[25] Tracy JE, Currier DP, Threlkeld AJ. Comparison of selected pulse frequencies from two different electrical stimulators on bio-structure flow in healthy subjects. Physical Therapy. 1988;**68**:1526-1532

[26] Vasileva D, Lubenova D, Mihova M, Dimitrova A, Grigorova-Petrova K. Influence of Kinesitherapy on Gaitin patients with ischemic stroke in the chronic period. Open Access Macedonian Journal of Medical Sciences. 2015;**3**:619-623

[27] Constantin B, Postolache P, Croitoru A, Nemes RM. Occupational bronchial asthma-clinical and epidemiological aspects. Journal of Environmental Protection and Ecology. 2015;**16**:517-520

[28] Mitchel M. Complexity: A Guided Tour. Oxford: Oxford University Press; 2009

8

Soliton-Like Solutions in the Problems of Vibrations of Nonlinear Mechanical Systems

Yury A. Rossikhin and Marina V. Shitikova

Abstract

In the given chapter, free vibrations of different nonlinear mechanical systems with one-degree-of-freedom, two-degree-of-freedom, and multiple-degree-of-freedoms are reviewed with the emphasis on the vibratory regimes which could go over into the aperiodic motions under certain conditions. Such unfavorable and even dangerous regimes of vibrations resulting in the irreversible process of energy exchange from its one type to another type are discussed in detail. The solutions describing such processes are found analytically in terms of functions, which are in frequent use in the theory of solitons.

Keywords: soliton-like solution, nonlinear mechanical systems, free vibrations, method of multiple time scales, suspension bridge

1. Introduction

It is known [1] that the periodical transfer of energy from one type to another is made possible during vibrational processes occurring in nonlinear mechanical systems. This phenomenon is called *energy exchange* [2, 3].

Investigations on the energy exchange originate from the chapter [4], wherein the authors studied small nonlinear vibrations of a two-degree-of-freedom (2dof) system consisting of a load suspended on a linearly elastic spring and executing pendulum vibrations and vibrations along the spring's axis in the same vertical plane. In spite of the apparent simplicity of that system, it realistically explains some phenomena occurring during vibrations of more complex nonlinear systems and in particular describes all types of energy exchange from pendulum vibratory motions into oscillatory motions along the spring's axis, and vice versa: the periodic and aperiodic energy interchange, as well as stationary regimes during which the energy exchange is absent.

The energy-exchange mechanism in a similar nonlinear 2dof system has been studied in [5, 6]. The system was made up of two loads, one of which was suspended on a linearly elastic spring and executed vertical vibrations, and the other was suspended on an unstretched rod and executed pendulum vibrations in the same vertical plane. Reviews devoted to nonlinear vibrations of 2dof systems can be found in [2, 3].

However, the energy transfer is observed during free vibrations of different nonlinear mechanical systems: possessing one-degree-of-freedom (1dof), two- (2dof), and more degrees-of-freedom (multiple-dof), and as well as having infinite number of degrees-of-freedom (deformable solids). *The internal resonance* is realized when magnitudes of natural frequencies of two natural modes belonging to the different types of vibrations of the system (*partial subsystems*) are approximately equal to each other or one of them two to three times larger than the other. This phenomenon is particular evident in modern engineering structures which are very light and flexible due to the application of present-day materials, resulting in finite displacements of individual structural elements as well as of the structure as a whole. Among such constructions are suspension-combined systems: suspension and cable-stayed bridges, suspension roofs in large public and industrial buildings, and so on. Suspension-combined systems and suspension bridges, in particular, are distinguished by high esthetic merits, and many of them are referred to the most remarkable up-to-date engineering structures. For example, "Golden Gate" suspension bridge in San Francisco with the span of 1281 m, cable-stayed bridge in Cologne with the span of 690 m, suspension roofing of Olympic sport complex in Moscow, and many others.

The majority of papers devoted to the dynamic behavior of suspension-combined systems studies free nonlinear vibrations of suspension bridges with a thin-walled stiffening girder [7–11]. Different dynamic loads (wind, seismic excitation, moving loads, etc.) after the completion of acting on a suspended structure setup prolonged free nonlinear vibrations of this structure, in so doing both vertical and flexural-torsional vibrations could be excited. One of the most unfavorable nonlinear effects, which is observed in suspension systems during free vibrations, is just the "energy exchange" from one type of vibratory motions into the other under the conditions of the internal resonance.

The intensity and frequency of energy exchange between strongly coupled modes essentially depend on an absolute level of the initial amplitudes [7, 8, 11, 12] which is governed by the value of the initial mechanical energy of the system.

However, the qualitative character of the energy exchange is dependent on the relative level of initial amplitudes which is independent of the system's initial energy and is defined as the ratio of the initial amplitudes of the two interacting modes [9]. It has been found in [9] that in accordance with a value of that level, three types of an energy-exchange mechanism exist: *two-sided energy exchange* (a periodic energy exchange from one subsystem to another), *one-sided energy exchange* (one subsystem completely or partially transfers the energy to another), and energy exchange does not occur (*stationary vibrations*). Among the three types of the behavior of the mechanical system, the second one may occur to be the most unfavorable. As for the behavior of a suspension bridge, then the most hazardous type is the irreversible transfer of the energy of vertical vibrations into the energy of its torsional vibrations in the case of a bisymmetrical stiffening girder or into the energy of flexural-torsional vibrations in the case of a mono-symmetrical girder. This is due to the fact that suspension bridges possess a rather higher flexural rigidity than torsional one, that is, they perceive better than those dynamic loads that result in vertical vibrations.

Solutions describing the one-sided energy transfer occurring in mechanical systems we shall call as *soliton-like solutions*, since the functions entering in such solutions are widely met in the theory of solitons [13, 14].

In this chapter, it is shown that solutions of such a type exist both in 1dof systems and in systems possessing two- and more degrees-of-freedom.

2. A one-degree-of-freedom system

The phenomenon of energy transfer, when one type of the energy completely and irreversibly goes into another type of the energy as time passes, can be observed on such a simple object as a mathematical pendulum (**Figure 1**).

In order to demonstrate this, let us consider the expression for the total mechanical energy of the mathematical pendulum which is combined from the kinetic energy

$$T = \frac{1}{2}mv^2 = \frac{1}{2}ml^2(\dot{\varphi})^2 \tag{1}$$

and the potential energy (**Figure 1**)

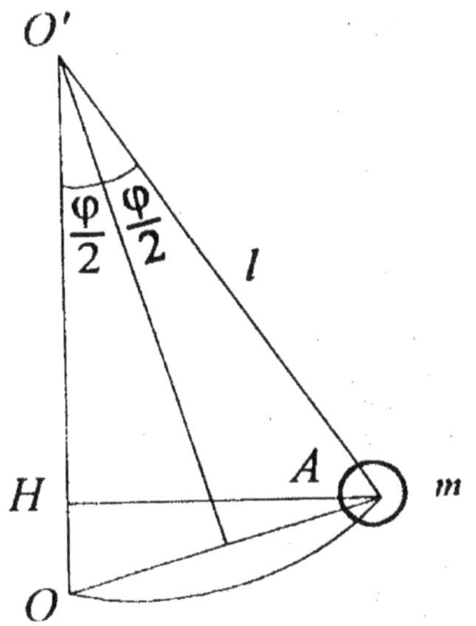

Figure 1. A mathematical pendulum.

$$\Pi = mg(OH) = 2mgl\sin^2\left(\frac{\varphi}{2}\right) \tag{2}$$

and has the form

$$E = T + \Pi = \frac{1}{2}ml^2(\dot{\varphi})^2 + 2mgl\sin^2\left(\frac{\varphi}{2}\right), \tag{3}$$

where an overdot denotes a time derivative, l is the string length, g is the gravity acceleration, m is the load mass, $v = l\dot{\varphi}$ is its velocity, and φ is the angle of the string's deflection from the vertical.

Rewrite Eq. (3) in the dimensionless form

$$\frac{(\dot{\varphi})^2}{\omega_0^2} + 4\sin^2\left(\frac{\varphi}{2}\right) = \frac{E}{E_0}, \tag{4}$$

where $E_0 = \frac{1}{2}m\omega_0^2 l^2$ and $\omega_0 = \sqrt{g/l}$.

Consider the case of motion of the mathematical pendulum when its energy E is exactly equal to $4E_0$. Then, the law of conservation of energy Eq. (4) gives the simple relationship [15].

$$\frac{(\dot{\varphi})^2}{\omega_0^2} = 4\cos^2\left(\frac{\varphi}{2}\right) \tag{5a}$$

or

$$\dot{\varphi} = 2\omega_0\cos\left(\frac{\varphi}{2}\right). \tag{5b}$$

Dividing the variables in Eq. (5b), integrating separately the right and left parts of the relationship obtained, and considering that $\varphi = 0$ at $t = 0$ yield

$$\ln\left[\tan\left(\frac{\pi - \varphi}{4}\right)\right] = -\omega_0 t \tag{6a}$$

or

$$\varphi = \pi - 4\arctan(e^{-\omega_0 t}). \tag{6b}$$

Differentiating Eq. (6b) over t, we find

$$\dot{\varphi} = \frac{2\omega_0}{\cosh(\omega_0 t)} \tag{7}$$

Reference to Eqs. (6) and (7) shows that if the mathematical pendulum begins its motion from the extreme low position, then at $t \to \infty$ its velocity $\dot{\varphi} \to 0$, in so doing does not vanish

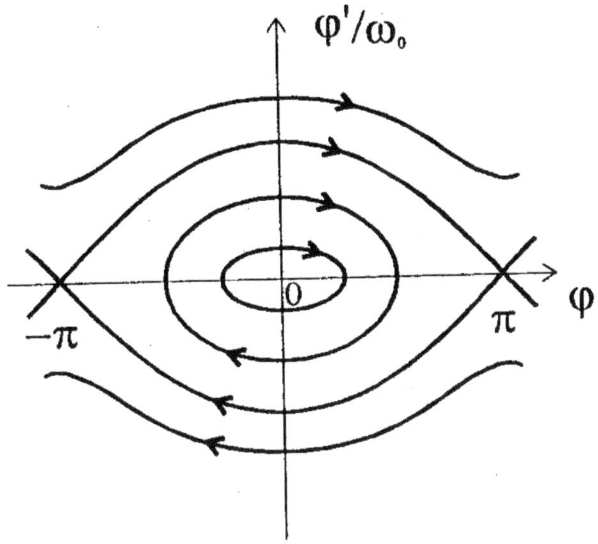

Figure 2. Phase portrait describing vibrations of mathematical pendulum.

anywhere, and the angle $\varphi \to \pi$, that is, the pendulum, tends to take the upper position of equilibrium which is an unstable one. As this takes place, the kinetic energy completely transforms into the potential energy. This solution is the soliton-like one, since the functions *arctan* and *ch* are frequently met in soliton solutions.

If one represents the phase trajectories of the pendulum motion on the phase plane $\dot{\varphi}/\omega_0 - \varphi$ at different magnitudes of the energy E, then solution (6) will correspond to the phase trajectory which is called as a *separatrix*. This line divides closed trajectories from nonclosed ones (**Figure 2**). Closed and nonclosed trajectories are consistent with the solutions for the periodic transfer of the potential and kinetic energies into each other, in doing so in the first case, the pendulum will vibrate, and in the second one, it will rotate around the point of suspension.

3. A two-degree-of-freedom system

3.1. Governing equations

Now, consider a 2dof system presented in **Figure 3**. The kinetic T and potential Π energies of such a system have the form

$$T = \frac{1}{2}(m_1 + m_2)\dot{y}^2 - m_2 l\, \dot{y}\dot{\varphi}\sin\varphi + \frac{1}{2}m_2 l^2 \dot{\varphi}^2, \tag{8a}$$

$$\Pi = \frac{1}{2}k(y + y_{cm})^2 - m_1 g(y + y_{cm}) - m_2 g(y + y_{cm} + l\cos\varphi), \tag{8b}$$

where $y_{cm} = (m_1 + m_2)g/k$, k is the elastic spring rigidity, m_1 and m_2 are the masses of the first and second loads, respectively, y is the vertical displacement of the first load, and φ is the angle of the pendulum's deflection.

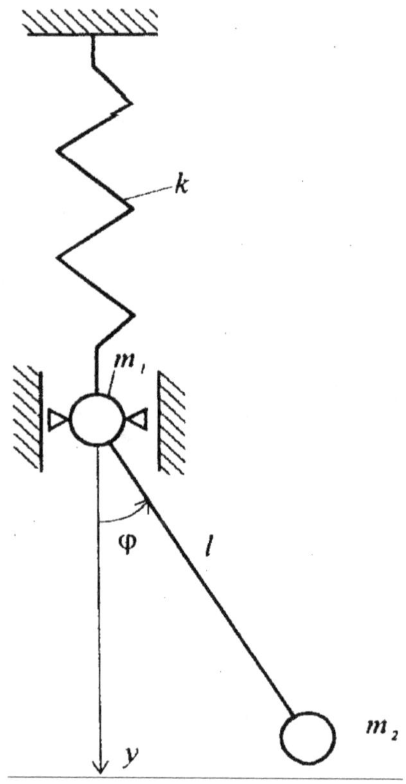

Figure 3. Scheme of a 2dof mechanical system.

Applying Lagrange equations of the second kind [15]

$$\frac{d}{dt}\left(\frac{\partial T}{\partial \dot{\lambda}}\right)-\frac{\partial T}{\partial \lambda}=-\frac{\partial \Pi}{\partial \lambda} \qquad (\lambda = y, \varphi)$$

and considering Eq. (8), the system's equations of motion in the dimensionless form within an accuracy of the values of the second order of smallness with respect to y and φ can be written as follows:

$$\ddot{y}^* + \omega^{*2} y^* - a\varphi\ddot{\varphi} - a\dot{\varphi}^2 = 0, \tag{9a}$$

$$\ddot{\varphi} + \Omega^{*2}\varphi - b\varphi\ddot{y}^* = 0, \tag{9b}$$

where

$$\omega^{*2} = \omega^2 y_0 g^{-1}, \quad \Omega^{*2} = \Omega y_0 g^{-1}, \quad \omega^2 = k(m_1+m_2)^{-1}, \quad \Omega^2 = g l^{-1}$$

$$y^* = y\,y_0^{-1}, \quad t^* = t\sqrt{g\,y_0^{-1}}, \quad y_0 = m_1 k^{-1} g, \quad a = \frac{m_2}{m_1+m_2}\cdot\frac{l}{y_0}, \quad b = \frac{y_0}{l}.$$

Suppose that the linear natural frequency ω^* is twice as large than the linear natural frequency Ω^*, that is,

$$\omega^* = 2\Omega^* \tag{10a}$$

or the linear natural frequency ω^* and the linear natural frequency Ω^* are equal to each other, that is,

$$\omega^* = \Omega^* . \tag{10b}$$

It is said that the system is being under the conditions of *the two-to-one internal resonance* or *the one-to-one internal resonance* if the condition Eq. (10a) or (10b) is fulfilled, respectively [2].

For analyzing nonlinear vibrations of the systems subjected to the internal resonance (10), assume that the amplitudes of vibrations are small but finite values and weakly vary with time. Then, perturbation technique could be used to construct the solution of the set of Eq. (9), and, particularly, the method of multiple time scales [16].

3.2. Method of solution

An approximate solution of Eq. (9) can be represented by an expansion in terms of different time scales limiting by the values of the third order of smallness in ε

$$y^*(t) = \varepsilon y_1(T_0,T_1,T_2,\ldots) + \varepsilon^2 y_2(T_0,T_1,T_2,\ldots) + \varepsilon^3 y_3(T_0,T_1,T_2,\ldots) + \ldots , \tag{11a}$$

$$\varphi(t) = \varepsilon\varphi_1(T_0,T_1,T_2,\ldots) + \varepsilon^2\varphi_2(T_0,T_1,T_2,\ldots) + \varepsilon^3\varphi_3(T_0,T_1,T_2,\ldots) + \ldots , \tag{11b}$$

where $T_n = \varepsilon^n t$ (n = 0,1,2…), and ε is a small parameter.

Substituting Eq. (11) into Eq. (9), considering that

$$\frac{d}{dt} = D_0 + \varepsilon D_1 + \varepsilon^2 D_2 + \ldots, \qquad D_n = \frac{\partial}{\partial T_n} \quad (n = 0,1,2,\ldots)$$

$$\frac{d^2}{dt^2} = \left(D_0 + \varepsilon D_1 + \varepsilon^2 D_2 + \ldots\right)^2 = D_0^2 + 2\varepsilon D_0 D_1 + \varepsilon^2 (D_0 D_2 + D_1^2) + \ldots$$

and equating the coefficients of like powers of ε, one obtains, to order ε,

$$D_0^2 y_1 + \omega^{*2} y_1 = 0, \qquad D_0^2\varphi_1 + \Omega^{*2}\varphi_1 = 0; \tag{12}$$

to order ε^2,

$$\begin{aligned} &D_0^2 y_2 + \omega^{*2} y_2 = -2D_0 D_1 y_1 + a\varphi_1 D_0^2 \varphi_1 + a(D_0\varphi_1)^2, \\ &D_0^2\varphi_2 + \Omega^{*2}\varphi_2 = -2D_0 D_1 \varphi_1 + b\varphi_1 D_0^2 y_1; \end{aligned} \tag{13}$$

to order ε^3,

$$
\begin{aligned}
D_0^2 y_3 + \omega^{*2} y_3 &= -2D_0D_1y_2 - (D_1^2 + 2D_0D_2)y_1 + a\varphi_2 D_0^2\varphi_1 \\
&\quad +2aD_0\varphi_1(D_1\varphi_1 + D_0\varphi_2) + a\varphi_1(D_0^2\varphi_2 + 2D_0D_1\varphi_1), \\
D_0^2\varphi_3 + \Omega^{*2}\varphi_3 &= -2D_0D_1\varphi_2 - (D_1^2 + 2D_0D_2)\varphi_1 \\
&\quad +b\varphi_2 D_0^2 y_1 + b\varphi_1(D_0^2 y_2 + 2D_0D_1y_1)
\end{aligned}
\tag{14}
$$

The solution of Eq. (12) could be sought in the form

$$
\begin{aligned}
y_1 &= A_1(T_1)\exp(i\omega^* T_0) + \overline{A}_1(T_1)\exp(-i\omega^* T_0), \\
\varphi_1 &= A_2(T_1)\exp(i\Omega^* T_0) + \overline{A}_2(T_1)\exp(-i\Omega^* T_0),
\end{aligned}
\tag{15}
$$

where A_1 and A_2 are unknown complex functions, while $\overline{A}_1$ and $\overline{A}_2$ are the complex conjugates of A_1 and A_2, respectively.

3.2.1. *The case of a two-to-one internal resonance*

Substituting Eq. (15) into the right-hand sides of Eq. (13) yields

$$
\begin{aligned}
D_0^2 y_2 + \omega^{*2} y_2 &= -2i\omega^* D_1A_1\exp(i\omega^* T_0) - 2aA_2^2\Omega^{*2}\exp(2i\Omega^* T_0) + cc, \\
D_0^2\varphi_2 + \Omega^{*2}\varphi_2 &= -2i\Omega^* D_1A_2\exp(i\Omega^* T_0) - bA_1A_2\omega^{*2}\exp[i(\omega^* + \Omega^*)T_0] - \\
&\quad - bA_1\overline{A}_2\omega^{*2}\exp[i(\omega^* - \Omega^*)T_0] + cc,
\end{aligned}
\tag{16}
$$

where cc denotes complex conjugate parts of the preceding terms.

The functions $\exp(i\omega^* T_0)$, $\exp(2i\Omega^* T_0) = \exp(i\omega^* T_0)$, $\exp(i\Omega^* T_0)$, $\exp[i(\omega^* - \Omega^*)T_0] = \exp(i\Omega^* T_0)$ entering into the right-hand sides of Eq. (16) produce secular terms in the expression for y_2 and φ_2, that is, the terms of the type of $T_0 e^{i\omega^* T_0}$ and $T_0 e^{i\Omega^* T_0}$. Since secular terms increase without any limits as time goes on, then there is a need to eliminate them by equating the coefficients standing at the enumerated functions to zero. As a result, we obtain

$$
i\omega^* D_1A_1 + aA_2^2\Omega^{*2} = 0, \tag{17a}
$$

$$
2i\Omega^* D_1A_2 + bA_1\overline{A}_2\omega^{*2} = 0. \tag{17b}
$$

Multiply Eq. (17a) by $\overline{A}_1$ and Eq. (17b) by $\overline{A}_2$ and find the complex conjugate equations. Two mutually conjugated equations first add to each other and then subtract one from another. As a result of such a procedure, we obtain more convenient set of four equations:

$$i\omega^*\left(\bar{A}_1 D_1 A_1 - A_1 D_1 \bar{A}_1\right) + a\Omega^{*2}\left(\bar{A}_1 A_2^2 + A_1 \bar{A}_2^2\right) = 0,$$
$$i\omega^*\left(\bar{A}_1 D_1 A_1 + A_1 D_1 \bar{A}_1\right) + a\Omega^{*2}\left(\bar{A}_1 A_2^2 - A_1 \bar{A}_2^2\right) = 0,$$
$$2i\Omega^*\left(\bar{A}_2 D_1 A_2 - A_2 D_1 \bar{A}_2\right) + b\omega^{*2}\left(A_1 \bar{A}_2^2 + \bar{A}_1 A_2^2\right) = 0,$$
$$2i\Omega^*\left(\bar{A}_2 D_1 A_2 + A_2 D_1 \bar{A}_2\right) + b\omega^{*2}\left(A_1 \bar{A}_2^2 - \bar{A}_1 A_2^2\right) = 0.$$

Representing the functions A_1 and A_2 in a polar form

$$A_1 = a_1(T_1)\exp[i\varphi_1(T_1)], \quad A_2 = a_2(T_1)\exp[i\varphi_2(T_1)], \tag{18}$$

we can rewrite the set of four differential equations as

$$\left(a_1^2\right)^\bullet = -\frac{1}{2}a\omega^* a_1 a_2^2 \sin\delta, \tag{19a}$$

$$\left(a_2^2\right)^\bullet = 2b\omega^* a_1 a_2^2 \sin\delta, \tag{19b}$$

$$\dot{\varphi}_1 = \frac{1}{4}a\omega^* a_2^2 a_1^{-1} \cos\delta, \tag{19c}$$

$$\dot{\varphi}_2 = b\omega^* a_1 \cos\delta, \tag{19d}$$

where an overdot denotes differentiation with respect to T_1, and $\delta = 2\varphi_2 - \varphi_1$.

Eliminating the value $\omega^* a_1 a_2^2 \sin\delta$ from Eqs. (19a) and (19b) and integrating the net relationship with respect to T_1 yield

$$a_1^2 + \frac{1}{4}ab^{-1}a_2^2 = E_0, \tag{20}$$

where E_0 is the initial magnitude of the system's energy, which represents the law of conservation of the total mechanical energy of the system under consideration. Expression (20) is the first integral of the set of Eq. (19).

Introducing a new function $\xi(T_1)$ $(0 \le \xi \le 1)$ such that

$$a_1^2 = E_0\xi(T_1), \quad a_2^2 = 4ba^{-1}E_0[1 - \xi(T_1)], \tag{21}$$

and substituting Eq. (21) in Eq. (19a), we have

$$\dot{\xi} = -B\sqrt{\xi}(1-\xi)\sin\delta, \tag{22}$$

where $B = 2\sqrt{E_0}\,\omega^* b$.

Doubling both sides of Eq. (19d) and subtracting from the net relationship Eq. (19c) with due account for Eq. (21) and

$$\dot{\delta} = 2\dot{\varphi}_2 - \dot{\varphi}_1,$$

we obtain

$$\dot{\delta} = B\frac{3\xi - 1}{2\sqrt{\xi}}\cos\delta . \tag{23}$$

Putting

$$\dot{\delta} = \frac{d\delta}{d\xi}\dot{\xi} \tag{24}$$

and substituting Eq. (24) into Eqs. (22) and (23), we are led to the equation

$$\frac{d\cos\delta}{d\xi} + \frac{1-3\xi}{2\xi(1-\xi)}\cos\delta = 0 . \tag{25}$$

Separating the variables in Eq. (25) and integrating the equation obtained yield

$$\cos\delta = G_0\xi^{-1/2}(1-\xi)^{-1} \tag{26a}$$

or

$$G(\xi,\delta) = \sqrt{\xi}(1-\xi)\cos\delta = G_0(\xi_0,\delta_0), \tag{26b}$$

where $G_0(\xi_0,\delta_0) = \sqrt{\xi_0}(1-\xi_0)\cos\delta_0$ is an arbitrary constant determined from the initial conditions, and ξ_0 and δ_0 are the initial magnitudes of the values ξ and δ, respectively. Note that relationship (26b) is the other first integral of the set of Eq. (19).

Finely, let us eliminate the value δ from Eqs. (26a) and (22), resulting in

$$\dot{\xi} = -B\xi^{1/2}(1-\xi)\sqrt{1 - \frac{G_0^2}{\xi(1-\xi)^2}} . \tag{27}$$

Separating the variables in Eq. (27) and integrating the net expression, we obtain implicitly the desired function $\xi(T_1)$

$$\int_{\xi_0}^{\xi} \frac{d\xi}{\left[\xi(1-\xi)^2 - G_0^2\right]^{1/2}} = -BT_1, \tag{28}$$

where ξ_0 is the value defining the relative level in the initial amplitudes.

The integral in Eq. (28) can be transformed into an incomplete integral of the first kind, which is tabulated in [17].

At $G_0 = 0$, the integral in Eq. (28) can be calculated, in so doing, it possesses two magnitudes. Really, changing the variable $\sqrt{\xi} = x$ in the integral in Eq. (28) at $G_0 = 0$, we have the first magnitude

$$\int_{\xi_0}^{\xi} \frac{d\xi}{\sqrt{\xi}(\xi-1)} = 2\int_{\sqrt{\xi_0}}^{\sqrt{\xi}} \frac{dx}{x^2-1} = \int_{\sqrt{\xi_0}}^{\sqrt{\xi}} \frac{dx}{x-1} - \int_{\sqrt{\xi_0}}^{\sqrt{\xi}} \frac{dx}{x+1} = \ln\left(\frac{\sqrt{\xi}-1}{\sqrt{\xi}+1}\right)\Bigg|_{\xi_0}^{\xi} \tag{29a}$$

and the second magnitude

$$\int_{\xi_0}^{\xi} \frac{d\xi}{\sqrt{\xi}(1-\xi)} = 2\int_{\sqrt{\xi_0}}^{\sqrt{\xi}} \frac{dx}{1-x^2} = \int_{\sqrt{\xi_0}}^{\sqrt{\xi}} \frac{dx}{1-x} - \int_{\sqrt{\xi_0}}^{\sqrt{\xi}} \frac{dx}{1+x} = \ln\left|\frac{1+\sqrt{\xi}}{1-\sqrt{\xi}}\right|\Bigg|_{\xi_0}^{\xi}. \tag{29b}$$

Considering Eq. (29), the solutions of Eq. (28) may be written in the following form:

the first solution

$$\ln\left|\frac{(\sqrt{\xi}-1)(\sqrt{\xi_0}+1)}{(\sqrt{\xi}+1)(\sqrt{\xi_0}-1)}\right| = -BT_1 \tag{30a}$$

or

$$\sqrt{\xi} = \frac{(1+\sqrt{\xi_0}) - (1-\sqrt{\xi_0})e^{-BT_1}}{(1+\sqrt{\xi_0}) + (1-\sqrt{\xi_0})e^{-BT_1}}, \tag{30b}$$

and the second solution

$$\ln\left|\frac{(1+\sqrt{\xi})(1-\sqrt{\xi_0})}{(1-\sqrt{\xi})(1+\sqrt{\xi_0})}\right| = -BT_1 \tag{31a}$$

or

$$\sqrt{\xi} = \frac{(1+\sqrt{\xi_0})e^{-BT_1} - (1-\sqrt{\xi_0})}{(1+\sqrt{\xi_0})e^{-BT_1} + (1-\sqrt{\xi_0})} \tag{31b}$$

Solutions (30b) and (31b) at $\xi_0 \neq 0$ and 1 describe the motions corresponding to the one-sided energy exchange between pendulum's vibrations and vertical vibration of the load. As this takes place, $\xi \to 1$ and $\xi \to 0$ in the first and second solutions, respectively, with the increase in time T_1. In other words, in the first solution, the energy of vibrations of the pendulum completely transforms into the energy of vertical vibrations of the load, but in the second solution, quite the reverse, the energy of vertical vibrations of the load completely goes into the energy of vibrations of the pendulum.

In the first solution, the process of energy transfer occurs over an infinitely large time interval, which resembles the phenomenon of the transfer of the kinetic energy into potential one, which is described by the soliton-like solution (6b) for the mathematical pendulum.

In the second solution, the process of energy transfer occurs during a finite instant of the time from 0 till T_1^*, where

$$T_1{*} = -\frac{1}{B}\ln\left(\frac{1-\sqrt{\xi_0}}{1+\sqrt{\xi_0}}\right).$$

According to our classification, both of them are the soliton-like solutions. At $\xi_0 = 0$ from (30b), we obtain the known soliton-like solution in the form of a single kink [13].

$$\sqrt{\xi} = \tanh\left(\frac{1}{2}BT_1\right). \qquad (32)$$

Physically speaking, this solution kink is responsible to the one-sided energy exchange when the energy of the pendulum vibration completely transforms with time into the energy of the vertical vibrations which energy was equal to zero at the initial moment of time, so the pendulum vibrations give way to the vertical vibrations.

In order to understand the physical meaning of the first integral (26b), let us introduce into consideration the phase plane $\delta-\xi$ and analyze on this plane the phase fluid flow that interprets the motion of the mechanical system in hand. The velocity vector **V** of the phase fluid particles motion has the components $v_\xi = \dot{\xi}$ and $v_\delta = \dot{\delta}$. From Eqs. (22) and (23), it follows that

$$v_\xi = B\frac{\partial G}{\partial \delta}, \qquad v_\delta = -B\frac{\partial G}{\partial \xi} \qquad (33)$$

Writing the equation of a streamline of the phase fluid $\frac{d\xi}{v_\xi} = \frac{d\delta}{v_\delta}$ and substituting Eq. (33) in it, we obtain that the function $G(\xi,\delta)$ defined by the relationship (26b) is *the stream function of the phase fluid*. In other words, Eq. (26b) at different magnitudes of ξ_0 and δ_0 governs a family of the streamlines of the phase fluid. Since the phase fluid is incompressible (div **V** = 0) and its flow is steady and solenoidal (rot **V**≠0), then streamlines of the phase fluid will coincide with trajectories of the phase fluid particles motion.

Streamlines constructed according to the relationship

$$\sqrt{\xi}(1-\xi)\cos\delta = \sqrt{\xi_0}(1-\xi_0)\cos\delta_0 \qquad (34)$$

at different magnitudes of ξ_0 and δ_0 are presented in **Figure 4**, where digits near the curves denote the magnitudes of the value $G_0(\xi_0,\delta_0) = \sqrt{\xi_0}(1-\xi_0)\cos\delta_0$. Reference to **Figure 4** shows that all phase trajectories are closed lines located around the perimeter of the rectangle

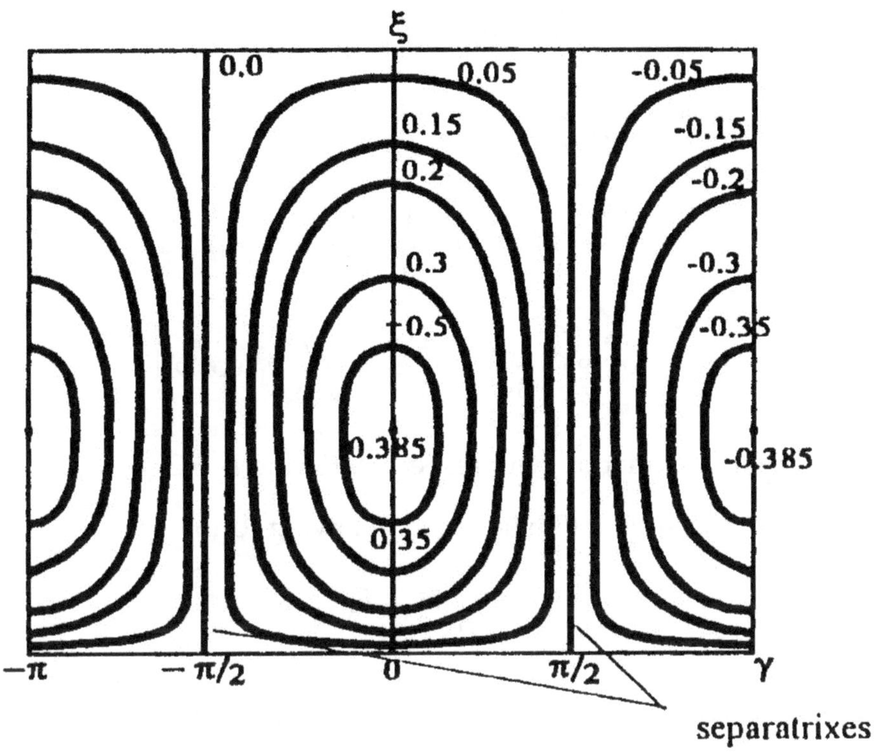

Figure 4. Phase portrait in the case of the two-to-one internal resonance $\omega^* = 2\Omega^*$.

bounded by the lines $\xi = 0$, $\xi = 1$, $\delta = \pm\pi/2 + \pi n$ $(n = 0,1,2,\ldots)$. The flow in each rectangle is isolated. On all four rectangle sides, $G_0 = 0$ and inside it the value G_0 preserves its sign. On the closed streamlines, a two-sided energy exchange takes place between the partial subsystems. Along the lines $\delta = \pm\pi/2 \pm \pi n$, a one-sided energy interchange occurs corresponding to pure amplitude-modulated aperiodic motions, in so doing on the lines with ascending flow of the phase fluid particles (an arrow is directed upwards), the aperiodic regime is described by Eq. (30b), and on the lines with descending flow (an arrow is directed downwards), the aperiodic regime is governed by Eq. (31b). On the line $\xi = 1$, there exists the boundary phase-modulated regime. The transition of fluid elements from the points with the coordinates $\xi = 0$, $\delta = \pi/2 + \pi n$ to the points $\xi = 0$, $\delta = -\pi/2 + \pi n$ proceeds instantly. The points with coordinates $\xi = 1/3$, $\delta = \pi n$ correspond to the stable stationary regimes.

3.2.2. *The case of a one-to-one internal resonance*

To construct the solution in the case of a one-to-one internal resonance (10b), it will suffice to restrict consideration to the terms of the order of ε^3 and to consider the amplitudes A_1 and A_2 as functions of T_1 and T_2.

The resonance (10b) is weaker than (10a), since in order to eliminate circular terms arising in the second approximation, it would suffice to consider the functions A_1 and A_2 dependent on T_2 only [18]. Under such an assumption, the set of equations providing the absence of circular terms in the expressions for y_3 and φ_3 has the form

$$\left(a_1^2\right)^{\bullet} = ab\omega^* a_1^2 a_2^2 \sin\delta \ , \tag{35a}$$

$$\left(a_2^2\right)^{\bullet} = -b^2\omega^* a_1^2 a_2^2 \sin\delta \ , \tag{35b}$$

$$\dot{\varphi}_1 = -ab\omega^* a_2^2\left(\frac{1}{3}+\frac{1}{2}\cos\delta\right) , \tag{35c}$$

$$\dot{\varphi}_2 = -ab\omega^*\left(\frac{1}{3}a_1^2 - \frac{4}{3}a_2^2 + \frac{1}{2}\frac{b}{a}a_1^2\cos\delta\right), \tag{35d}$$

where overdots denote differentiation with respect to T_2, and $\delta = 2(\varphi_2 - \varphi_1)$.

The two first integrals of the system (35) have the following form:

$$a_1^2 + \frac{a}{b}a_2^2 = E_0 , \tag{36}$$

$$G(\xi,\delta) = \xi(1-\xi)\cos\delta - \frac{a}{3b}\xi^2 - \frac{5}{3}(1-\xi)^2 = G_0(\xi_0,\delta_0) , \tag{37}$$

in so doing

$$\dot{\xi} = -\omega^* b^2 E_0 \frac{\partial G}{\partial \delta}, \qquad \dot{\delta} = \omega^* b^2 E_0 \frac{\partial G}{\partial \xi} ,$$

where $G_0(\xi_0,\delta_0) = \xi_0(1-\xi_0)\cos\delta_0 - \frac{a}{3b}\xi_0^2 - \frac{5}{3}(1-\xi_0)^2$, the function ξ (T_2) is connected with a_1^2 and a_2^2 by the relationships

$$a_1^2 = E_0\xi(T_2), \qquad a_2^2 = ba^{-1}\left[1-\xi(T_2)\right],$$

and the rest of the values have the same meaning as in the abovementioned case (10a).

Streamlines constructed according to Eq. (37) at different magnitudes of ξ_0 and δ_0 are presented in **Figure 5** when $a/b = 5$ and $\omega^{*2} = b$. Magnitudes of the value $G_0(\xi_0,\delta_0)$ that correspond to the streamlines are indicated by digits near the curves; the flow direction of the phase fluid elements is shown by arrows on the streamlines. Reference to **Figure 5** shows that there exist two types of the streamlines, namely (1) nonclosed which correspond to the periodic change of amplitudes and the aperiodic change of phases and (2) closed ones which correspond to the periodic change of both amplitudes and phases. The alignment of the circulation zones resembles that of Von Karman vortex streets with a symmetric arrangement. The adjacent circulation zones osculate at the *saddle points* with the coordinates $\xi_0 = 0.5$, $\delta_0 = \pi \pm 2\pi n \quad (n = 0,1,2\ldots)$ and $G_0 = -1.0833$, wherein *the unstable stationary regime* occurs.

On the boundary lines of these zones (separatrixes), the value $G_0 = -13/12$, and the analytical solution corresponding to the soliton-like regime has the form

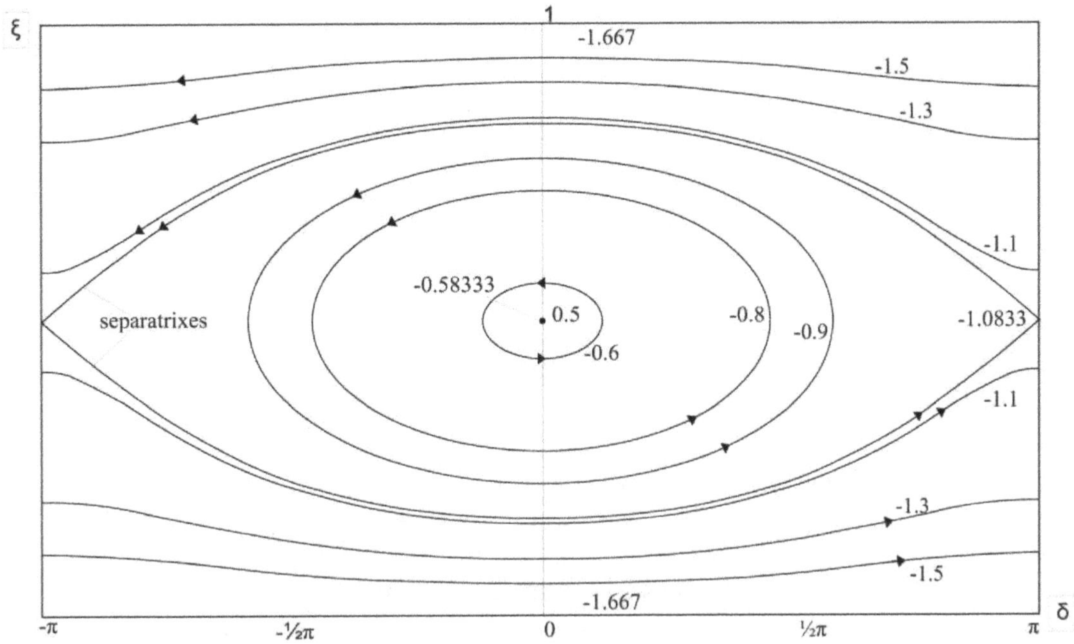

Figure 5. Phase portrait in the case of the one-to-one internal resonance.

$$\ln\left\|\frac{2\sqrt{-0.1154(\xi-0.5)^2+0.3397}+0.2308}{\xi-0.5}\right\|_{\xi_0}^{\xi}=\pm 0.083E_0T_2, \tag{38}$$

where the sign "+" fits to the initial magnitudes $0.5<\xi_0\le 0.8397$, $-\pi\pm 2\pi n<\delta_0\le 2\pi n$ $0.16032\le\xi_0<0.5$, and $-2\pi\pm 2\pi n<\delta_0\le -\pi\pm 2\pi n$, but the sign "–" conforms to the initial magnitudes $0.5<\xi_0\le 0.8397$, $-2\pi\pm 2\pi n<\delta_0\le -\pi\pm 2\pi n$ and $0.16032\le\xi_0<0.5$, $-\pi\pm 2\pi n<\delta_0\le\pm 2\pi n$.

The upper branch of the separatrix describes the partial irreversible energy transfer from the vertical vibrations to the pendulum vibrations, but the lower branch, on the contrary, is in compliance with partial irreversible transfer of the energy of the pendulum vibrations to the energy of the vertical vibrations.

The points with coordinates $\xi_0=0.5$, $\delta_0=\pm 2\pi n$, $G_0=-0.5833$ (*points like a center*) corresponding to *the stable stationary regime* are located inside closed streamlines.

4. System with an infinite number of degrees-of-freedom

Similar solutions corresponding to the one-sided energy interchange could be obtained for more complex nonlinear systems that describe dynamic behavior of real structures, as an

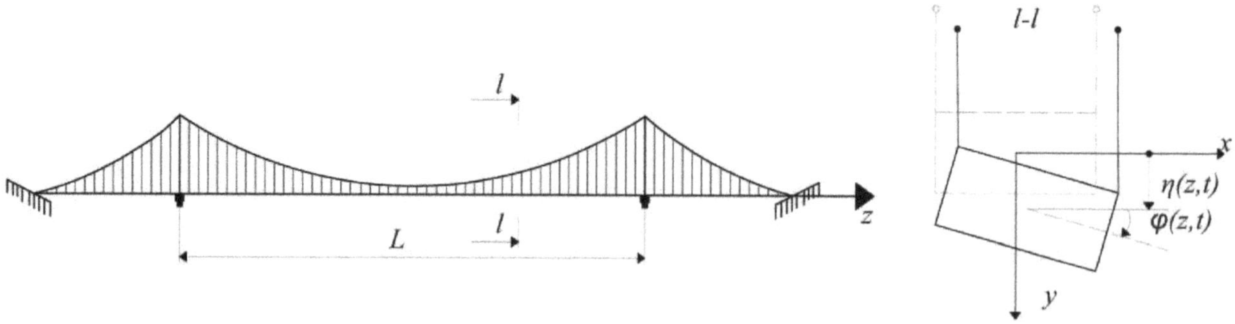

Figure 6. Scheme of a suspension bridge.

example, for systems with an infinite number of degree-of-freedom. Among such systems are suspension bridges, the scheme of one of them is shown in **Figure 6**.

The suspension bridge scheme presents a bisymmetrical thin-walled stiffening girder, which is connected with two suspended cables by virtue of vertical suspensions. The cables are thrown over the pilons and are tensioned by anchor mechanisms. The suspensions are considered as inextensible and uniformly distributed along the stiffening girder. The cables are parabolic, and the contour of the girder's cross section is undeformable. The cross section $l-l$ in **Figure 6** illustrates the displacements of the girder's contour during vibratory motions of the suspension system. Reference to this scheme shows that the girder's contour translates as a rigid body vertically (in the y-axis direction) on the value of $\eta(z,t)$ and rotates with respect to the girder's axis (the z-axis) through the angle of $\varphi(z,t)$. The origin of the frame of references is in the center of gravity of the cross section.

It is known for suspension bridges [8] that some natural modes belonging to different types of vibrations could be coupled with each other, that is, the excitation of one natural mode gives rise to another one. Two modes interact more often than not, although the possibility for the interaction of a greater number of modes is not ruled out.

If only two modes predominate in the vibrational process, namely the vertical n-th mode with linear natural frequency ω_{0n} and the torsional m-th mode with the natural frequency Ω_{0m}, such that the modes interaction is observed under the conditions (10a) or (10b), then the functions $\eta(z,t)$ and $\varphi(z,t)$ can be approximately defined as

$$\eta(z,t) \sim v_n(z)\, x_{1n}(t), \quad \varphi(z,t) \sim \Theta_m(z)\, x_{2m}(t), \tag{39}$$

where x_{1n} and x_{2m} are the generalized displacements, and $v_n(z)$ and $\Theta_m(z)$ are natural shapes of the two interacting modes of vibrations.

The resolving system of equations in a dimensionless form is written as [7, 8]

$$\begin{aligned} &\ddot{x}_{1n} + \omega_{0n}^2 x_{1n} + a_{11}^n x_{1n}^2 + a_{22}^{nm} x_{2m}^2 + \left(b_{11}^n x_{1n}^2 + b_{22}^{nm} x_{2m}^2\right) x_{1n} = 0, \\ &\ddot{x}_{2m} + \Omega_{0m}^2 x_{2m} + a_{12}^{nm} x_{1n} x_{2m} + \left(c_{11}^{nm} x_{1n}^2 + c_{22}^m x_{2m}^2\right) x_{2m} = 0, \end{aligned} \tag{40}$$

where the coefficients a_{ij}, b_{ij}, and c_{ij} $(i=1,2,\ j=2)$ are defined in [7]. Subsequently, for the ease of presentation, the indices n and m will be omitted.

An approximate solution of Eq. (40) for small but finite amplitudes could be written as an expansion in terms of different time scales in the following form [16]:

$$\begin{aligned} x_1(t) &= \varepsilon x_{11}(T_0,T_1,T_2,\ldots)+\varepsilon^2 x_{12}(T_0,T_1,T_2,\ldots)+\ldots, \\ x_2(t) &= \varepsilon x_{21}(T_0,T_1,T_2,\ldots)+\varepsilon^2 x_{22}(T_0,T_1,T_2,\ldots)+\ldots \end{aligned} \tag{41}$$

The number of the independent time scales needed depends on the order to which the expansion is carried out. Here, $T_0=t$ is the first scale characterizing motions with the natural frequencies ω_0 and Ω_0, and T_n are slow scales characterizing the modulations of the amplitudes and phases.

Substituting Eq. (41) into Eq. (40) and equating the coefficients of like powers of ε, we obtain on each step a set of two linear equations. On the first step, it is convenient to seek the solution in the form:

$$\begin{aligned} x_{11} &= A_1(T_1,T_2)\exp(i\omega_0 T_0)+\overline{A}_1(T_1,T_2)\exp(-i\omega_0 T_0), \\ x_{21} &= A_2(T_1,T_2)\exp(i\omega_0 T_0)+\overline{A}_2(T_1,T_2)\exp(-i\omega_0 T_0), \end{aligned} \tag{42}$$

where and are unknown complex functions, and $\overline{A}_1$ and $\overline{A}_2$ are the complex conjugates of A_1 and A_2, respectively.

Substituting Eq. (42) into the set of equations obtained on the first step and using the second step to eliminate secular terms, as well as representing the functions A_1 and A_2 in the polar form $A_1=a_1\exp(i\varphi_1)$, $A_2=a_2\exp(i\varphi_2)$, we are led to the following system of equations for the case of the two-to-one internal resonance (10a):

$$\begin{aligned} \dot{\xi} &= -b(1-\xi)\sqrt{\xi}\,\sin\gamma, \\ \dot{\gamma} &= -\frac{1}{2}b(1-3\xi)\xi^{-1/2}\cos\gamma, \end{aligned} \tag{43}$$

where $\xi=\xi(T_1)$ is an unknown function, $\gamma=2\varphi_2-\varphi_1$, $b=a_{12}\Omega_0^{-1}\sqrt{E_0}$, $E_0=a_1^2+a_{22}\Omega_0(a_{12}\omega_0)^{-1}a_2^2$ is the system's initial energy, $a_1=\sqrt{E_0\xi}$, $a_2=\sqrt{E_0 a_{12}\omega_0(a_{22}\Omega_0)^{-1}(1-\xi)}$, and an overdot denotes differentiation with respect to T_1.

Representing $\dot{\gamma}=\dot{\xi}d\gamma/d\xi$ and considering Eq. (43) yield

$$\frac{d\cos\gamma}{d\xi}+\frac{1}{2}\frac{1-3\xi}{\xi(1-\xi)}\cos\gamma=0. \tag{44}$$

The solution to Eq. (44) has the form

$$G_1(\xi,\gamma)=\xi^{1/2}(1-\xi)\cos\gamma=G_1^0, \tag{45}$$

where G_1^0 is an arbitrary constant determined from the initial conditions. Note that relationship (45) is similar to the first integral (26b) for a two-degree-of-freedom system.

In the case of the one-to-one internal resonance (10b), we seek the solution in the form of Eq. (42) also. Using the procedure for the elimination of secular terms, we obtain the following set of equations:

$$\begin{aligned}
&\dot{\xi}=\frac{1}{2}\Gamma_2 E_0\xi(1-\xi)\sin\gamma,\\
&\frac{1}{2}\dot{\gamma}=\frac{1}{4}\Gamma_2 E_0(1-2\xi)\cos\gamma-(\lambda_1-\lambda_3)E_0\xi-\frac{\Gamma_2}{\Gamma_1}(\lambda_2-\lambda_4)E_0(1-\xi),
\end{aligned} \tag{46}$$

where $\xi=\xi(T_2)$ is an unknown function, $\gamma=2(\varphi_2-\varphi_1)$, $E_0=a_1^2+\Gamma_1\Gamma_2^{-1}a_2^2$ is the system's initial energy, $a_1=\sqrt{E_0\xi}$, $a_2=\sqrt{\Gamma_2\Gamma_1^{-1}E_0(1-\xi)}$, an overdot denotes differentiation with respect to T_2, and the coefficients λ_i and Γ_j $(i=1,\ldots,4;\ j=1,2)$ dependent upon the system parameters [8].

Representing $\dot{\gamma}=\dot{\xi}d\gamma/d\xi$ and using Eq. (46) yield

$$\frac{d\cos\gamma}{d\xi}+\frac{1-2\xi}{\xi(1-\xi)}\cos\gamma-\frac{4(\lambda_1-\lambda_3)}{\Gamma_2(1-\xi)}-\frac{4(\lambda_2-\lambda_4)}{\Gamma_1\xi}=0. \tag{47}$$

The solution to Eq. (47) has the form

$$G_2(\xi,\gamma)=\xi(1-\xi)\cos\gamma-2(\lambda_1-\lambda_3)\Gamma_2^{-1}\xi^2+2(\lambda_2-\lambda_4)\Gamma_1^{-1}(1-\xi)^2=G_2^0, \tag{48}$$

where G_2^0 is an arbitrary constant determined from the initial conditions.

Eliminating the variable γ in Eq. (48) and in the second equation of (46) and integrating over T_2 yield

$$\frac{1}{\sqrt{m_1m_2}}\int_{\xi_0}^{\xi}\frac{d\xi}{\sqrt{(\xi^2+p_1\xi+q_1)(\xi^2+p_2\xi+q_2)}}=\frac{E_0}{2\Gamma_1}T_2, \tag{49}$$

where ξ_0 is a value determined by the relative level of the initial amplitudes, and the quantities m_i, p_i, and q_i are the coefficients [11]. The integral in Eq. (49) can be transformed to an incomplete elliptic integral of the first kind [17].

4.1. Soliton-like solutions

As examples, the nonlinear free vibrations of the Golden Gate Bridge in San Francisco are considered. All geometrical data, as well as natural frequency spectra and mode shapes for this one of the most beautiful suspension bridges, are available in [19].

It can be shown that under the relationship among the natural frequencies $\omega_6^s = 2\Omega_1^s = 2.66$ rad/s (a two-to-one internal resonance between the sixth symmetrical mode of vertical vibrations and the first symmetrical mode of torsional vibrations), one can obtain the analytical solution in the form of a single kink (32), where B should be replaced by the coefficient b defined by the system's parameters according to Eq. (43). The physical sense of this solution kink is that it is responsible for the one-sided energy exchange when the energy of the torsional vibrations completely transforms into the energy of the vertical vibrations with time, so that the torsional vibrations initiate the vertical vibrations [20].

Under the relationships among the natural frequencies, $\omega_5^s = \Omega_3^s = 2.61$ rad/s and $\omega_3^s = \Omega_1^s = 1.33$ rad/s (a one-to-one internal resonance), the analytical solutions may be found by solving Eq. (49), respectively, as [20]

$$\begin{aligned} &\ln\left|2\xi^{-1}\sqrt{-0.13\xi^2-0.195\xi+0.019}+0.27\xi^{-1}-1.4\right|_{\xi_0}^{\xi} = 0.431E_0T_2, \\ &\ln\left|2(1-\xi)^{-1}\sqrt{-0.066(1-\xi)^2+0.037(1-\xi)+0.004}+0.132(1-\xi)^{-1}+0.536\right|_{\xi_0}^{\xi} \\ &\qquad = -0.036E_0T_2, \end{aligned} \tag{50}$$

where $\Big|_{\xi_0}^{\xi}$ denotes the evaluation at the upper and lower limits of integration.

In the first case of Eq. (50), the coefficients q_1 and q_2 in the integral (49) become zero, and the analytical solution corresponding to the separatrix $G_2 = -0.354$ describes a one-sided energy transfer from the vertical vibration to the torsional vibration (a low aperiodic regime), which leads in time to the conversion of the flexural-torsional vibrations to the predominantly torsional vibrations. This regime is the most unfavorable and dangerous for suspension bridges.

In the second case of Eq. (50), the analytical solution corresponding to the separatrix $G_2 = 0.487$ describes a one-sided energy transfer from the torsional vibration to the vertical vibration (an upper aperiodic regime), so that the flexural-torsional vibrations evolve into the predominantly vertical vibrations with time.

The solutions obtained may be interpreted on the phase plane $\xi - \gamma$ by virtue of streamlines of the phase fluid which is demonstrated in **Figures 5** and **7** for solutions (32) and (50), respectively. Digits near curves indicate the magnitudes of the values G_1 and G_2 corresponding to the streamlines.

The analysis of the phase portraits in terms of the variables ξ and γ for various oscillatory regimes demonstrates that they contain both closed and nonclosed streamlines which are separated by the curves separatrixes. Along the separatrixes, one succeeds in finding analytical solutions that are inherently soliton-like solutions in the theory of vibrations and describe the complete one-sided energy transfer from one subsystem to another.

Note that soliton-like solutions could be found also in an analytical form for the case of free damped vibrations of a suspension bridge, when damping features of the system are described

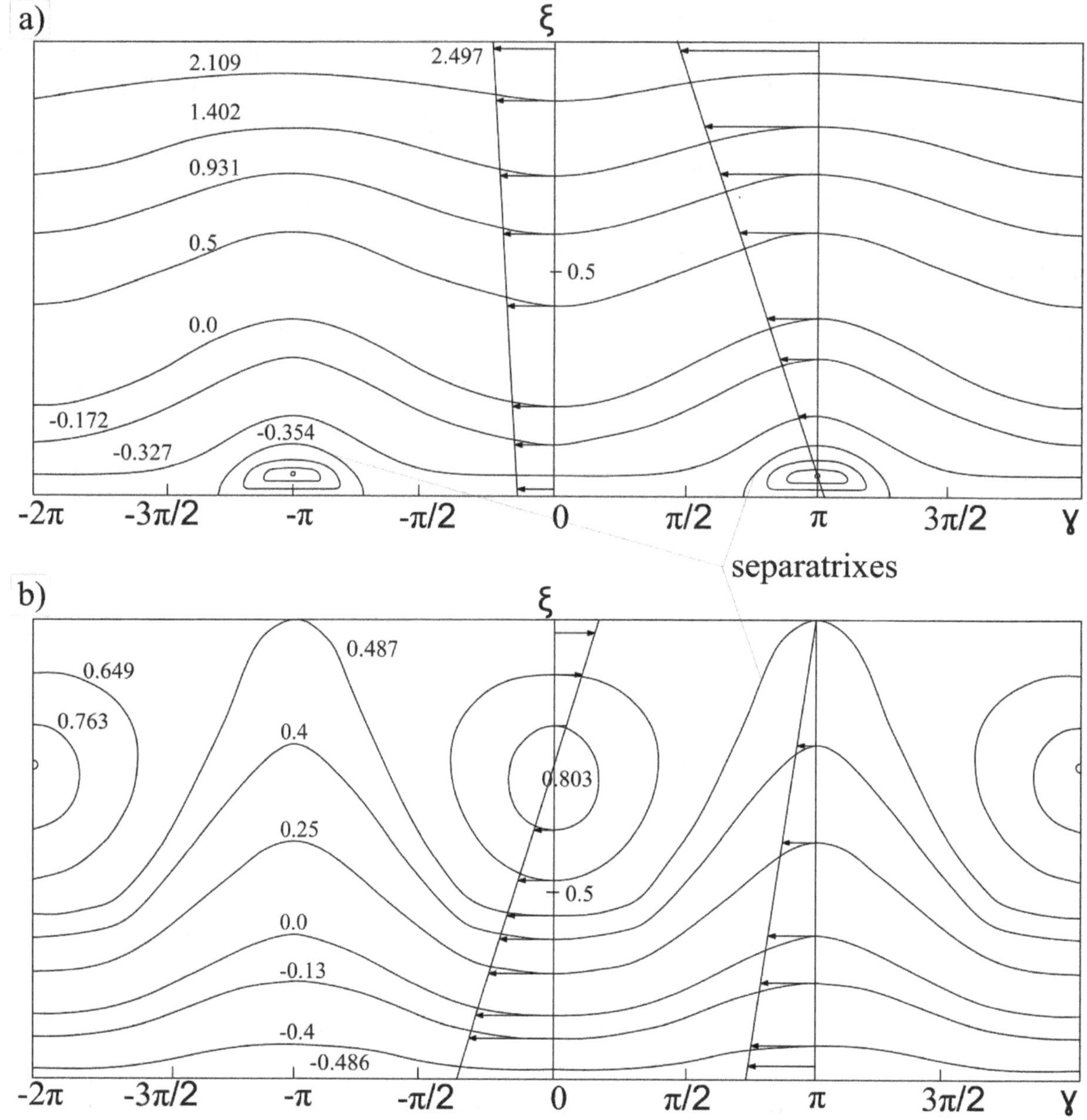

Figure 7. Phase portraits: (a) $\omega_5^s = \Omega_3^s$, and (b) $\omega_3^s = \Omega_1^s$

by ordinary first-order time derivative [21] or defined by a fractional derivative with a fractional parameter (the order of the fractional derivative) changing from zero to one [22].

5. Conclusions

From the review presented, the following conclusions could be deduced. In all considered vibratory systems—1dof, 2dof, and multi-dof—under certain conditions, there exist solutions that describe irreversible processes of energy transfer from its one type to another. Such solutions are called *soliton-like solutions* and could be written in an analytical form.

On the phase plane, these solutions correspond to streamlines which separate closed lines of phase fluid flow from nonclosed ones. These lines are called *separatrixes*.

Since soliton-like solution may describe unfavorable vibratory regimes of real mechanical systems, then they should be investigated systematically by virtue of mathematical models of these systems, in order to avoid, wherever possible, such dangerous vibratory regimes when designing and constructing real structures. A thorough analysis of internal resonances in thin plates and cylindrical shells could be found in [23, 24] and [25, 26], respectively.

Soliton-like solutions in the cases of combinational internal resonances for systems with an infinite number of degrees-of-freedom, when more than two natural modes of vibration are coupled, could be found in sight as well, and such examples for nonlinear plates and cylindrical shells are presented in [27, 28] respectively.

Acknowledgements

The research described in this publication was made possible in part by the Ministry of Education and Science of the Russian Federation under Project # 9.5138.2017/8.9.

Author details

Yury A. Rossikhin and Marina V. Shitikova*

*Address all correspondence to: mvs@vgasu.vrn.ru

Voronezh State Technical University, Voronezh, Russian Federation

References

[1] Mandel'shtam L. Lecture Notes on Theory of Vibrations (in Russian). Moscow: Nauka; 1972

[2] Nayfeh AH, Balachandran B. Modal interactions in dynamical and structural systems. Applied Mechanics Reviews. 1989;**42**:175-201

[3] Sado D. Energy transfer in two-degree-of-freedom vibrating systems—A survey. Mechanika Teoretyczna i Stosowana. 1993;**31**:151-173

[4] Vitt AA, Gorelik GA. Vibrations of an elastic pendulum as an example of vibrations of two parametrically coupled linear systems (in Russian). Journal of Technical Physics. 1933;**2–3**: 294-307

[5] Sado D. Analysis of vibration of two-degree of freedom system with inertial coupling. Machine Dynamics Problems. 1984;**1**:67-77

[6] Shitikova MV. Modelling of free nonlinear vibrational processes in suspension bridges by a two-mass system (in Russian). In: Advanced Methods of Static and Dynamic Analysis of Structures 1. Voronezh: Voronezh Civil Engineering Institute; 1992. pp. 147-153

[7] Abdel-Ghaffar AM, Rubin LI. Nonlinear free vibrations of suspension bridges: Theory and application. ASCE Journal of Engineering Mechanics. 1983;**109**:313-345

[8] Rossikhin YA, Shitikova MV. Nonlinear free spatial vibrations of combined suspension systems. Applied Mathematics and Mechanics. 1990;**54**:825-832

[9] Rossikhin YA, Shitikova MV. Effect of initial conditions on the behavior of vibrational processes in a combined suspended system. Mechanics of Solids. 1991;**26**:143-154

[10] Rossikhin YA, Shitikova MV. Effect of viscosity on the vibrational processes in a combined suspension system. Mechanics of Solids. 1995;**30**:157-166

[11] Rossikhin YA, Shitikova MV. Analysis of nonlinear free vibrations of suspension bridges. Journal of Sound and Vibration. 1995;**186**:369-393

[12] Goldenblat II. Dynamic Stability of Structures (in Russian). Moscow: Stroy Izdat; 1948

[13] Dodd R, Eilbeck J, Gibbon J, Morris H. Solitons and Non-linear Wave Equations. London: Academic Press; 1982. 670 p

[14] Filippov AT. Manifold Soliton (in Russian). "Kvant" Library Series. Moscow: Nauka; 1990. 287 p

[15] Appell PE. Theoretical Mechanics (in Russian). Moscow: Fizmatlit; 1960. 516 p

[16] Nayfeh AH. Perturbation Methods. New York: Wiley; 1973. 450 p

[17] Abramowitz M, Stegan I, editors. Handbook of Mathematical Functions with Formulas, Graphs, and Tables. Washington: National Bureau of Standards; 1964. 558 p

[18] Rossikhin YA, Shitikova MV. Analysis of nonlinear vibrations of a two-degree-of-freedom mechanical systems with damping modeled by a fractional derivative. Journal of Engineering Mathematics. 2000;**37**:343-362

[19] Abdel-Ghaffar AM, Scanlan RH. Ambient vibration studies of Golden Gate Bridge. I: Suspended structure. ASCE Journal of Engineering Mechanics. 1985;**111**:463-482

[20] Rossikhin YA, Shitikova MV. Soliton-like solution to the equations of free nonlinear vibrations of suspension bridges. In: Proceedings of the 1993 International Symposium on Nonlinear Theory and Its Applications; December 5–10, 1993; Hawaii. pp. 705-708

[21] Rossikhin YA, Shitikova MV. Soliton-like solutions to the nonlinear damped vibrations of a suspension bridge under an internal resonance. In: Proceedings of the 2nd European Nonlinear Oscillations Conference; September 9–13, 1996; Prague. Vol. 2, pp. 203-206

[22] Rossikhin YA, Shitikova MV. Application of fractional calculus for analysis of nonlinear damped vibrations of suspension bridges. Journal of Engineering Mechanics. 1998;**124**: 1029-1036

[23] Rossikhin YA, Shitikova MV. Analysis of free non-linear vibrations of a viscoelastic plate under the conditions of different internal resonances. International Journal of Non-Linear Mechanics. 2006;**41**:313-325

[24] Rossikhin YA, Shitikova MV, Ngenzi JC. A new approach for studying nonlinear dynamic response of a thin plate with internal resonance in a fractional viscoelastic medium. Shock and Vibration. 2015;**2015**:795606

[25] Rossikhin YA, Shitikova MV. Nonlinear dynamic response of a fractionally damped cylindrical shell with a three-to-one internal resonance. Applied Mathematics and Computation. 2015;**257**:498-525

[26] Rossikhin YA, Shitikova MV. A new approach for studying nonlinear dynamic response of a thin fractionally damped cylindrical shell with internal resonances of the order of ε. In: Altenbach H, Mikhasev GI, editors. Shell and Membrane Theories in Mechanics and Biology: From Macro- to Nanoscale Structures. Advanced Structural Materials. Berlin-Heidelberg: Springer; 2015. Chapter 17. pp. 301-321

[27] Rossikhin YA, Shitikova MV, Ngenzi JC. Fractional calculus application in problems of non-linear vibrations of thin plates with combinational internal resonances. Procedia Engineering. 2016;**144**:849-858

[28] Rossikhin YA, Shitikova MV. Analysis of non-linear vibrations of a fractionally damped cylindrical shell under the conditions of combinational internal resonance. In: Mastorakis N et al., editors. Computational Problems in Science and Engineering. Lecture Notes in Electrical Engineering. Vol. 343. Berlin-Heidelberg: Springer; 2015. pp. 59-107

9

Nonlinear Aeroelastic Response of Highly Flexible Flying Wing Due to Different Gust Loads

Ehsan Izadpanahi and Pezhman Mardanpour

Abstract

Nonlinear aeroelastic responses of a flying wing aircraft due to different gust profiles are investigated. Three different gust profiles are obtained considering light, moderate, and severe turbulence. A flying wing configuration is designed for the purpose of this investigation. The structural properties of the wings are obtained using VABS software, and then the flying wing is simulated with Nonlinear Aeroelastic Trim and Stability of HALE Aircraft (NATASHA) computer program. The results of time domain analysis are reported for the cases when engine is placed at the root of the wing and close to the area of maximum flutter speed. It has been found that the flying wing experiences limit cycle oscillation, when the engines are mounted at the root of the aircraft, for all three gust profiles. However, when the engines are placed at the area of maximum flutter speed, the oscillations die out. In addition, the real and imaginary part of eigenvalues and the unstable mode shape of the aircraft are reported.

Keywords: gust response, flying wing, nonlinear time domain analysis, flutter analysis, gust suppression

1. Introduction

Very flexible high-aspect-ratio wings are widely used in the design of high altitude long endurance (HALE) aircrafts. These wings due to their characteristics may subject to large deformation, which causes geometric nonlinearities. As a result, conducting the nonlinear aeroelastic analysis is necessary when it comes to the design of very flexible configurations [1–3]. In addition, time-dependent external excitation including gust [4–7] and blast [8–12] can lead to instability even if the aircraft is flying below the stability boundary. Therefore, the determination of nonlinear aeroelastic responses to time-dependent excitation is a crucial topic for the design of very flexible flying wings.

Gust loads can result in large deformations in the case of a highly flexible aircraft. The flight dynamic characteristics and gust response of highly flexible aircraft were investigated by Patil and Taylor [13]. It was reported that the non-uniform gust creates higher responses in a case of high-aspect-ratio flying wing compared to uniform gusts. In addition, the nonlinear gust response of a highly flexible aircraft was reported by Patil [14], which he found that the time domain response matches with frequency domain response presented in the work by Patil and Taylor [13]. Ricciardi et al. [15] investigated the accuracy of the Pratt method for unconventional HALE aircraft. The Pratt method and transient method were used to analyze the gust response on the joined-wing and flying-wing model. It was found that Pratt method is only useful for the preliminary design of the joined-wing model. However, when it comes to the design of flying-wing model, the Pratt method is inadequate. Yi et al. [16] compared a theoretical and experimental approach of a flexible high-aspect-ratio wing exposed to a harmonic gust. It was found that a very flexible wing experiences different gust response characteristics under different load conditions and the responses are difficult to evaluate using linear analysis.

On the other hand, finding ways to suppress the responses of a highly flexible configuration due to time-dependent excitations is a challenging aspect of design. Tang et al. [17] conducted an experimental and theoretical study to investigate the effect of store span location and its pitch stiffness on the flutter velocity and LCO. A delta wing for the purpose of experimentation was chosen. In addition, the von-Karman plate theory, three-dimensional vortex lattice model, and slender body aerodynamic theory were used for modeling the wing structure and determining the aerodynamic loads, respectively. It was reported that the experimental investigation and theoretical studies were in good agreement, and they showed that the structural natural frequency of the wing/store declines as the store moves from the root to the tip of the wing. They concluded that mounting the store at the leading edge of the wing tip leads to a higher critical flutter velocity. Moreover, Mardanpour et al. [18] found that the maximum flutter speed happened for engine placement at 60% of span forward the reference line. It was reported that the body-freedom flutter mode was unaffected by the engine location except for cases in which the engine was mounted at the wing tip and near the reference line.

Fazelzadeh et al. [19] investigated the effects of a nonlinear active control system on the flutter vibration of a wing/store exposed to a random gust disturbance. It was found that the control system is effective in suppressing the flutter vibration. In addition, Mardanpour et al. [7, 20] reported that the gust response of a very flexible high-aspect-ratio wing can be suppressed by changing the location of the engine. It was found that placing the engine close to 75% of the span forward of the reference line increases the flutter speed and also leads to suppression of the LCO due to gust loads.

In this chapter, the effect of engine placement on nonlinear aeroelastic gust response of a flying wing aircraft is investigated using three gust profiles with different gust intensities. The gust profiles are obtained utilizing different magnitude of turbulence at 10,000 m of altitude [21]. A flying wing aircraft is simulated for this study. The wings are designed using the structural properties which were obtained utilizing VABS software for NACA0012 airfoil. The computer program Nonlinear Aeroelastic Trim and Stability of High Altitude Long Endurance Aircraft (NATASHA) [3, 22] is used to simulate the nonlinear behavior of the flying wing aircraft. NATASHA is a powerful tool for the simulation of nonlinear behavior of HALE aircraft. It uses the nonlinear composite beam theory [23] that accommodates the modeling of high-aspect-ratio wings and the aerodynamic theory of Peters et al. [24] to model the aerodynamic forces and the p method to evaluate the aeroelastic stability. NATASHA has been verified and validated against experimental and theoretical studies many times [25, 26]. The nonlinear responses of the aircraft are obtained for the cases when the engines are mounted at the root of wings and at the area of maximum flutter speed (i.e., 60% of span forward of reference line).

2. Theory

2.1. Nonlinear composite beam theory

The equations of motion, which are presented in Eq. (1), are based on force, moment, angular velocity, and velocity with nonlinearities of second order. These variables can be expressed in the bases of the deformed and undeformed frames, $B(x_1, t)$ and $b(x_1)$, respectively, see **Figure 1**.

$$\begin{aligned} &F'_B + \tilde{K}_B F_B + f_B = \dot{P}_B + \tilde{\Omega}_B P_B \\ &M'_B + \tilde{K}_B M_B + (\tilde{e}_1 + \tilde{\gamma}) F_B + m_B = \dot{H}_B + \tilde{\Omega}_B H_B + \tilde{V}_B P_B \end{aligned} \tag{1}$$

In this set of equations, F_B and M_B represent the column matrices of cross-sectional stress and moment resultant; V_B and Ω_B define column matrices of cross-sectional frame velocity and angular velocity; P_B and H_B indicate the column matrices of cross-sectional linear and angular

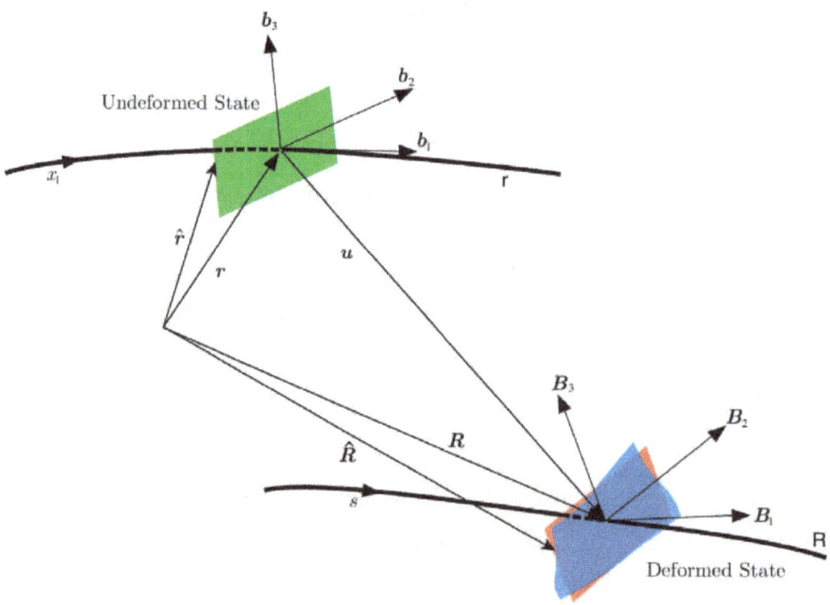

Figure 1. Sketch of beam kinematics.

momentum measures; $\tilde{K}_B$ is Column matrix of deformed beam's curvature and twist. All of the abovementioned variables measure in $\mathbf{B}_i$ basis. The structural and the inertial constitutive equations relate the stress resultants and moments to the generalized strains and velocities as follows:

$$\begin{Bmatrix} \gamma \\ \kappa \end{Bmatrix} = \begin{bmatrix} R & S \\ S^T & T \end{bmatrix} \begin{Bmatrix} F_B \\ M_B \end{Bmatrix} \tag{2}$$

$$\begin{Bmatrix} P_B \\ H_B \end{Bmatrix} = \begin{bmatrix} \mu\Delta & -\mu\tilde{\xi} \\ \mu\tilde{\xi} & I \end{bmatrix} \begin{Bmatrix} V_B \\ \Omega_B \end{Bmatrix} \tag{3}$$

here, R, S, and T represent 3×3 partitions of the cross-sectional flexibility matrix; μ is the mass per unit length; Δ is the 3×3 identity matrix; I defines the 3×3 cross-sectional inertia matrix; ξ is $\lfloor 0 \quad \xi_2 \quad \xi_3 \rfloor^T$ in which ξ_2 and ξ_3 represent the position coordinates of the cross-sectional mass center with respect to the reference line. Finally, strain- and velocity-displacement equations are utilized to derive the intrinsic kinematical partial differential Equations [23].

$$\begin{aligned} V_B' + \tilde{K}_B V_B + (\tilde{e}_1 + \tilde{\gamma})\Omega_B &= \dot{\gamma} \\ \Omega_B' + \tilde{K}_B \Omega_B &= \dot{\kappa} \end{aligned} \tag{4}$$

In these equations, the tilde $(\tilde{\ })$ represents the antisymmetric 3×3 matrix associated with the column matrix over which the tilde is placed, $(\dot{\ })$ defines the partial derivative with respect to time, and $(\ )'$ is the partial derivative with respect to the axial coordinate, x_1. More details about these equations can be found in Ref. [27]. In order to solve these first-order, partial differential equations, one may eliminate γ and κ using Eq. (2) and P_B and H_B using Eq. (3), and also 12 boundary conditions are required, in terms of force (F_B), moment (M_B), velocity (V_B), and angular velocity (Ω_B). Displacement and rotation variables do not appear in this formulation, and singularities due to finite rotations are avoided. The position and the orientation can be obtained as postprocessing operations by integrating

$$\begin{aligned} r_i' &= C^{ib} e_1 \\ r_i + u_i' &= C^{iB}(e_1 + \gamma) \end{aligned} \tag{5}$$

and

$$\begin{aligned} \left(C^{bi}\right)' &= -\tilde{k} C^{bi} \\ \left(C^{Bi}\right)' &= -\left(\tilde{k} + \tilde{\kappa}\right) C^{Bi} \end{aligned} \tag{6}$$

2.2. Finite state-induced model of Peters et al.

The aerodynamic model of Peters et al. [24] is utilized in this study. This finite state model is a state-space, thin-airfoil, inviscid, incompressible approximation of an infinite-state representation of the aerodynamic loads. By using known airfoil parameters, it can consider induced flow in the wake and apparent mass effects. In addition, it can accommodate large motion of the airfoil as well as deflection of a small trailing-edge flap. Available studies in literature [24–26] indicate that although this model cannot simulate the three-dimensional effects associated with the wing tip, it can accurately approximate the aerodynamic loads acting on high-aspect-ratio wings. The lift, drag, and pitching moment at the quarter-chord are given by

$$L_{\text{aero}} = \rho b\left[\left(c_{l_0} + c_{l_\beta}\beta\right)V_T V_{a_2} - c_{l_\alpha}\dot{V}_{a_3}b/2 - c_{l_\alpha}V_{a_2}(V_{a_3} + \lambda_0 - \Omega_{a_1}b/2) - c_{d_0}V_T V_{a_3}\right] \quad (7)$$

$$D_{\text{aero}} = \rho b\left[-\left(c_{l_0} + c_{l_\beta}\beta\right)V_T V_{a_3} + c_{l_\alpha}(V_{a_3} + \lambda_0)^2 - c_{d_0}V_T V_{a_2}\right] \quad (8)$$

$$M_{\text{aero}} = 2\rho b\left[\left(c_{m_0} + c_{m_\beta}\beta\right)V_T - c_{m_\alpha}V_T V_{a_3} - bc_{l_\alpha}/8V_{a_2}\Omega_{a_1} - b^2 c_{l_\alpha}\dot{\Omega}_{a_1}/32 + bc_{l_\alpha}\dot{V}_{a_3}/8\right] \quad (9)$$

Where,

$$V_T = \sqrt{V_{a_2}^2 + V_{a_3}^2}. \quad (10)$$

$$\sin\alpha = \frac{-V_{a_3}}{V_T} \quad (11)$$

$$\alpha_{\text{rot}} = \frac{\Omega_{a_1}b/2}{V_T} \quad (12)$$

and β is the angle of flap deflection, V_{a_2} and V_{a_3} denote the measure numbers of V_a. The effect of unsteady wake (induced flow) and apparent mass included as λ_0 and acceleration terms in the force and moment equation, which λ_0 can be calculated using the induced flow model of Peters et al. [24]:

$$[A_{\text{induced flow}}]\{\dot{\lambda}\} + \left(\frac{V_T}{b}\right)\{\lambda\} = \left(-\dot{V}_{a_3} + \frac{b}{2}\dot{\Omega}_{a_1}\right)\{c_{\text{induced flow}}\} \quad (13)$$

$$\lambda_0 = \frac{1}{2}\{b_{\text{induced flow}}\}^T\{\lambda\} \quad (14)$$

here, λ defines the column matrix of induced flow states, and $[A_{\text{induced flow}}]$, $\{c_{\text{induced flow}}\}$, $\{b_{\text{induced flow}}\}$ represent constant matrices, which are derived in Ref. [24].

2.3. Gust airloads model

The gust airloads are taken into account separately from the aerodynamic forces of the flight dynamic velocities. The unsteady gust model measures the chordwise variation of the gust field on the deformed state of the wing. Here, an interpretation of the Peters and Johnson [28] theory that considers these effects is provided. The total induced flow is ω^B, defining the vertical gust velocity in the deformed beam frame

$$\overline{L} = \omega_0 + \frac{1}{2}\omega_1 + \frac{1}{2}\left(\dot{\omega}_0 + \frac{1}{2}\dot{\omega}_1\right)\frac{b}{V_T} \tag{15}$$

here, $\overline{L}$ denotes the velocity-normalized lift coefficient presented by Peters and Johnson [28]; ω_n is the coefficient of the nth Chebychev polynomial mode shape. ω^B can approximated as

$$\omega^B = \sum_0^N \omega_n T_n \tag{16}$$

where T_n is the nth order Chebyshev polynomial. The gust force can be provided as

$$f_{gust} = \begin{Bmatrix} 0 \\ -\rho b C_{l\alpha}(V_3 + \omega_0)\overline{L} \\ \rho b C_{l\alpha} V_2 \overline{L} \end{Bmatrix} \tag{17}$$

and the gust contribution to the induced flow can be presented as

$$\lambda_{0_{gust}} = \dot{\omega}_0 + \frac{1}{2}\dot{\omega}_1 \tag{18}$$

2.4. Aeroelastic system

By unifying the aerodynamic equations with the structural equations, the aeroelastic system is constructed

$$[A]\{\dot{x}\} + \{B(x)\} = \{f_{cont}\} + \{f_{gust}\} \tag{19}$$

here, $\{x\}$, $\{f_{cont}\}$, and $\{f_{gust}\}$ define the vectors of all of the aeroelastic variables, the flight controls, and gust loads, respectively. The resulting nonlinear ordinary differential equations are then linearized about a static equilibrium state, which is obtained by nonlinear algebraic equations. Utilizing the Newton-Raphson procedure, NATASHA solves these equations to obtain the steady-state trim solution [3]. The stability of the structure can be analyzed by linearizing this system of nonlinear aeroelastic equations about the resulting trim state, which leads to a standard eigenvalue problem. The linearized system is represented as

$$[A]\left\{\dot{\widehat{x}}\right\} + [B]\{\widehat{x}\} = \left\{\widehat{f}_{cont}\right\} + \left\{f_{gust}\right\} \tag{20}$$

where $\widehat{()}$ is the perturbation about the steady-state values.

2.5. Transient gust response

The dynamic aeroelastic equations are solved in time to obtain the transient gust response. A central difference scheme in time-marching algorithm is used with a high-frequency damping. The linearized system in time can be written as follows:

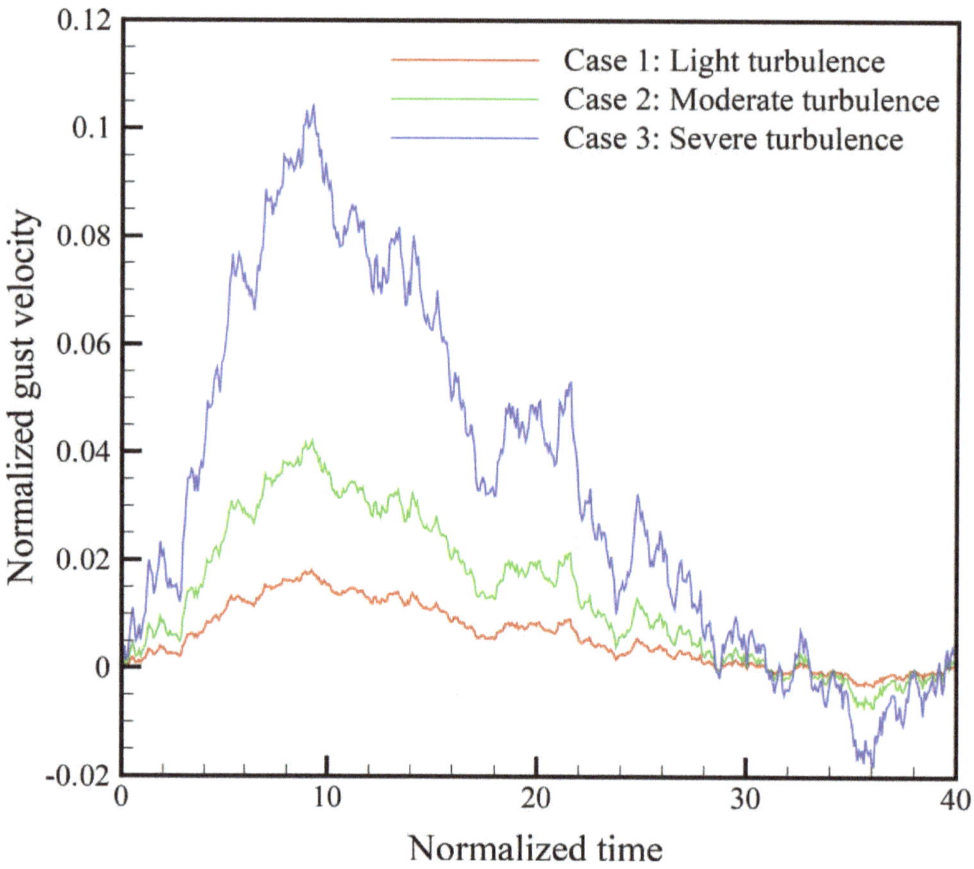

Figure 2. Gust velocity profile versus time.

$$\frac{1}{\delta t}[A]\left(\left\{\widehat{x}^{t+\delta t}\right\}-\left\{\widehat{x}^{t}\right\}\right)+\frac{1}{2}[B]\left((1+\varsigma)\left\{\widehat{x}^{t+\delta t}\right\}+(1-\varsigma)\left\{\widehat{x}^{t}\right\}\right)=\left\{\widehat{f}_{cont}\right\}+\left\{f_{gust}\right\} \tag{21}$$

here, δt and ς are the time step and the high-frequency-damping parameter, respectively. Utilizing ς approximately equal to 0.01 provides a good time-marching algorithm, which the results are close to the central difference method.

The gust profiles are presented in **Figure 2**. These profiles presented in **Figure 2** are generated by passing the Gaussian white noise through the Dryden spectrum model.

3. Case study

A very flexible high-aspect-ratio flying wing (see **Figure 3**) is designed in order to investigate the effects of different gust loads. The properties of the flying wing are presented in **Table 1**. The wings are aft swept 15°, and each wing has 20 elements. The fuselage is considered as a rigid body which contains four elements. The weight of each element of fuselage is five times of the weight of the elements of the wings. The aircraft has two engines with the mass of 10 kg.

Figure 3. A schematic 3D view of a very flexible high-aspect-ratio wing.

Property	Value
Span	16
Number of elements	20
Sweep angle	15
R	$\begin{bmatrix} 9.06 \times 10^{-9} & 0 & 0 \\ 0 & 3.50 \times 10^{-8} & 7.22 \times 10^{-13} \\ 0 & 7.22 \times 10^{-13} & 1.18 \times 10^{-6} \end{bmatrix}$
S	$\begin{bmatrix} 0 & 2.63 \times 10^{-12} & 7.57 \times 10^{-11} \\ -3.01 \times 10^{-12} & 0 & 0 \\ -1.02 \times 10^{-6} & 0 & 0 \end{bmatrix}$
T	$\begin{bmatrix} 4.33 \times 10^{-6} & 0 & 0 \\ 0 & 5.53 \times 10^{-6} & 2.42 \times 10^{-14} \\ 0 & 2.42 \times 10^{-14} & 8.43 \times 10^{-8} \end{bmatrix}$
I	$\begin{bmatrix} 4.78 \times 10^{-1} & 0 & 0 \\ 0 & 7.2 \times 10^{-3} & -1.04 \times 10^{-10} \\ 0 & -1.04 \times 10^{-10} & 4.71 \times 10^{-1} \end{bmatrix}$
ξ	$\begin{bmatrix} 0 \\ 8.98 \times 10^{-4} \\ -4.76 \times 10^{-7} \end{bmatrix}$
Mass per unit length	4.38
Chord, c	1
Offset of aerodynamic center from reference line, e	0.125
c_{l_α}	2π
c_{l_δ}	1
c_{d_0}	0.01
c_{m_0}	0.0
c_{m_α}	−0.08
Gravity, g	9.8
Air Density, ρ	0.4135

Table 1. Properties of wing in SI units.

4. Results and discussion

In this section, two cases are considered. First, when the engine mounted at the root of the wing and the second case when the engines are located at 60% of the span forward of the reference line. For each case, the eigenvalues, the unstable mode shape of the aircraft, and the nonlinear time domain responses to the gust profiles are reported. The velocity results are normalized with the aircraft cruise speed of 50 m/s. The wing tip deflections also normalized with the length of the entire flying wing (i.e., 35.2 m), and the time is normalized with the period of oscillation of the flying wing at the flutter boundary when the engines are located at the root (i.e., 0.129 s).

4.1. Engine at the root

When the engines are located at the root of the flying wing, the wings experience a flutter at the speed of 48.9 m/s with a frequency of 7.7 rad/s. The real and imaginary parts of the eigenvalues are shown in **Figure 4**. In addition, the mode shape of the unstable mode is shown in **Figure 5**. The mode shape seems to contain first and second free-free bending mode.

Figures 6–11 illustrate the results of time domain analysis when the engine is mounted at the root of the flying wing for different gust profiles in which Case 1, Case 2, and Case 3 indicate the results when the flying wing is exposed to light, moderate, and severe turbulence, respectively. It is found that the tip deflection increases in all directions when the gust load changes from light to severe turbulence. The same also happens for velocities. The velocity of the wing tip in different directions increases.

4.2. Engine at 60% of span forward of reference line

In another case, the engines are mounted at 60% of span forward of reference line. Mardanpour et al. [18] reported that this area coincides with the area of maximum flutter

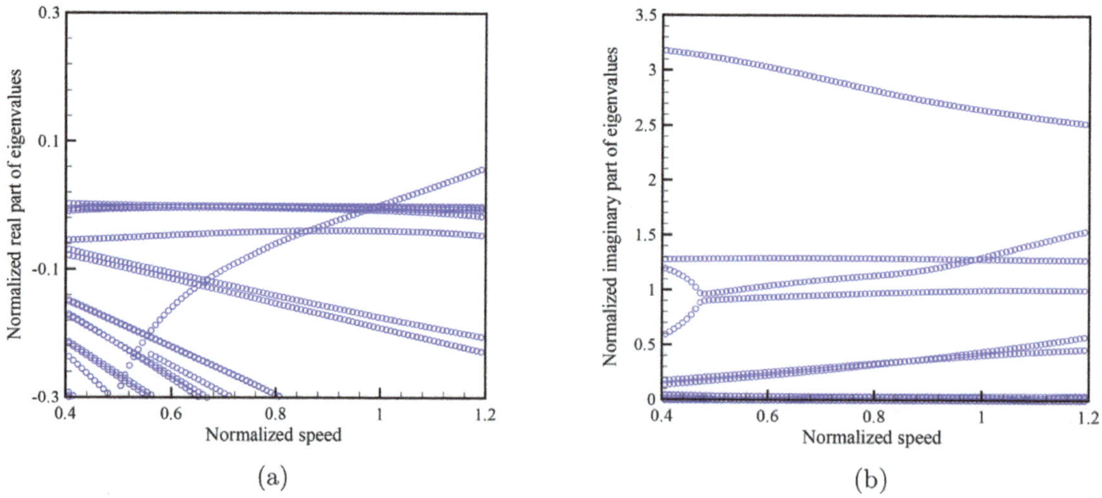

Figure 4. (a) Real part of eigenvalues and (b) imaginary part of eigenvalues.

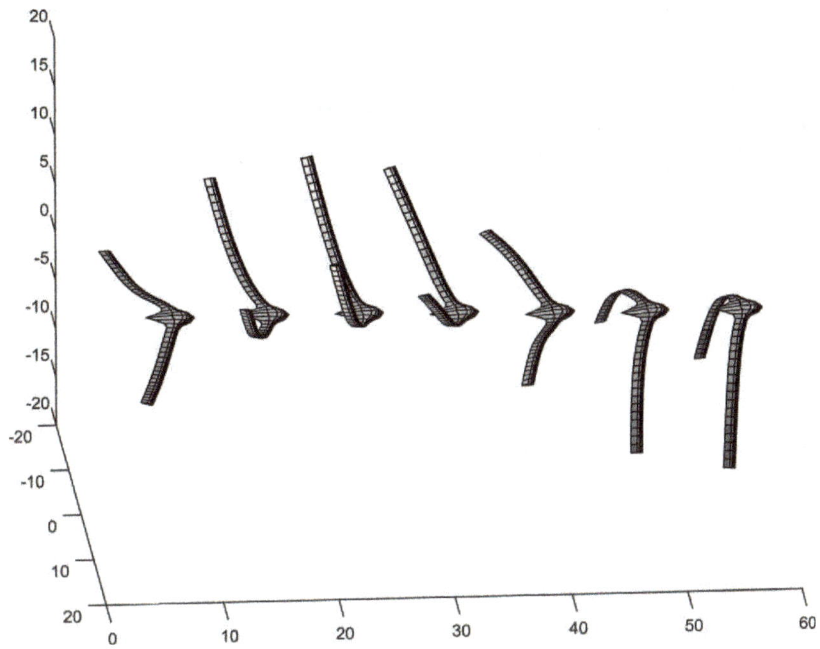

Figure 5. Unstable mode of the flying wing.

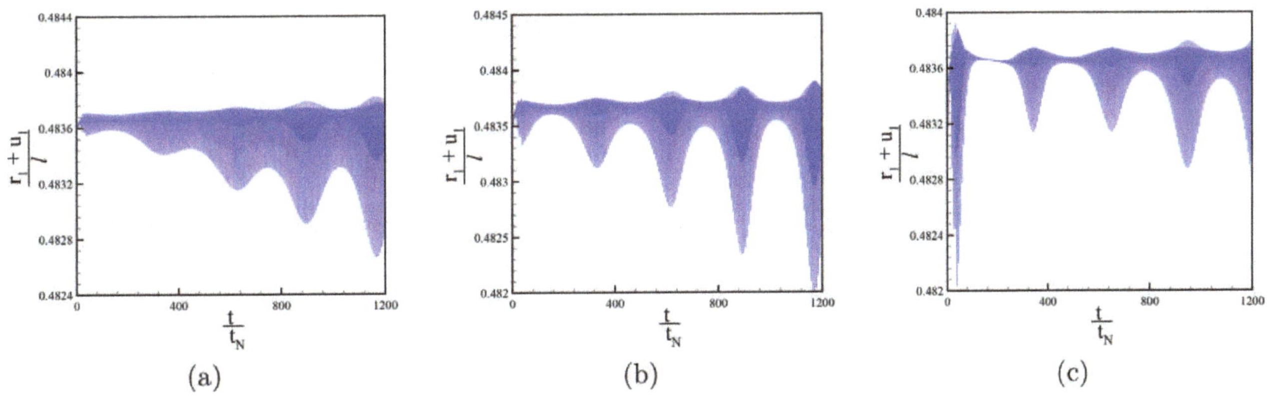

Figure 6. Normalized wing tip position $\frac{r_1+u_1}{l}$ versus normalized time $\frac{t}{t_N}$. (a) Case 1, (b) Case 2, and (c) Case 3.

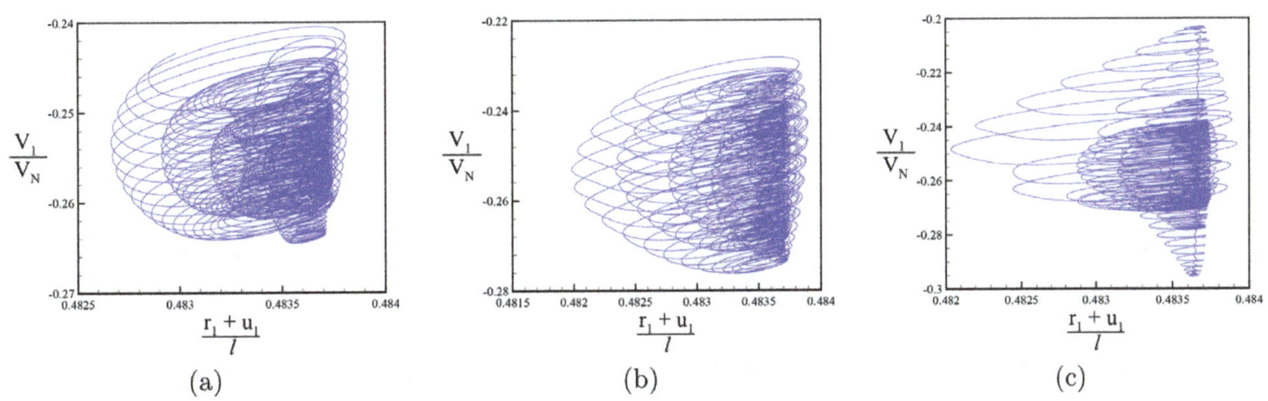

Figure 7. Normalized velocity vector of wing tip $\frac{V_1}{V_N}$ versus normalized wing tip position $\frac{r_1+u_1}{l}$. (a) Case 1, (b) Case 2, and (c) Case 3.

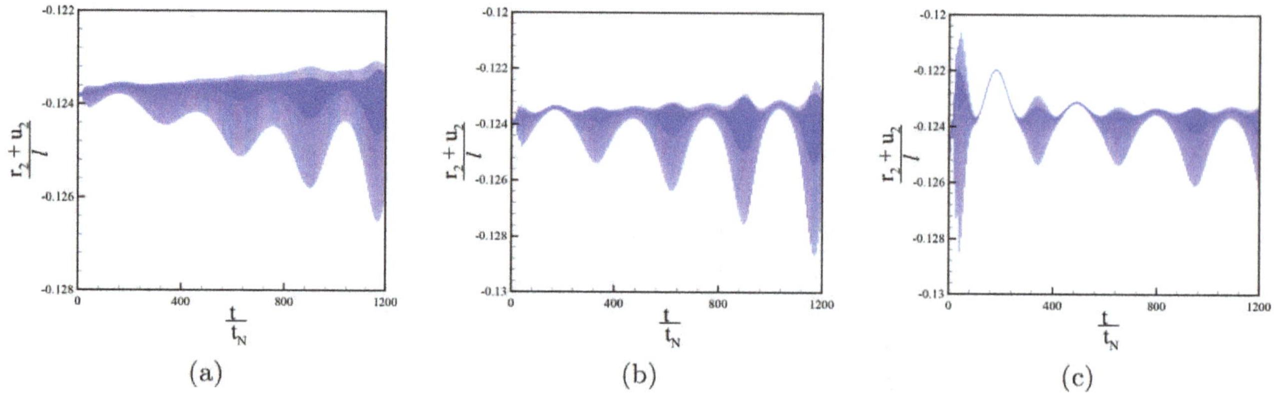

Figure 8. Normalized wing tip position $\frac{r_2+u_2}{l}$ versus normalized time $\frac{t}{t_N}$. (a) Case 1, (b) Case 2, and (c) Case 3.

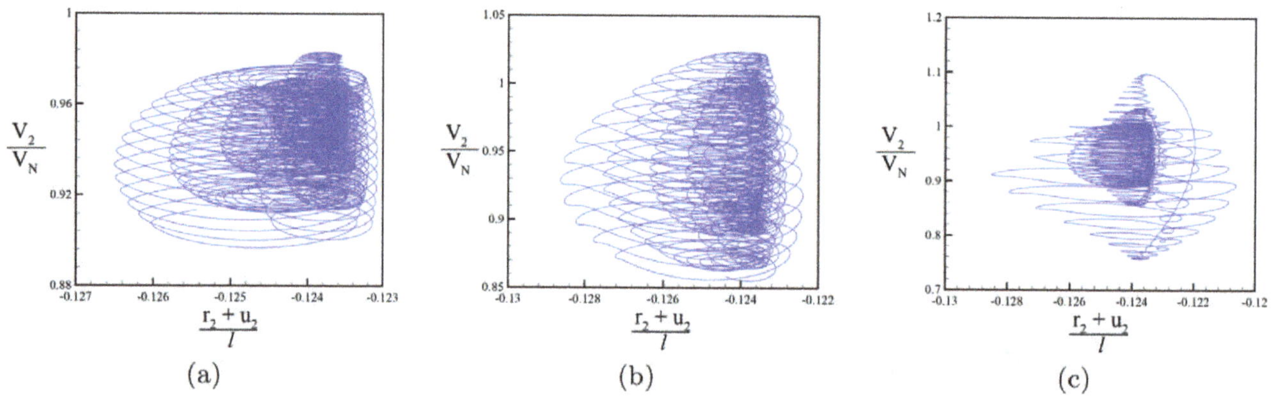

Figure 9. Normalized velocity vector of wing tip $\frac{V_2}{V_N}$ versus normalized wing tip position $\frac{r_2+u_2}{l}$. (a) Case 1, (b) Case 2, and (c) Case 3.

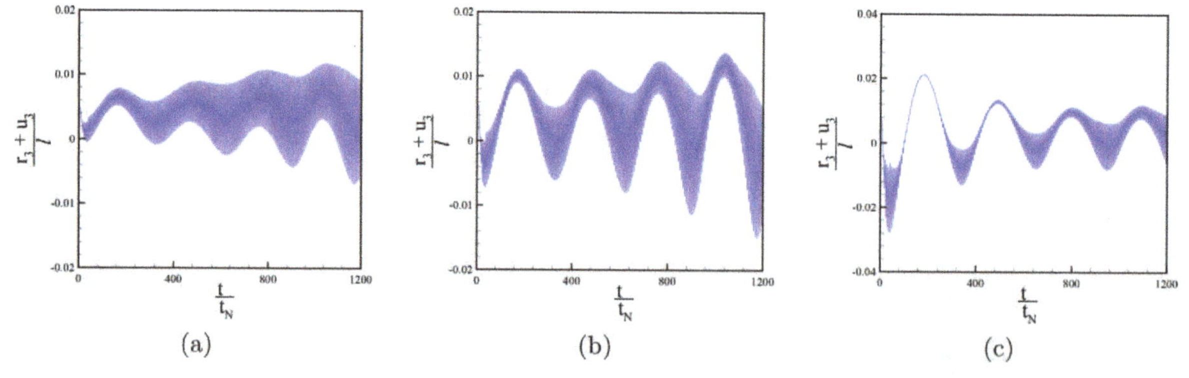

Figure 10. Normalized wing tip position $\frac{r_3+u_3}{l}$ versus normalized time $\frac{t}{t_N}$. (a) Case 1, (b) Case 2, and (c) Case 3.

speed. It is found that the flying wing becomes unstable at the speed of 75.6 m/s. The real and imaginary parts of the eigenvalues are shown in **Figure 12**, and the mode shape of the unstable mode is displayed in **Figure 13**. Apparently, the mode shape only contains the first symmetric free-free bending mode.

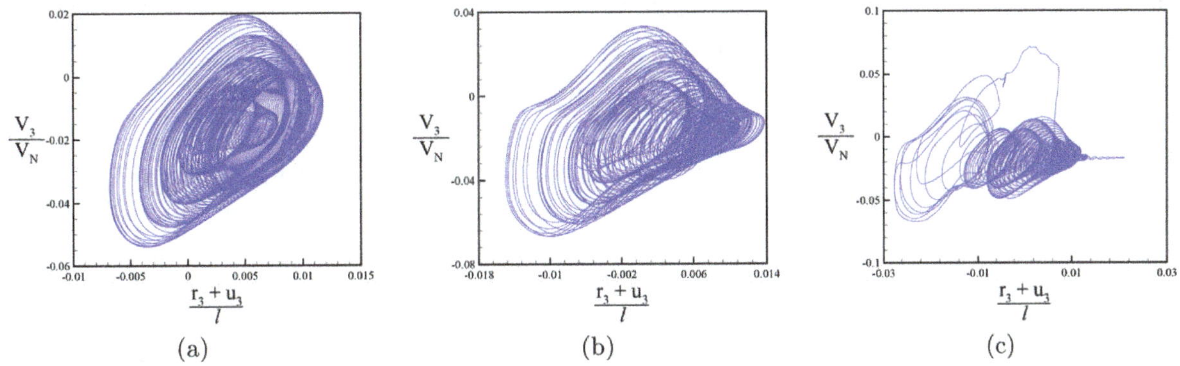

Figure 11. Normalized velocity vector of wing tip $\frac{V_3}{V_N}$ versus normalized wing tip position $\frac{r_3+u_3}{l}$. (a) Case 1, (b) Case 2, and (c) Case 3.

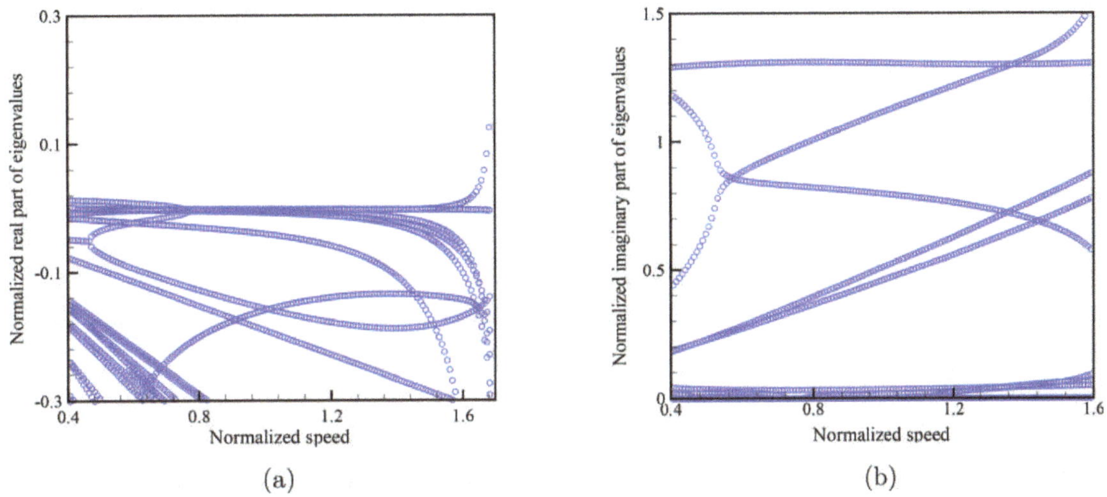

Figure 12. (a) Real part of eigenvalues and (b) imaginary part of eigenvalues.

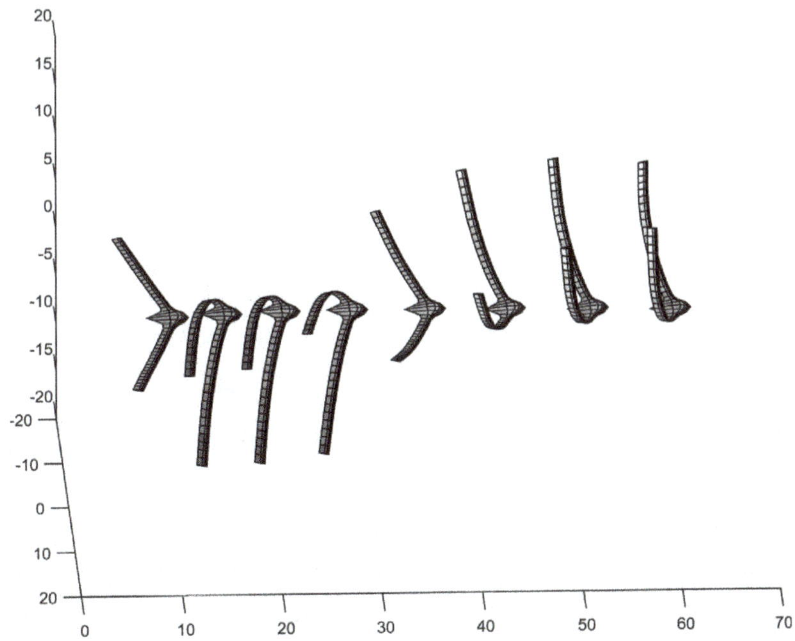

Figure 13. Unstable mode of the flying wing.

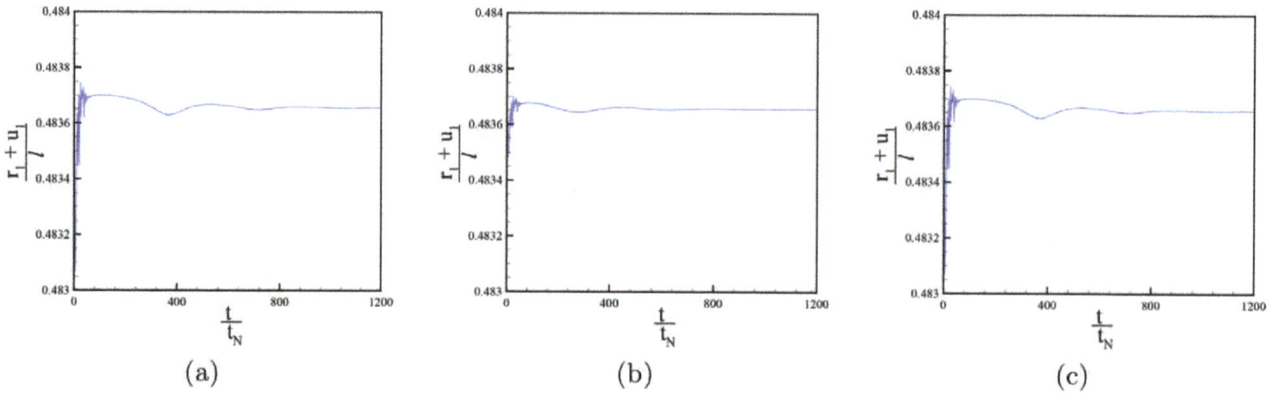

Figure 14. Normalized wing tip position $\frac{r_1+u_1}{l}$ versus normalized time $\frac{t}{t_N}$. (a) Case 1, (b) Case 2, and (c) Case 3.

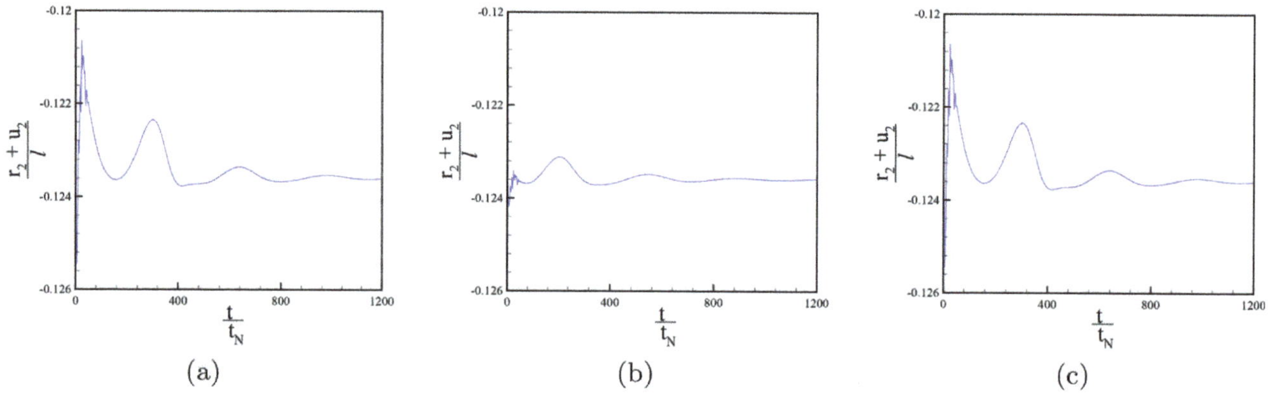

Figure 15. Normalized wing tip position $\frac{r_2+u_2}{l}$ versus normalized time $\frac{t}{t_N}$. (a) Case 1, (b) Case 2, and (c) Case 3.

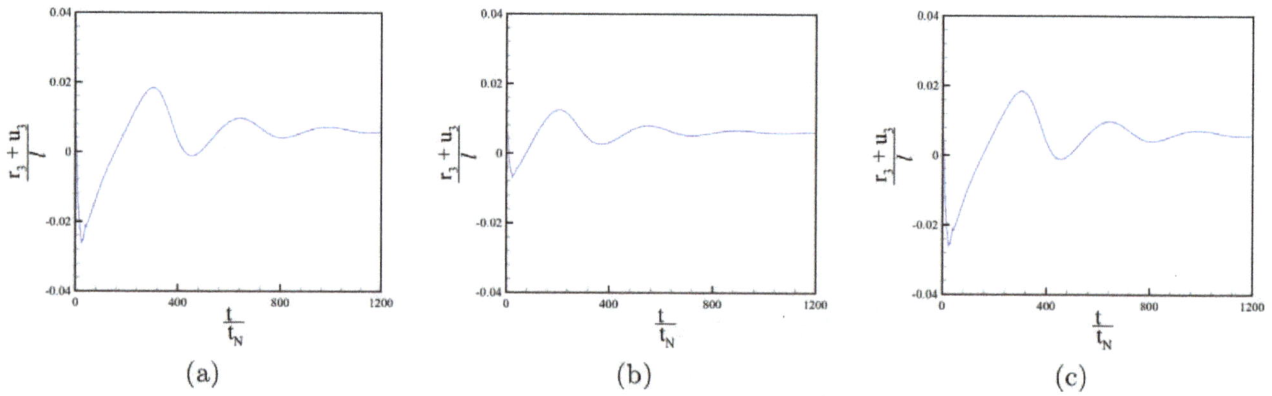

Figure 16. Normalized wing tip position $\frac{r_3+u_3}{l}$ versus normalized time $\frac{t}{t_N}$. (a) Case 1, (b) Case 2, and (c) Case 3.

Figures 14–16 show the results of time domain analysis when the engine is located at the area of maximum flutter speed (i.e., 60% of span forward of reference line). The results are reported for three different gust profiles. The results for this arrangement indicate that all the excitations from gust loads with different strength ranges from light to severe loads die out and the wing remains stable.

5. Conclusion

The nonlinear aeroelastic responses of a flying wing aircraft are investigated when the aircraft is exposed to different gust profiles with different intensities. The aircraft is designed with two aluminum wings with NACA 0012 airfoil. The properties of the wings are obtained using VABS software. The properties are then used in geometrically exact beam formulation, which is coupled with two-dimensional finite state aerodynamic model of Peters. The flutter characteristics for two configurations of the aircraft (i.e., engines at the root of the wings and engines at 60% of span forward of the reference line) as well as the eigenvalues and mode shape of the unstable modes for each configuration are studied. The flutter results are in agreement with the previous conclusion by Mardanpour et al. [18], which shows a higher flutter speed when the engines are mounted at 60% of span forward of reference line.

Three different gust profiles are then produced by passing white noise through Dryden gust model. The gust loads with light, moderate, and severe intensities are applied to the aircraft in time domain when the aircraft is cruising at 50 m/s. The results indicate that when the engines are mounted at the root of the wings, large oscillations exist, which their amplitude increases as the intensity of the gust loads increases. On the contrary, for all of the gust loads, when the engines are located at 60% of span forward of the reference line, the oscillations suppress. Previous study on gust alleviation by Mardanpour et al. [7, 20] for a cantilever wing also showed the suppression of gust responses when the engines are mounted at the area of maximum flutter speed.

Nomenclature

a deformed beam aerodynamic frame of reference

b undeformed beam cross-sectional frame of reference

B deformed beam cross-sectional frame of reference

$\mathbf{b}_i$ unit vectors in undeformed beam cross-sectional frame of reference ($i = 1, 2, 3$)

$\mathbf{B}_i$ unit vectors of deformed beam cross-sectional frame of reference ($i = 1, 2, 3$)

c chord

$c_{m\beta}$ pitch moment coefficient w.r.t. flap deflection (β)

$c_{l\alpha}$ lift coefficient w.r.t. angle of attack (α)

$c_{l\beta}$ lift coefficient w.r.t. flap deflection (β)

e_1 column matrix $\lfloor 1 \quad 0 \quad 0 \rfloor^T$

e offset of aerodynamic center from the origin of frame of reference along $\mathbf{b}_2$

f	column matrix of distributed applied force measures in $\mathbf{B}_i$ basis
F	column matrix of internal force measures in $\mathbf{B}_i$ basis
$\mathbf{g}$	gravitational vector in $\mathbf{B}_i$ basis
H	column matrix of cross-sectional angular momentum measures in $\mathbf{B}_i$ basis
i	inertial frame of reference
$\mathbf{i}_i$	unit vectors for inertial frame of reference ($i = 1, 2, 3$)
I	cross-sectional inertia matrix
k	column matrix of undeformed beam initial curvature and twist measures in $\mathbf{b}_i$ basis
K	column matrix of deformed beam curvature and twist measures in $\mathbf{B}_i$ basis
l	wing length
$\overline{L}$	velocity-normalized lift coefficient
m	column matrix of distributed applied moment measures in $\mathbf{B}_i$ basis
M	column matrix of internal moment measures in $\mathbf{B}_i$ basis
P	column matrix of cross-sectional linear momentum measures in $\mathbf{B}_i$ basis
r	column matrix of position vector measures in $\mathbf{b}_i$ basis
u	column matrix of displacement vector measures in $\mathbf{b}_i$ basis
U_∞	free stream velocity
V	column matrix of velocity measures in $\mathbf{B}_i$ basis
x_1	axial coordinate of beam
β	trailing edge flap angle
Δ	identity matrix
γ	column matrix of 1D-generalized force strain measures
κ	column matrix of elastic twist and curvature measures (1D-generalized moment strain measures)
η	dimensionless position of the engine along the span
λ	column matrix of induced flow states
Λ	sweep angle
μ	mass per unit length
ξ	column matrix of center of mass offset from the frame of reference origin in $\mathbf{b}_i$ basis

ψ column matrix of small incremental rotations

ω induced flow velocity

Ω column matrix of cross-sectional angular velocity measures in $\mathbf{B}_i$ basis

$(\ )'$ partial derivative of ( ) with respect to x_1

$\dot{(\ )}$ partial derivative of ( ) with respect to time

$\widehat{(\ )}$ nodal variable

Author details

Ehsan Izadpanahi and Pezhman Mardanpour*

*Address all correspondence to: Pezhman.Mardanpour@FIU.edu

Department of Mechanical and Materials Engineering, Florida International University, Miami, FL, USA

References

[1] Patil MJ, Hodges DH. On the importance of aerodynamic and structural geometrical nonlinearities in aeroelastic behavior of high-aspect-ratio wings. Journal of Fluids and Structures. 2004;**19**(7):905-915

[2] Patil MJ, Hodges DH, Cesnik CES. Nonlinear aeroelasticity and flight dynamics of high-altitude long-endurance aircraft. Journal of Aircraft. 2001;**38**(1):88-94

[3] Patil MJ, Hodges DH. Flight dynamics of highly flexible flying wings. Journal of Aircraft. 2006;**43**(6):1790-1799

[4] Tang D, Dowell EH. Experimental and theoretical study of gust response for a wing–store model with freeplay. Journal of Sound and Vibration. 2006;**295**(3):659-684

[5] Lau ASH, Haeri S, Kim JW. The effect of wavy leading edges on aerofoil–gust interaction noise. Journal of Sound and Vibration. 2013;**332**(24):6234-6253

[6] Xiang J, Yining W, Li D. Energy harvesting from the discrete gust response of a piezoaeroelastic wing: Modeling and performance evaluation. Journal of Sound and Vibration. 2015;**343**:176-193

[7] Mardanpour P, Izadpanahi E, Rastkar S, Hodges DH. Effects of engine placement on nonlinear aeroelastic gust response of high-aspect-ratio wings. In: AIAA Modeling and Simulation Technologies Conference, Page 0576. 2017

[8] Türkmen HS, Mecitoğlu Z. Dynamic response of a stiffened laminated composite plate subjected to blast load. Journal of Sound and Vibration. 1999;**221**(3):371-389

[9] Na S, Librescu L. Dynamic response of adaptive cantilevers carrying external stores and subjected to blast loading. Journal of Sound and Vibration. 2000;**231**(4):1039-1055

[10] Marzocca P, Librescu L, Chiocchia G. Aeroelastic response of a 2-d airfoil in a compressible flow field and exposed to blast loading. Aerospace Science and Technology. 2002;**6**(4): 259-272

[11] Librescu L, Na S, Marzocca P, Chung C, Kwak MK. Active aeroelastic control of 2-d wing-ap systems operating in an incompressible flowfield and impacted by a blast pulse. Journal of Sound and Vibration. 2005;**283**(3):685-706

[12] Kazancı Z, Mecitoğlu Z. Nonlinear dynamic behavior of simply supported laminated composite plates subjected to blast load. Journal of Sound and Vibration. 2008;**317**(3):883-897

[13] Patil MJ, Taylor DJ. Gust response of highly exible aircraft. In Proceedings of the 47th Structures, Structural Dynamics, and Materials Conference, Newport, Rhode Island, Reston, Virginia, May 2006. AIAA. AIAA-2006-1638

[14] Patil MJ. Nonlinear gust response of highly exible aircraft. In: Proc. 48th AIAA, Structural Dynamics, and Materials Conf, Pages 2007–2103. 2007

[15] Ricciardi AP, Patil MJ, Canfield RA, Lindsley N. Evaluation of quasi-static gust loads certification methods for high-altitude long-endurance aircraft. Journal of Aircraft. 2013; **50**(2):457-468

[16] Yi L, Xie C, Yang C, Cheng J. Gust response analysis and wind tunnel test for a high-aspect ratio wing. Chinese Journal of Aeronautics. 2016;**29**(1):91-103

[17] Tang D, Attar P, Dowell EH. Flutter/limit cycle oscillation analysis and experiment for wing-store model. AIAA Journal. 2006;**44**(7):1662-1675

[18] Mardanpour P, Hodges DH, Neuhart R, Graybeal N. Engine placement effect on nonlinear trim and stability of flying wing aircraft. Journal of Aircraft. 2013;**50**(6):1716-1725

[19] Fazelzadeh SA, Azadi M, Azadi E. Suppression of nonlinear aeroelastic vibration of a wing/store under gust effects using an adaptive-robust controller. Journal of Vibration and Control. 2017;**23**(7):1206-1217

[20] Mardanpour P, Izadpanahi E, Rastkar S, Hodges DH. Nonlinear aeroelastic gust suppression and engine placement. Journal of Aircraft. 2017:1-4

[21] Military Standard. Flying qualities of piloted aircraft. In: Mil-Std-1797a Ed. 1990

[22] Chang C-S, Hodges DH, Patil MJ. Flight dynamics of highly flexible aircraft. Journal of Aircraft. Mar.-Apr. 2008;**45**(2):538-545

[23] Hodges DH. Geometrically-exact, intrinsic theory for dynamics of curved and twisted anisotropic beams. AIAA Journal. June 2003;**41**(6):1131-1137

[24] Peters DA, Karunamoorthy S, Cao W-M. Finite state induced flow models; part I: two-dimensional thin airfoil. Journal of Aircraft. Mar.-Apr. 1995;**32**(2):313-322

[25] Sotoudeh Z, Hodges DH, Chang CS. Validation studies for aeroelastic trim and stability analysis of highly flexible aircraft. Journal of Aircraft. 2010;**47**(4):1240-1247

[26] Mardanpour P, Hodges DH, Neuhart R, Graybeal N. Effect of engine placement on aeroelastic trim and stability of flying wing aircraft. In Proceedings of the 53rd AIAA/ASME/ASCE/AHS/ASC Structures, Structural Dynamics and Materials Conference, Honolulu, Hawaii, Reston, Virginia, April 23–26, 2012. AIAA. AIAA Paper 2012-1634

[27] Hodges DH. Nonlinear Composite Beam Theory. Reston, Virginia: AIAA; 2006

[28] Peters DA, Johnson MJ. Finite-state airloads for deformable airfoils on fixed and rotating wings. In: Symposium on Aeroelasticity and Fluid/Structure Interaction, Proceedings of the Winter Annual Meeting, AD Volume 44, Pages 1–28. ASME, November 6–11. 1994

10

Nonlinear Response on External Electric Field and Nonlinear Generalization of Fluctuation-Dissipation Theorem for Levy Flights

Valeriy E. Arkhincheev and Lubsan V. Budazapov

Abstract

As well known, the fluctuation-dissipation theorem (FDT) establishes the relation between two different physical phenomena: the fluctuations and the dissipation. The fluctuations or the stochastic motion are determined by random stochastic forces. The dissipation or the directed motion is determined by regular forces. Nevertheless in the linear case, they are related by the FDT. One of the first and well-known examples of the FDT is Einstein's relation between diffusion coefficient and mobility of particle. It has been shown that a particle's velocity depends on electrical field in a nonlinear way in arbitrary weak fields due to anomalous super-diffusion character of Levy flight. The relation between two different critical indexes, describing Levy flight diffusion and dependence of current on electric field, has been established. This relation is the generalization of fluctuation-dissipation theorem for such a nonlinear Levy flight case. The physical interpretation of these results is given.

Keywords: fluctuation-dissipation theorem, nonlinear response, generalization of Einstein relation, random walks, Levy flights, diffusion on self-similar clusters

1. Introduction

As well known, the fluctuation-dissipation theorem (FDT) establishes the relation between two different physical phenomena: the fluctuations and the dissipation. The fluctuations or the stochastic motion are determined by random stochastic forces. The dissipation or the directed motion is determined by regular forces. Nevertheless in the linear case, they are related by the fluctuation-dissipation theorem (FDT). One of the first and well-known examples of this FDT is Einstein's relation between diffusion coefficient D and mobility of particle η:

$$qD = \eta \ kT \tag{1}$$

Here T is the temperature of the system, k is Boltzmann's constant, and q is the charge of the particle.

We first recall the well-known Einstein's arguments [1]. Let the diffusion current be J_d and the field current be J_f in the system. In the equilibrium state, the diffusion current is compensated by the field current:

$$J_d + J_f = 0 \tag{2}$$

and the particles are in the equilibrium state and are described by Boltzmann's distribution function:

$$N_{eq} = N_0 \exp\left(-\frac{U}{kT}\right) \tag{3}$$

where U is the potential energy, T is the temperature, and k is Boltzmann's constant, N_0 is the initial number of particles. Let us consider in more details the assumptions, which are used. There are three assumptions:

i. Boltzmann's statistics

ii. Fick's law for the diffusion current:

$$J_d = -D\nabla n \tag{4}$$

It also means that the root-mean-square displacement depends on time in a linear way and it is characterized by diffusion coefficient D:

$$< X^2(t) >\sim D \ t \tag{5}$$

iii. Ohm's law, which describes a linear dependence on electric field

$$J_f = n\eta E \tag{6}$$

Consequently, if one of these above assumptions does not hold, then we expect that Einstein's relation is broken and the new generalized relation will be appeared.

Subsequently, we consider the case, when diffusion has an anomalous power character:

$$< X^2(t) >\sim t^k \tag{7}$$

These anomalous stochastic processes were intensively studied [2]. The value $k = 1$ corresponds to the usual ordinary diffusion, the value of exponent $k < 1$ corresponds to the sub-diffusion case, and the value of exponent $k > 1$ corresponds to the super-diffusion or Levy flights case. Usually, anomalous sub-diffusion random walks were observed in disordered materials as fractals and percolations clusters [3–5]. Another anomalous super-diffusion, that is, Levy flights, was observed in the chaotic dynamics problem [6–10].

In this chapter, the Levy flights diffusion in an external weak electric field is considered. The problem consists of that the diffusion coefficient for Levy flight, which is determined in a usual way, has an infinite value:

$$D = \lim_{t \to \infty} \frac{< X^2(t) >}{t} \to \infty$$

It occurs due to the possibility of diffusing particle to move for an arbitrary distances at every step. So, if we apply the usual Einstein relation (1), then we obtain the infinite value for a mobility of particle η at arbitrary weak fields:

$$\eta \to \infty$$

But it is not possible to have infinite value of mobility from the physical point of view. What does it means? We believe that it means Einstein's relation in its usual form does not apply. Furthermore, we show that instead of linear response—Ohm's law—another new nonlinear response is appeared in the studied problem. Namely, the drift velocity depends on a weak electric field in a nonlinear way:

$$V \sim E^{\nu} \tag{8}$$

Here, ν is the critical exponent of new nonlinearity. The relation between the exponent of nonlinearity ν and the exponent of anomalous super-diffusion μ has been established:

$$\nu = \mu - 1 \tag{9}$$

It is necessary to emphasize that this nonlinearity occurs in arbitrary weak fields and it was a consequence of the anomalous Levy super-diffusion. In other words, Ohm's law (the linear response to a field) holds in the case of usual diffusion and Ohm's law does not apply at all for case of Levy flight super-diffusion.

This chapter was organized as follows. In Section 2, the preliminary generalization of Einstein's relation for a Levy flights was obtained. The qualitative estimations for drift velocity in two cases of super diffusion and usual diffusion were obtained too in Section 2. In Section 3, the one-dimensional discrete Levy flight diffusion was studied. The stable non-Gaussian distribution was deduced. The problem of Levy random walks in an external electric field or anisotropic Levy diffusion was studied in Section 4. The numerical simulations of Levy flights in an electric field were presented in Section 5. In Section 6, obtained new results for particle mobility were represented in the scaling form. The fluctuation-dissipation theorem for Levy flight case was rewritten in the scaling form also in Section 7. Section 8 concludes the chapter and the discussion of results was given in this section.

2. Qualitative estimation and generalization of Einstein relation for Levy flight case

Let us briefly remind the Levy flights diffusion. A feature of Levy flight random walks consists of the possibility for a diffusing particle to move on arbitrary large distances at every step, so that the root-mean-square displacement appears to be infinite. The numerical simulation of Levy hops diffusion has shown that the points, visited during Levy flights diffusion, have formed spatially well-defined clusters. "For more in-depth consideration it makes easy to see that each of clusters consists of a collection of clusters, in turn, so a structure of self -similar clusters was appeared due to Levy flights" [6]. The probability distribution function P in (k, t)-representation is

$$P(k,t) = \exp\left(-A|k|^{\mu} t\right) \tag{10}$$

where A and μ are positive quantities, $1 < \mu < 2$. Such distributions are called as stable Levy distributions. For more information about diffusion, see also [7, 8].

Let us check the above three assumptions for Einstein's relation—formulae (2–4) in the case of anomalous Levy flights super-diffusion. The first assumption about Gibbs-Boltzmann's statistics keeps the same, because the type of statistics—Gibbs- Boltzmann's classical statistics—was determined by the statistical properties of the system in the equilibrium and it does not depend on the kinetic properties of the system. (The kinetic phenomena as relaxation and diffusion describe the processes or ways, which lead to the equilibrium state, only.) So we use Gibbs-Boltzmann's distribution function too. But the second assumption about Fick's law for diffusion current is broken. The diffusion current has another form in the Levy flights case, and we write it in a general operator form:

$$J_d = -\widehat{K} n \tag{11}$$

Here, n is the concentration of diffusing particles, the operator $\widehat{K}$ in the k- representation is equal to

$$\left(\widehat{K}\right)_k = ik|k|^{\mu-2} \tag{12}$$

And in the r- presentation, it is equal to

$$\widehat{K} = \vec{\nabla}\ \left|\Delta^2\right|^{\mu-2/4} \tag{13}$$

where Δ is the Laplace operator and K is the fractional order operator—see, for example [10]. And finally, we use the general form for the field current instead of linear Ohm's law approximation as

$$\vec{J_f} = n\,\vec{V} \tag{14}$$

where $\vec{V}$ is the drift velocity. In the general case, this velocity depends on electric field in an arbitrary way: a linear or may be a nonlinear way. Repeating the same reasons for equilibrium stated as above, we obtain the general formula for the drift velocity:

$$\vec{V} = \frac{\widehat{K} N_{eq}}{N_{eq}} = \exp\left(\frac{U}{kT}\right)\ \widehat{K} \exp\left(-\frac{U}{kT}\right) \tag{15}$$

In the case of the anomalous diffusion, we obtain

$$\widehat{K} = \vec{\nabla} \left|\Delta^2\right|^{\mu-2/4}$$

By taking a definition for the derivative of the fractional order in the form of the set [10]:

$$\vec{V} = \exp\left(\frac{U}{kT}\right) \lim_{\varepsilon \to 0} \left(\Delta^2 + \varepsilon\right)^{(\mu-2)/4} \vec{p} \exp\left(\frac{U}{kT}\right) \tag{16}$$

we recover that the drift velocity depends on the homogeneous electric field $U = -q\vec{E}\ \ \vec{r}$ in a nonlinear way:

$$\vec{V} \propto \vec{E^{\nu}} \tag{17}$$

It should be emphasized that this nonlinearity occurs in arbitrarily weak fields, and it was a result of the unusual anomalous character of Levy flights diffusion. The exponent of this nonlinearity relates with the critical exponent of the Levy hop diffusion as above (9): $\nu = \mu - 1$. We consider this relation between two critical exponents, which describe the nonlinear mobility on one hand and the anomalous super-diffusion on the other hand, as generalization of FDT for Levy flight diffusion case.

2.1. Qualitative estimations for particle velocity

Subsequently, we want to confirm the result (17), which was obtained from the phenomenological approach, in another way. For this aim, we consider the problem of diffusion in an electric field in more details. When we introduce the electric field into the diffusion problem, then the new "field" length, governed by external electric field, was appeared:

$$L_E = \frac{kT}{qE} \tag{18}$$

To understand physical sense of this new "field" length and to make necessary estimations for drift velocity, let us imagine that the medium was partitioned into the boxes of size L_E [11]. Further, we proceed the particle motion inside of this box. After leaving this box, the particle goes along the electric field direction with the probability W_+, which is approximately equal to the unity, ($W_+ \propto 1$) and the particle leaves this box with the approximately zero probability W_- in the opposite direction ($W_- \propto 0$). It means that at these scales of L_E, the directed motion prevails over the random diffusion motion. So we can estimate the particle velocity as follows:

$$V = \frac{L_E}{T_E} \tag{19}$$

where T_E is a diffusion time for a length L_E.

In the case of usual diffusion, this diffusion time equals to: $T_E = \frac{L_E^2}{D}$ and we obtain Ohm's law with well-known Einstein relation between diffusion and mobility:

$$\vec{V} = \frac{q^2 D \vec{E}}{kT} \tag{20}$$

In the case of Levy flight, the diffusion time is proportional to powers of "field" length: $T_E \propto L_E^{\mu}$ according to Eq. (10). Repeating the same estimation, we obtain the same nonlinear relation as formula (17):

$$\vec{V} = \frac{q^2 \tilde{D}}{kT} \left|\frac{qE}{kT}\right|^{\mu-2} \vec{E} \tag{21}$$

Here, $\tilde{D}$ is the constant diffusion coefficient for Levy hop diffusion. Correspondingly for a case of two different diffusion regimes, we obtain two different laws for drift velocity: nonlinear behavior (21) and Ohm's law (20).

We want to stress that these preliminary generalizations of Einstein's relation in Section 2, see formulae (17, 21), only reveal the possibility of new nonlinear behavior for drift velocity in the anomalous super-diffusion case. To prove this result in an exact way, we need to study the microscopic model.

3. The Levy flight diffusion

To prove the fluctuation-dissipation for Levy flights diffusion case, let us consider the one-dimensional Levy flights diffusion in more details. Briefly, we remind how the Levy stable law (10) for distribution function has been obtained. Let us denote the probability of particle to occupy l- site after n steps as $P_n(l)$ and the probability of hops on length l at every step as $f(l)$. So we obtain the following master equation for a discrete case:

$$P_{n+1}(l) = \sum_{m=-\infty}^{\infty} f(|l-m|) P_n(m) \tag{22}$$

Here, l and m are integer numbers, which describe positions of sites. In the case of usual diffusion, when the particle hops on the nearest (left or right) sites only, this function $f(l)$ is equal to

$$2f(l) = \delta_{l,b} + \delta_{l,-b} \tag{23}$$

where $\delta_{l,b}$ is the Kronecker's delta symbol. And the known main equation describing diffusion on the nearest sites is received:

$$P_{n+1}(l) = \frac{1}{2}(P_n(l+1) + P_n(l-1)) \tag{24}$$

To simulate a Levy flight, the following Weierstrass function has been used as $f(l)$

$$f(l) = \sum_{n=0}^{\infty} a^{-n}\left(\delta_{l,-b^n} + \delta_{l,b^n}\right) \tag{25}$$

Here, parameter b is a length of hop, parameter $\frac{1}{a}$ is a possibility to make hop of length b (e.g., a possibility to hop for a distance b^2 is equal to $\frac{1}{a^2}$ and so on). The value of parameter a is confined by the bound values:$b < a < b^2$, consequently

$$1 < \mu = \frac{\ln a}{\ln b} < 2 \tag{26}$$

Let us shortly discuss the physical picture of Levy flight diffusion. Due to power distribution of hops over the lengths according to (25), the diffusing particle prefers to hop at nearest sites due to the biggest probability for nearest sites, to create the cluster from the nearest visited sites. But there is a small possibility to make a long hop from time to time. After this long hop, the new cluster of another nearest visited sites has formed at new place. So finally, the structure of self-similar clusters appears [6]. So we can say that Levy diffusion is the random walks along self-similar clusters.

Then the structural function for such random walks is equal to

$$\lambda(k) = \int f(l)\exp(ikl)dl = \sum_{n=0}^{\infty} a^{-n}\cos(kb^n) \tag{27}$$

Note too that the structural function of Levy flight satisfies the functional equation:

$$\lambda(k) = a\lambda(kb) + \cos(k) \tag{28}$$

Therefore, for $k \to 0$, it has a power behavior:

$$\lambda(k) \approx k^{\mu}, \text{where } \mu = \ln a / \ln b \tag{29}$$

Exactly, the nonanalytic power behavior for k has been established by means of Mellin's transformation or by formulas of Poisson's set summation. In more detail, see [7].

4. Introduction of field in the Levy flight problem and nonlinear response on electric field

Let us introduce an anisotropy into the random walk on self-similar clusters, formed during Levy flights diffusion. By virtue of specific nature of Levy hops, a particle can move for an arbitrary distance b^n at every step. For this reason, a small anisotropy $(1+\alpha)$ for small displacements s (with $\alpha = \frac{qEs}{kT}$) has an exponential large value at large distances b^n as $(1+\alpha)^{b^n}$. Since at each step, a diffusing particle certainly leaves a site, so the sum of probability of motion along the electric field direction W_+ and probability of motion in opposite direction W_- must be equal to 1:

$$W_- + W_+ = 1 \tag{30}$$

Hence, we obtain the following expressions for these probabilities:

$$W_\pm = \frac{(1 \pm \alpha)^{b^n}}{(1+\alpha)^{b^n} + (1-\alpha)^{b^n}} \tag{31}$$

Therefore, the structural function λ (k; E) for Levy flights diffusion in an electrical field is equal to

$$2\lambda(k, E) = \sum a^{-n}[\cos(kb^n) + i\sin(kb^n)(W_+ - W_-)] \tag{32}$$

As well as for the usual ordinary diffusion, the second member with anisotropy for small $k \to 0$ contains the expression for the drift velocity:

$$V = i\frac{\partial\lambda(k, E)}{\partial k}\bigg|_{k\to 0} = \sum \left(\frac{b}{a}\right)^n \frac{(1+\alpha)^{b^n} - (1-\alpha)^{b^n}}{(1+\alpha)^{b^n} + (1-\alpha)^{b^n}} \approx \sum \left(\frac{b}{a}\right)^n th(\alpha b^n) \tag{33}$$

here $th(x)$ is the hyperbolic tangent.

It is easy to see that the drift velocity satisfies the following functional equation:

$$V(\alpha) = \frac{b}{a}V(\alpha b) + cth(\alpha) \tag{34}$$

It means that at weak fields $\alpha \to 0$, the velocity depends on the electric field in a power-like way:

$$V(\alpha) \propto \alpha^\nu \tag{35}$$

with exponent $\nu = (\mu - 1)$.

To calculate the velocity by exact way, we used Poisson's formula:

$$\sum_{n=0}^{\infty} f(n) = \frac{1}{2}f(0) + \int f(t)dt + 2\sum_{m=1}^{\infty} f(t)\cos(2\pi mt)dt \tag{36}$$

After calculations, we obtain the formula for the velocity:

$$V(\alpha) = \frac{\alpha}{2} + \alpha^{(\mu-1)}\left[\sum_{m=-\infty}^{\infty}\int th(z)z^{-\gamma_m}dz\right] + \int_0^\alpha th(z)z^{-\gamma_m}dz] \tag{37}$$

where a power exponent is equal to $\gamma_m = \mu + 2\pi\, mi/\ln b$. It is easy to see that for arbitrary weak fields α, the first term has been neglected in comparison with the second term in the brackets. Thus, in arbitrary weak electric fields, the nonlinear dependence on electrical field of velocity (35) has appeared.

5. Numerical simulations

Subsequently, the results of numerical simulations of Levy random walks were reported. Let us briefly explain the algorithm of simulations. Probabilities of left and right walks are determined as probabilities to have a random value from [0, 0.5] and [0.5, 1] correspondingly. The anisotropy of random walks is simulated by the decreasing length of [0, 0.5] for quantity W_- anti-parallel field and increasing [0.5, 1] for quantity W_+ in parallel field case. The simulations are made at different values of parameters a and b. As the probability a^{-n} decreases rather rapidly, so we can confine finite members in the sum (10). For example, at a = 50, b = 10, n = 6 and a = 6, b = 3, n = 12. But we proceed that at every hop, the sum of all probabilities with finite numbers of hops is equal to 1, that is, particle does not stay in the site.

The results of random walks, **Figure 1**, are in accordance with the known results [2].

The step-like dependence of rms as a function of time is easy to understand as follows. The particle diffuses at nearest sites mainly, making the cluster from visited sites, and with a small probability hops at big distance (at next step) and again diffuses at nearest sites and so on.

The electric field leads to the particle drift. The dependence of the average displacement $< X >$ as function of the time is represented in **Figure 2** at different values of anisotropy. From linear dependence, it is easy to find the particle velocity by standard way: V = <X>/N. The value of the nonlinear dependence index is determined from numerical simulation data as.

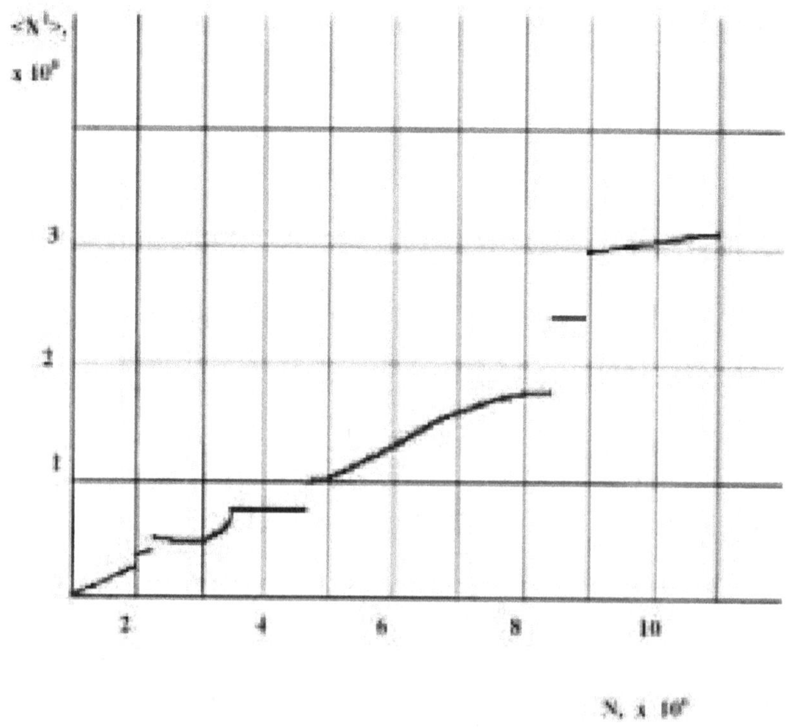

Figure 1. Typical dependence of RMS for Levy flight.

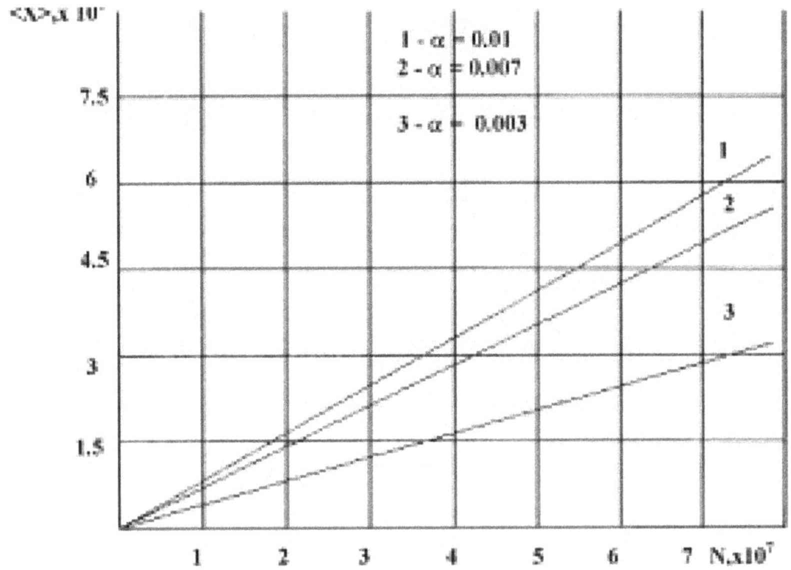

Figure 2. Dependence of the average displacement <X(t) > on number of hops N.

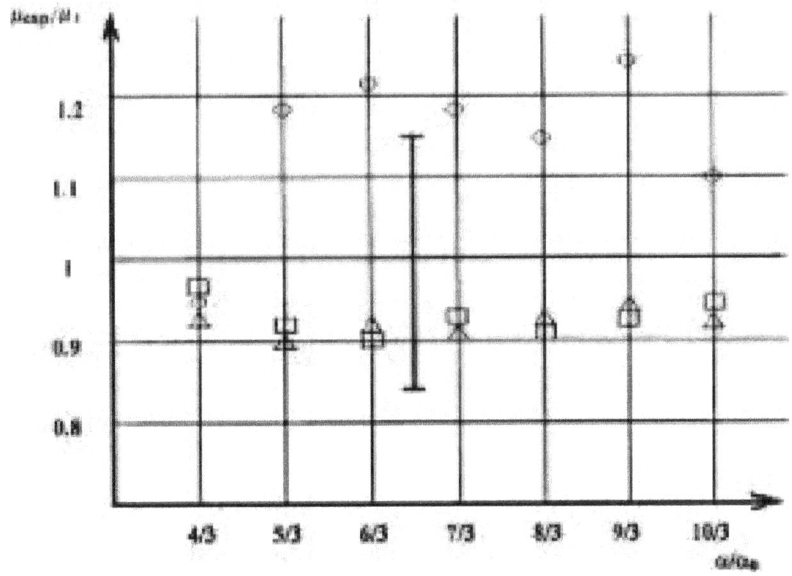

Figure 3. Dependence of relation $\mu_{\exp}/\mu_{theor}$ at different values of anisotropy.

$$\mu_{\exp} = 1 + \ln(V/V_0)/\ln(\alpha/\alpha_0)$$

These results are represented in **Figure 3**. The main distortion in the simulations is due to the random character of walks, and it was checked in the calculations from values of average displacement at zero fields.

6. Transition from nonlinear response to Ohm's law

6.1. Transition from Levy super-diffusion to ordinary diffusion

In this section, we additionally introduce the usual diffusion on the nearest neighboring sites in the process of random Levy walks. It gives us the possibility to proceed the transition from Levy super-diffusion to the usual diffusion. For this aim, the finite hop length ξ at every step has been introduced. So we construct the complex random walks, in which the Levy super-diffusion alternates with the usual diffusion. Accordingly, the distribution function for lengths of hops has the following form:

$$f(l,\xi) = \sum_{n=0}^{\infty} a^{-n}\left(\delta_{l,-(b^n+\xi)} + \delta_{l,b^n+\xi}\right) \tag{38}$$

Hence, the structural function for complex random walks with Levy diffusion and ordinary diffusion is equal to

$$\lambda(k,\xi) = \sum a^{-n} \cos\left(kb^n + k\xi\right) \tag{39}$$

In the case of complex alternative diffusion, the main contribution to the root-mean-square displacement was provided by Levy flights on long times, corresponding to big scales. Correspondingly, on small times and at small scales, the main contribution was provided by the usual diffusion. In the limit of the small lengths of hops $b << \xi$, we obtain the formula for structural function, which corresponds to the usual diffusion:

$$\lim_{b\to 0} \lambda(k,\xi) = \frac{a}{a-1} \cos\left(k\xi\right) \tag{40}$$

We consider this transition $b << \xi$ as transition from the discrete medium to the continuous medium with heterogeneity length as ξ. It is easy to check that the usual diffusion equation has followed from this structural function as a result.

6.2. The drift in the case of both ordinary and Levy diffusion

Let us introduce the anisotropy into these complex random walks as described earlier, but now we replace the hop length b^n to the new quantity: $b^n + \xi$. After this replacement, we obtain the formula for the new structural function in an electric field with finite hop length:

$$2\lambda(k,\xi,E) = \sum_{n=0}^{\infty} a^{-n}[\cos\left(kb^n + k\xi\right) + i\sin\left(kb^n + k\xi\right)(W_+ - W_-)] \tag{41}$$

Accordingly, the velocity has been described by the following formula:

$$V = i\frac{\partial\lambda(k,\xi,E)}{\partial k}\bigg|_{k\to 0} = \sum\left(\frac{b^n+\xi}{a^n}\right)\tanh(\alpha b^n + \alpha\xi). \tag{42}$$

To calculate this sum in formula (42), Poisson's method of summation has been used again.

The following results have obtained. For weak electric fields ($\frac{qE\xi}{kT} << 1$), the velocity is a nonlinear function of the electric field:

$$V \sim E^{\mu-1} \tag{43}$$

and in the strong fields ($\frac{qE\xi}{kT} >> 1$), the velocity became a linear function in the field:

$$V \sim E\xi^{2-\mu} \tag{44}$$

Note that the particle velocity has two asymptotic regimes in accordance with the diffusion limits: Levy hops and usual ordinary diffusion. The Levy flight diffusion leads to the nonlinear response, and the usual diffusion leads to the linear Ohm's law. So the two different power dependencies of particle mobility (43, 44) were obtained for a specific distribution of hops as (38). But before this result was obtained without any assumptions about the nature of hops, only specific form of Levy diffusion current was used as (11). And now, we consider the specific distribution of hops (38) only as microscopic model. We believe that the same nonlinear result will be correct for another hops distribution over lengths.

7. Scaling for particle mobility

We want to remark that above results look similar to the phase transition theory results [12, 13]. First of all, we have the analog of correlation radius for phase transition L_c—in our case, this is the finite length of hop ξ. At scales, which bigger than ξ, we have anomalous super-diffusion and at scales, which are smaller than ξ, we have the usual diffusion. So this length ξ has a role of heterogeneous scale as correlation radius. Second as it is well known that if the correlation radius L_c trends to the infinity at the phase transition point (at threshold), then any characteristic scales in the phase transition theory at threshold are absent, so any response for external fields has the power behavior, which is described by the critical exponents of phase transition theory. Near threshold point, results of the phase transitions theory were easy to understand if they have the scaling form. So we want to present the above-obtained results in the general scaling automodel form too, using the finite hop length ξ instead of correlation length L_c.

So to clarify the obtained results, the expression for the particle mobility $\eta = \frac{V}{E}$ has been rewritten in the scaling form too:

$$\eta \propto \xi^{-\lambda} f\left(\frac{qE\xi}{kT}\right) \tag{45}$$

where λ is the critical exponent of scaling, and the scaling function $f(x)$ has the asymptotic power behavior:

$$f(x) = \begin{Bmatrix} 1, x << 1 \\ x^{\lambda}, x >> 1 \end{Bmatrix} \tag{46}$$

For our model of Levy flights diffusion, this scaling exponent λ is connected with the exponent of the super-diffusion as

$$\lambda = \mu - 2 \tag{47}$$

At the small scales $\xi << \frac{kT}{qE}$, where the usual diffusion dominates, the particle mobility depends on the homogeneity length ξ only (correlation radius in the phase transitions theory). At the large scales, where the Levy super-diffusion dominates, the mobility depends on the electric field E only or, in other words, the mobility became a function of the new "field" length $L_E = \frac{kT}{qE}$ with the same exponent λ (see formula (42) too).

7.1. The scaling form for fluctuation-dissipation theorem for nonlinear case

Usually, the Einstein relation between diffusion and conductivity was considered as a simple example of fluctuation-dissipation theorem (FDT), which was connected by the different characteristics of the considered system: the dissipation, described by the relaxation time τ (the particle mobility $\eta = \frac{q\tau}{m}$), and the fluctuation characteristic, described by the diffusion coefficient D:

$$qD = \frac{q\tau}{m} kT = \eta kT \tag{48}$$

We want to stress that this obtained nonlinearity (43) essentially differs from the usual nonlinearity, and our result means that the relation between the nonlinear mobility and the coefficient of diffusion existed in the new nonlinear form, when the mobility became as nonlinear function of the electric field

$$\eta(E) \sim E^{\lambda} \tag{49}$$

Here, λ is exponent of the nonlinear dependence of mobility. And new nonlinear generalized fluctuation-dissipation theorem relates the exponent of the nonlinear response λ with the exponent of the anomalous diffusion μ:

$$\lambda = \frac{d \ln \eta(E)}{d \ln E} = \mu - 2 \tag{50}$$

It seems that this investigated case was a first case when the fluctuation-dissipation theorem in the usual form of linear relation between two coefficients was broken. And instead of simple relation between linear coefficients, the new and more general relation between exponents of mobility and exponent of the super-diffusion appeared.

From this point of view, we believe that the case of usual diffusion or Einstein's relation between two coefficients of diffusion and mobility is the limiting case of new generalized

FDT between exponents of mobility of particle in an electric field and exponent of diffusion: $\lambda = 0 \ (\mu = 2)$.

8. Discussion

Let us discuss the results. All the above obtained both the nonanalytic behavior of structural function for small k $\to 0$ and the nonlinear electric field dependence of the velocity in arbitrarily weak fields which were the asymptotical results. We show that the current (velocity) depends on electric field in a nonlinear way due to the anomalous character of Levy flights and possibility to fly at arbitrary distances:

$$J(E) \sim E^{\nu} \tag{51}$$

Nonlinear properties of media intensively have been studied. Usually, the nonlinearity has been connected with the expansion of electric current for set in powers of the electric field and with consideration of the cubic nonlinearity [14]:

$$\vec{J} = \sigma \vec{E} + \chi |E|^2 \vec{E} + \ldots \tag{52}$$

But our result essentially differs from the results, obtained by this method. We show that in the investigated case of Levy super-diffusion, the nonlinear behavior appeared due to anomalous super-diffusion character and the electric current depends on electric field in a power nonlinear way. It means that Ohm's law or a linear term was absent in the field series expansion of the current (58) in the investigated case.

The generalization of fluctuation-dissipation theorem for a case of Levy flights diffusion was obtained. Instead of well-known Einstein's relation between diffusion coefficient D and mobility η, which is correct in linear Ohm's law case, the new relation between exponents, which describes the nonlinear response of system ν on the hand and anomalous Levy flight diffusion μ on the other hand, was obtained:

$$\nu = \mu - 1$$

It is interesting to note that from the above-obtained results, we understand what two results were contained in Einstein's relation (1). Firstly, we can say that Einstein recovers or proves the existence of Ohm's law (linear response) for any systems with usual diffusion, and secondly, he established the relation between diffusion coefficient and mobility of particle in a linear case.

As for "real" systems, the different theories with different predictions have been existed and numerical simulations have not given a clear answer yet: the non-monotonically dependence with time were founded [15, 16]. We hope that these results may be applied for real disordered systems and in particular also for the problem of hopping in the disordered systems, but we need to make further investigations for it [17].

Author details

Valeriy E. Arkhincheev[1,2*] and Lubsan V. Budazapov[3]

*Address all correspondence to: valeriy.arkhincheev@tdt.edu.vn

1 Theoretical Physics Research Group, Advanced Institute of Materials Science, Ton Duc Thang University, Ho Chi Minh City, Vietnam

2 Faculty of Applied Sciences, Ton Duc Thang University, Ho Chi Minh City, Vietnam

3 Buryat Research Institute of Agriculture, Ulan-Ude, Russian Federation

References

[1] Einstein A. Investigations on the Theory of Brownian Movement. New York: Dover; 1956

[2] Metzler R, Klafter J. The random walks guide to anomalous diffusion: A fractional dynamics approach. Physics Reports. 2000;**339**:1-77

[3] Metzler R, Klafter J. Anomalous stochastic processes in the fractional dynamics framework: Fokker-Planck equation, dispersive transport, and non-exponential relaxation. Advances in Chemical Physics. 2001;**116**:223-264

[4] Isichenko MB. Percolation, statistical topography and transport in random media. Reviews of Modern Physics. 1992;**64**:961-1043

[5] Arkhincheev VE, Baskin EM. Anomalous diffusion and drift in a comb model of percolation clusters. Soviet Physics - JETP. 1991;**73**:161-165

[6] Mandelbrot B. Fractals: Form, Chance and Dimension. San-Francisco: Freeman; 1977

[7] Hughes BD, Shlesinger MF, Montroll EW. Random walks with self-similar clusters. Proceedings of the National Academy of Sciences of the United States of America. 1981;**78**: 3287-3291

[8] Barkai E. Fractional Fokker-Planck equation, solution, and application. Physical Review E. 2001;**63**:046118

[9] Uchaikin VV. Anomalous diffusion and fractional stable distributions. JETP. 2003;**97**:810-825

[10] Arkhincheev VE, Baskin EM, Batiev EG. Diffusion and conductivity m percolation systems. Journal of Non-Crystalline Solids. 1987;**90**:21-24

[11] Batyev EG. Private communications

[12] Pokrovskii VL, Patashinskii AZ. Fluctuation Theory of Phase Transitions. Moscow: Nauka; 1982

[13] Stanley HE. Introduction to Phase Transitions and Critical Phenomena. USA: Oxford University Press; 1987

[14] Klimontovich Y. Nonlinear Brownian motion. Uspekhi Fiziologicheskikh Nauk. 1994;**164**: 811-844

[15] Overhoff H, Beyer W. Philosophical Magazine B. 1981;**43**:433-438

[16] Benjamin J, Adkins C, van Clever J. Journal of Physics C. 1981;**17**:559-564

[17] Arkhincheev VE. Hopping by Levy flights and nonlinear relation between diffusion and conductivity. SPIE Proceedings. 2004;**5471**:560-567

11

Controlling Equilibrium and Synchrony in Arrays of FitzHugh–Nagumo Type Oscillators

Elena Adomaitienė, Skaidra Bumelienė and Arūnas Tamaševičius

Abstract

We present a case study of the FitzHugh–Nagumo (FHN) type model with a strongly asymmetric activation function. The proposed model is an electronically rather than a biologically inspired approach. The asymmetric exponential model imitates the shape of spikes in real neurons better than the classical FHN model with a cubic van der Pol activation function. An array of mean-field coupled non-identical FHN type oscillators is considered. The effect of mutual synchronization (phase locking) of units, originally oscillating at their individual frequencies, is demonstrated. Several feedback control methods, including stable tracking filter technique, mean field nullifying, and repulsive coupling are shown either to stabilize unstable equilibrium states or to suppress synchrony of the coupled FHN oscillators. The stability of the equilibrium states is analyzed by employing the eigenvalues, obtained from the characteristic equation, and by using the diagonal minors of the Routh–Hurwitz matrix. Nonlinear differential equations are solved numerically.

Keywords: nonlinear dynamics, spiking neuron model, FitzHugh–Nagumo oscillator, arrays of coupled oscillators, equilibrium states, synchronization, control methods

1. Introduction

The stability of any either natural or artificial system is a valuable and desired property. Therefore, the control of dynamical systems, in particular stabilization of their unstable equilibrium (UEQ) states, is an important problem in basic science and engineering applications, if periodic or chaotic oscillations are unacceptable behaviors. Usual control methods, based on proportional feedback control [1, 2] require knowledge of a mathematical model of a dynamical system or at least the exact coordinates of the UEQ in the phase space for the reference point. However, in many real complex systems, especially in biology, physiology, economics, sociology, and chemistry neither the full reliable models nor the exact coordinates of the UEQ are a priori known. Moreover, the position of the UEQ may change with time because of external unknown and unpredictable forces. In these cases, adaptive, that is model-independent and reference-free methods, automatically tracing and stabilizing unknown UEQ, can be helpful [3–5].

Synchronization is a universal and very common phenomenon, widely observed in nature, science, engineering, and social life [6]. Coupled oscillators and their arrays, exhibiting synchrony, range from pendulum clocks to various biological populations. In many cases, synchronization plays a positive role. However, sometimes, it has an unfavorable impact. Strong synchronization of neurons in human brain is an example. It is assumed that synchrony of spiking neurons in a neuronal population causes the symptoms of the Parkinson's disease and essential tremor [7]. Therefore, development of the methods and practical techniques for controlling, more specifically, for suppressing synchrony of coupled oscillators, in general, and particularly with possible application to neuronal arrays, is of great importance [8–10].

A variety of adaptive feedback methods for stabilizing UEQ of nonlinear dynamical systems have been described in literature. Here, we mention only some of them, e.g., derivative control technique [11–13], stable filter technique [3, 4, 14–17], unstable filter technique [18–20], and combined filters techniques [21–23]. A comprehensive list and an overview of control methods developed to stabilize UEQ states can be found in [24]. We note that the above mentioned techniques deal with single unstable dynamical systems. Stabilization of a network of coupled oscillators has been considered in a recent paper [25].

Suppression of synchrony in arrays of oscillators by means of feedback methods has been described in many papers [7–10, 26–29]. More publications and discussion on the feedback techniques for control of synchrony are presented in [24, 25].

Another way to avoid synchrony in arrays of oscillators is a non-feedback method using external periodic drive at relatively high frequency (much higher than the natural frequency of the oscillators). In neurology, it is known as deep brain stimulation (DBS), applying about 150 Hz periodic pulses to certain brain areas [30]. It is a clinically approved therapy for patients with the Parkinson's disease symptoms. However, mechanism of the DBS is not fully understood. There are several papers considering the Hodgkin–Huxley and the FitzHugh–Nagumo models and demonstrating that high frequency forcing can stabilize the UEQ of the neuronal oscillators and thus inhibit spiking cells [31–33].

In this chapter, we present a case study of the FitzHugh–Nagumo (FHN) type model with a strongly asymmetric activation function (Section 2). An array of mean-field coupled non-identical FHN type oscillators is considered in Section 3. The effect of mutual synchronization (phase locking) of units originally oscillating at their individual frequencies is demonstrated. Several feedback control methods, including stable tracking filter technique (Section 4), mean field nullifying (Section 5), and repulsive coupling (Section 6) are shown either to stabilize UEQ states or to suppress synchrony of the coupled FHN type oscillators.

2. Single FHN type oscillator

An extremely simple electrical circuit, imitating a single spiking neuron, is sketched in **Figure 1**. The negative resistance R_n can be implemented by means of a negative impedance converter [34]. Typical train of spikes from its output is presented in **Figure 2**.

We apply the Kirchhoff's laws to electrical circuit in **Figure 1**, use the Shockley current–voltage characteristic for the diode, and introduce the following dimensionless quantities:

$$x = \frac{V_C}{V^*},\ y = \frac{\rho I_L}{V^*},\ t \to \frac{t}{\sqrt{LC}},\ \alpha = \frac{\rho}{|R_n|},\ \beta = \frac{r}{\rho},\ \gamma = \frac{I\rho}{V^*},\ \delta = \frac{I_S\rho}{V^*},\ \mu = \frac{qV^*}{nk_BT},\ \rho = \sqrt{\frac{L}{C}} \quad (1)$$

where $V^* = 1$ V, k_B is the Boltzmann constant, T is the absolute temperature (in K), q is the elementary charge, k_BT/q is the thermal potential (≈ 25 mV at room temperature, $T = 293$ K), n is a diode ideality factor, sometimes called emission coefficient (assumed value $n = 2$). Then, differential equations, convenient for analysis and numerical integration, are derived:

$$\begin{aligned} \dot{x} &= F(x) - y - \gamma, \\ \dot{y} &= x - \beta y. \end{aligned} \quad (2)$$

Activation function $F(x)$ in Eq. (2) is a strongly asymmetric one (**Figure 3**):

$$F(x) = \alpha x + \delta\left[\exp\left(-\mu x\right) - 1\right]. \quad (3)$$

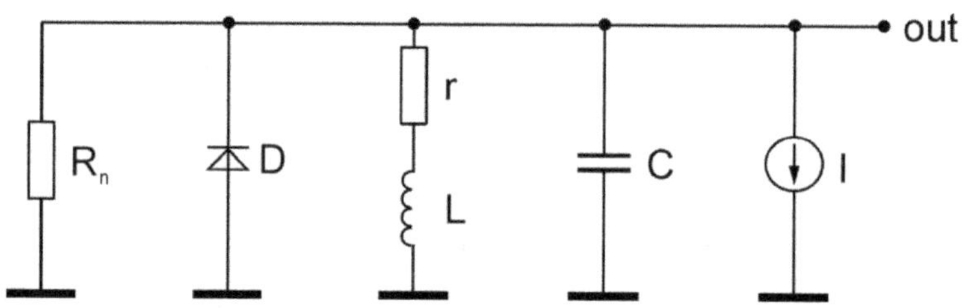

Figure 1. Circuit diagram of the electronic analog of spiking neuron. R_n is a negative resistance.

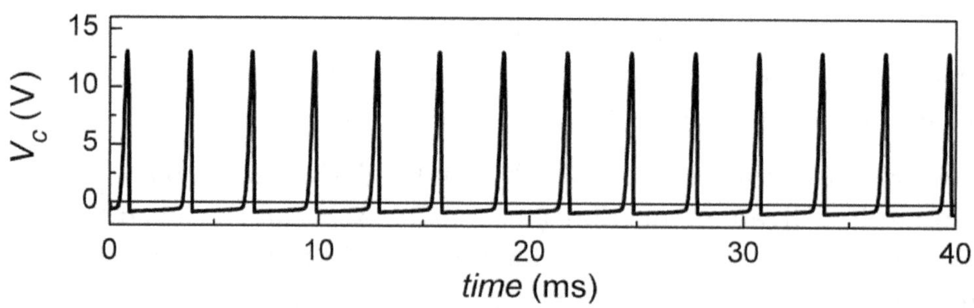

Figure 2. Voltage spikes from the circuit in Figure 1, generated by means of Electronics Workbench Professional software. $R_n = -680\ \Omega$, D is a semiconductor diode (BAV99 type) with saturation current $I_S = 10$ nA ($\delta = 10^{-5}$), $L = 100$ mH, $C = 100$ nF, ($\rho = 1$ kΩ), $r = 50\ \Omega$ ($\beta = 0.05$), $I = 1$ mA ($\gamma = 1$).

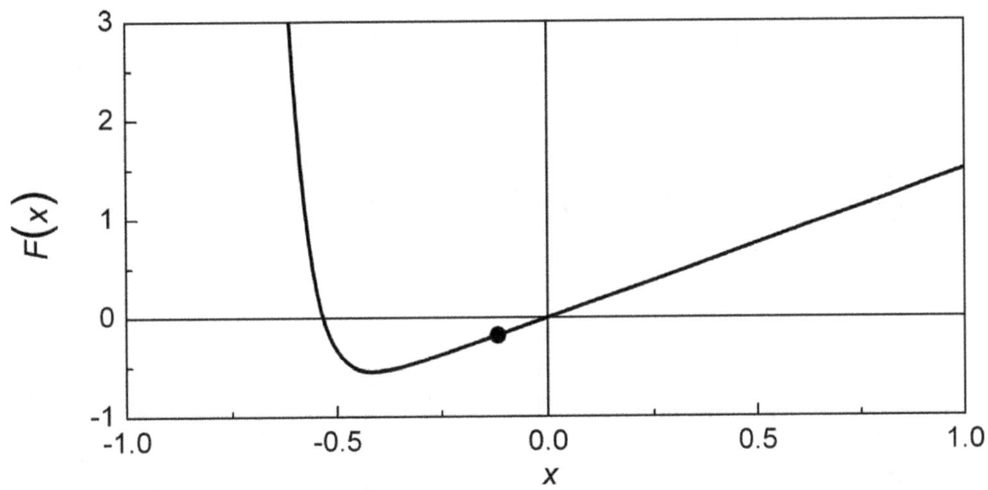

Figure 3. Activation function $F(x)$ from formula (3). α = 1.5, δ = 10^{-5}, and μ = 20. Black dot on the curve marks the equilibrium coordinate $x_0 = -0.12$ from formula (5) at $\beta = 0.1$ and $\gamma = 1$.

$F(x)$ essentially differs from the odd function $F_{FH}(x) = x - x^3/3$, introduced by FitzHugh [35] and used in many later papers, e.g., in [28]. It also differs from the asymmetric three-segment $[x < -1, -1 \le x \le 1, x > 1]$ piecewise linear function $F_{PL}(x) = \alpha x + d(x+1)H(-x-1) + g(x-1)H(x-1)$ suggested in [36], where $d >> g$ and $H(u)$ is the Heaviside unit step function, i.e., $H(u > 0) = 1$, $H(u \le 0) = 0$. In contrast to the $F_{PL}(x)$, the $F(x)$ is a smooth function, and therefore it seems a more realistic option.

For $\alpha\beta < 1$ and

$$\gamma << \frac{1-\alpha\beta}{\mu\beta}\ln\delta^{-1} \tag{4}$$

the equilibrium solution of Eq. (2) is given by the fixed point coordinates

$$x_0 = -\frac{\beta\gamma}{1-\alpha\beta}, \quad y_0 = -\frac{\gamma}{1-\alpha\beta}. \tag{5}$$

Due to the exponent in the activation function $F(x)$, strong inequality (4) practically can be replaced with a simple inequality:

$$\gamma \le \frac{1-\alpha\beta}{2\mu\beta}\ln\delta^{-1}. \tag{6}$$

Note empiric factor 2 is in the denominator. Eqs. (2), linearized around the fixed point (5), read

$$\begin{aligned} \dot{x} &= \alpha x - y, \\ \dot{y} &= x - \beta y. \end{aligned} \tag{7}$$

The corresponding characteristic equation is

$$\lambda^2 - (\alpha - \beta)\lambda + 1 - \alpha\beta = 0. \tag{8}$$

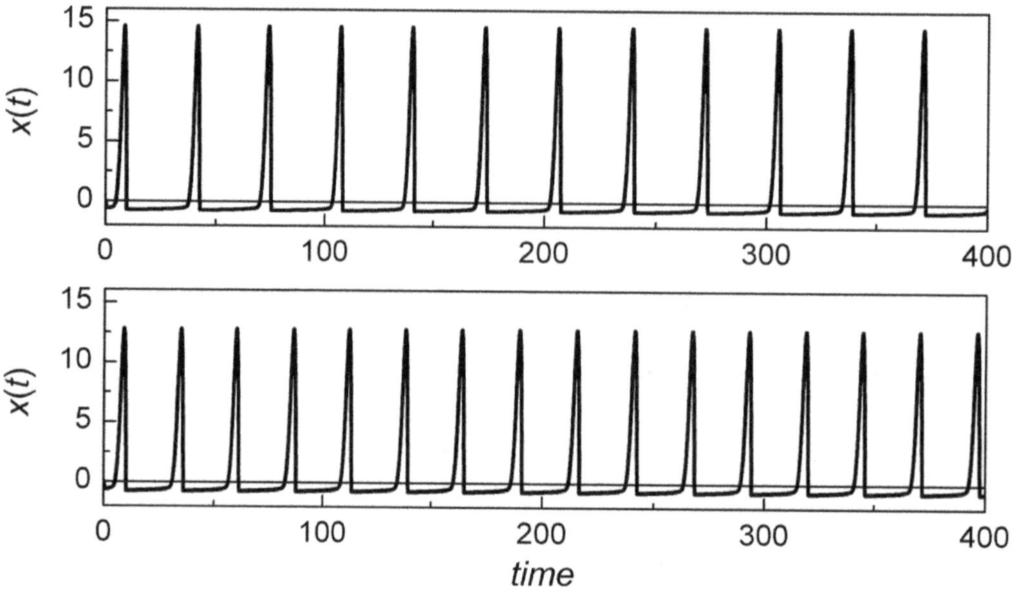

Figure 4. Waveforms $x(t)$ from Eq. (2) with $\alpha = 1.5$, $\gamma = 1$, $\delta = 10^{-5}$, and $\mu = 20$ for different damping β. (Top) $\beta = 0.05$ and (bottom) $\beta = 0.1$. Note, different inter-spike periods in the two plots.

It has two eigenvalues

$$\lambda_{1,2} = \frac{(\alpha - \beta)}{2} \pm \sqrt{\frac{(\alpha - \beta)^2}{4} - (1 - \alpha\beta)} = \frac{(\alpha - \beta)}{2} \pm \sqrt{\frac{(\alpha + \beta)^2}{4} - 1}. \tag{9}$$

If $\alpha > \beta$, then both real parts of the eigenvalues, $\mathrm{Re}\lambda_{1,2}$ are positive, proving that the equilibrium (x_0, y_0) is an unstable fixed point. If $\alpha + \beta > 2$, it is a node, and if $\alpha + \beta < 2$, it is a spiral.

Numerical solution of nonlinear equation Eq. (2) is presented in **Figure 4**.

3. Array of FHN type oscillators

An array of isolated (non-coupled) oscillators is given by

$$\begin{aligned} \dot{x}_i &= F(x_i) - y_i - \gamma, \\ \dot{y}_i &= x_i - \beta_i y_i, \end{aligned} \tag{10}$$

$$F(x_i) = \alpha x_i + \delta\left[\exp\left(-\mu x_i\right) - 1\right]. \tag{11}$$

Here and elsewhere $i = 1, 2, \ldots, N$. Note that the structure of function $F(x)$ and parameters α, δ, and μ are the same for all oscillators, whereas the damping parameters β_i in Eq. (10) are intentionally set different for each oscillator to make them slightly non-identical units.

Now we introduce interaction between oscillators. To be specific, we consider mean-field coupling, which is also called "star" coupling in electronics (**Figure 5**):

$$\dot{x}_i = F(x_i) - y_i - \gamma + k(\langle x \rangle - x_i), \quad \dot{y}_i = x_i - \beta_i y_i, \tag{12}$$

Here, $k = \rho/R^*$ is the strength of coupling and

$$\langle x \rangle = \frac{1}{N}\sum_{i=1}^{N} x_i. \tag{13}$$

Typical phase portraits for isolated and coupled (synchronized) oscillators are shown in **Figure 6**.

Intricate phase trajectories in **Figure 6 (left)** indicate that the oscillators are not synchronized, but oscillate at their individual frequencies, whereas simple closed loop in **Figure 6 (right)**

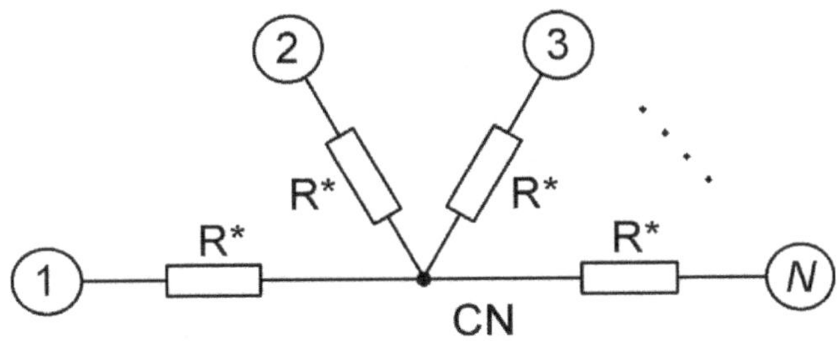

Figure 5. Diagram of mean-field coupled oscillators. R* are coupling resistors and CN is a coupling node.

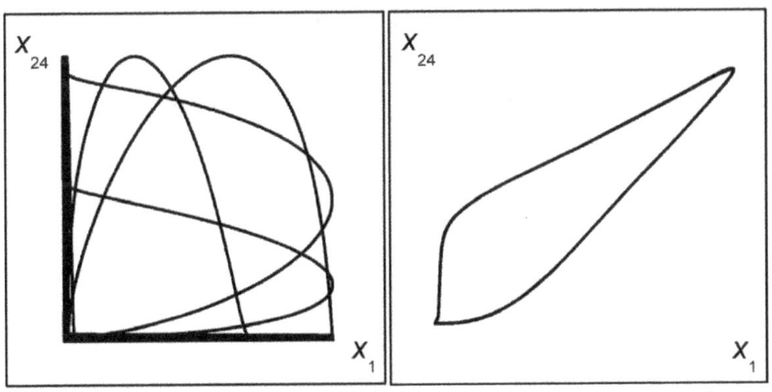

Figure 6. Phase portraits. $N = 24$, $\alpha = 1.5$, $\beta_i = 0.05 + 0.001i$, $\gamma = 1$, $\delta = 10^{-5}$, and $\mu = 20$. (Left) Isolated oscillators either from Eq. (10) or Eq. (12) with $k = 0$, and (right) coupled oscillators from Eq. (12) with $k = 1$.

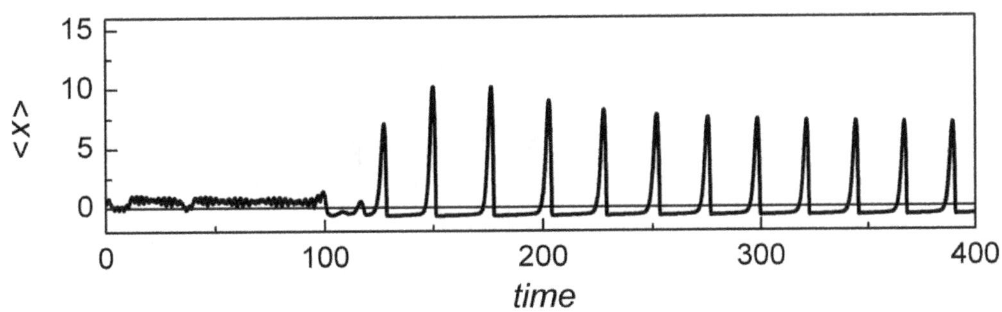

Figure 7. Waveform of the mean-field variable $\langle x \rangle$ from Eq. (12). $N = 24$, $\alpha = 1.5$, $\beta_i = 0.05 + 0.001i$, $\gamma = 1$, $\delta = 10^{-5}$, and $\mu = 20$. Coupling ($k = 1$) is switched on at $t = 100$.

proves that oscillators are in synchrony (phase-locked), i.e., oscillate at the same frequency. For synchronized oscillators, the phase difference is not necessarily zero (the phase portrait is not fine diagonal), but it does not change with time. The mean variable <x> for the two cases is shown in **Figure 7**. The amplitude of mean-field variable <x> is relatively low for isolated oscillators ($k = 0$), but becomes large for synchronized state ($k = 1$).

4. Stabilizing equilibrium states in array of oscillators

When an external capacitor is applied to the coupling node CN (**Figure 8**), the overall system becomes ($2N + 1$)-dimensional system:

$$\begin{aligned} \dot{x}_i &= F(x_i) - y_i - \gamma + k(z - x_i), \\ \dot{y}_i &= x_i - \beta_i y_i, \\ \dot{z} &= \omega_f(\langle x \rangle - z). \end{aligned} \tag{14}$$

Here, z is a dimensionless dynamical variable related to voltage across the external capacitor C_0, $z = V_{C0}/V^*$, the mean <x> is given by formula (13), and ω_f is the dimensionless cut-off frequency of the filter composed by R^* and C_0.

Analysis of the high-dimensional system is very complicated. Therefore, we consider a mean-field approach. We average all terms in Eq. (14) over all oscillators $i = 1, 2, \ldots, N$:

$$\begin{aligned} \langle \dot{x} \rangle &= \langle F \rangle - \langle y \rangle - \gamma + k(z - \langle x \rangle), \\ \langle \dot{y} \rangle &= \langle x \rangle - \langle \beta y \rangle, \\ \dot{z} &= \omega_f(\langle x \rangle - z). \end{aligned} \tag{15}$$

Here,

$$\langle x \rangle = \frac{1}{N}\sum_{i=1}^{N} x_i, \quad \langle y \rangle = \frac{1}{N}\sum_{i=1}^{N} y_i, \quad \langle F \rangle = \frac{1}{N}\sum_{i=1}^{N} F(x_i), \quad \langle \beta y \rangle = \frac{1}{N}\sum_{i=1}^{N} \beta_i y_i, \quad \omega_f = \frac{N\sqrt{LC}}{R^* C_0}. \tag{16}$$

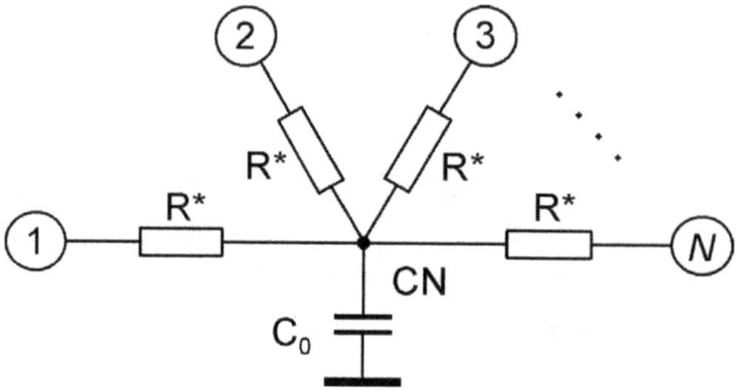

Figure 8. Diagram of mean-field coupled oscillators with a stabilizing capacitor C_0. Stable RC filter is composed of coupling resistors R^* and capacitor C_0 (see formulas (16)).

Eq. (15) is not suitable to describe full dynamics of the system. However, we can exploit it to find equilibrium coordinates. If inequality (6) is valid for all oscillators with different β_i, the steady-state equations read

$$\begin{aligned} 0 &= \langle x_0 \rangle - \langle y_0 \rangle - \gamma + k(z_0 - \langle x_0 \rangle), \\ 0 &= \langle x_0 \rangle - \langle \beta y_0 \rangle, \\ 0 &= \langle x_0 \rangle - z_0. \end{aligned} \tag{17}$$

Here,

$$\langle x_0 \rangle = \frac{1}{N}\sum_{i=1}^{N} x_{0i}, \quad \langle y_0 \rangle = \frac{1}{N}\sum_{i=1}^{N} y_{0i}, \quad \langle \beta y_0 \rangle = \frac{1}{N}\sum_{i=1}^{N} \beta_i y_{0i}, \quad \langle \beta \rangle = \frac{1}{N}\sum_{i=1}^{N} \beta_i. \tag{18}$$

There is a problem in Eq. (17) with the term $<\beta y_0>\langle \beta y_0 \rangle$. In general, $<\beta y_0> \neq <\beta> < y_0>$. However, if the ranges of the multiplicands β_i and y_{0i} in (18) are considerably different (in our case, the individual equilibrium coordinates y_{0i}, in comparison with β_i, much weaker depend on i), then $<\beta y_0> \approx <\beta> <y_0>$. Similarly to a single oscillator, considered in Section 2, for $\alpha\beta_i < 1$, the equilibrium coordinates are

$$\langle x_0 \rangle = -\frac{\langle \beta \rangle \gamma}{1 - \alpha\langle \beta \rangle}, \quad \langle y_0 \rangle = -\frac{\gamma}{1 - \alpha\langle \beta \rangle}, \quad z_0 = \langle x_0 \rangle. \tag{19}$$

Linearization of Eqs. (15) around the equilibrium coordinates yields:

$$\begin{aligned} \langle \dot{x} \rangle &= \alpha\langle x \rangle - \langle y \rangle + k(z - \langle x \rangle), \\ \langle \dot{y} \rangle &= \langle x \rangle - \langle \beta \rangle \langle y \rangle, \\ \dot{z} &= \omega_f(\langle x \rangle - z). \end{aligned} \tag{20}$$

The corresponding characteristic equation is

$$\lambda^3 + h_2\lambda^2 + h_1\lambda + h_0 = 0, \tag{21}$$

where

$$h_2 = -\alpha + \langle \beta \rangle + k + \omega_f, \quad h_1 = 1 - \alpha\langle \beta \rangle + \langle \beta \rangle k - (\alpha - \langle \beta \rangle)\omega_f, \quad h_0 = (1 - \alpha\langle \beta \rangle)\omega_f. \tag{22}$$

Numerical solution of Eq. (21) is presented in **Figure 9** for different values of the coupling parameter k. The equilibrium is stable, if the real parts of all three eigenvalues are negative, $\mathrm{Re}\lambda_{1,2,3} < 0$.

Necessary and sufficient conditions of stability can be found analytically from the Hurwitz matrix

$$H = \begin{pmatrix} h_2 & h_0 & 0 \\ 1 & h_1 & 0 \\ 0 & h_2 & h_0 \end{pmatrix}. \tag{23}$$

The Routh–Hurwitz stability criterion claims that the system is stable, if all diagonal minors of the matrix H are positive:

$$\Delta_1 = h_2 > 0, \quad \Delta_2 = h_2 h_1 - h_0 > 0, \quad \Delta_3 = h_0 \Delta_2 > 0. \tag{24}$$

The first minor is $\Delta_1 > 0$, if

$$k > k_1 = \alpha - \langle \beta \rangle - \omega_f. \tag{25}$$

For $\alpha = 1.5$, $\beta_i = 0.05 + 0.001i$, and $\omega_f = 0.1$, the threshold is $k_1 = 1.34$.

The second minor Δ_2 is more cumbersome and yields quadratic equation:

$$\langle \beta \rangle k^2 + dk + g = 0. \tag{26}$$

where

$$\begin{aligned} d &= 1 - 2\alpha\langle \beta \rangle + \langle \beta \rangle^2 - (\alpha - 2\langle \beta \rangle)\omega_f, \\ g &= -(\alpha - \langle \beta \rangle)\left[1 - \alpha\langle \beta \rangle - (\alpha - \langle \beta \rangle)\omega_f + \omega_f^2\right]. \end{aligned} \tag{27}$$

Eq. (26) has an analytical solution

$$k_{2,3} = -\frac{d}{2\langle \beta \rangle} \pm \sqrt{\frac{d^2}{4\langle \beta \rangle^2} - \frac{g}{\langle \beta \rangle}}, \tag{28}$$

which provides two different values. For $\alpha = 1.5$, $\beta_i = 0.05 + 0.001i$, and $\omega_f = 0.1$, the values are $k_2 = 1.44$ and $k_3 = -12.3$. Eventually, we evaluate the threshold $k_{th} = \max(k_1, k_2, k_3) = 1.44$. It is in a very good agreement with numerical value of k_{th} obtained from $\mathrm{Re}\lambda_{1,2,3}(k)$ in **Figure 9**.

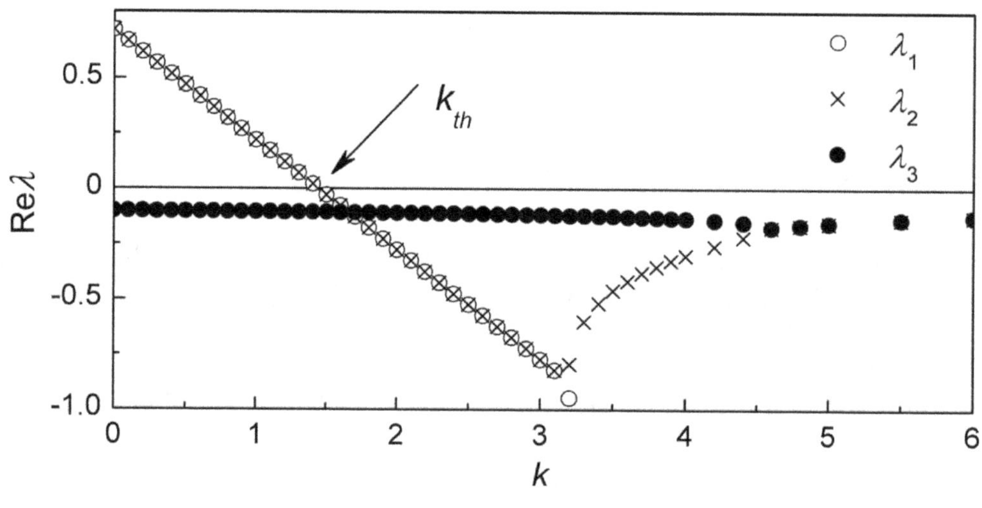

Figure 9. Real parts of the eigenvalues from Eq. (21). $N = 24$, $\alpha = 1.5$, $\beta_i = 0.05 + 0.001i$, $\langle\beta\rangle = 0.0625$, and $\omega_f = 0.1$. Arrow in the plot indicates the threshold coupling parameter $k_{th} = 1.44$, where the largest eigenvalues become negative.

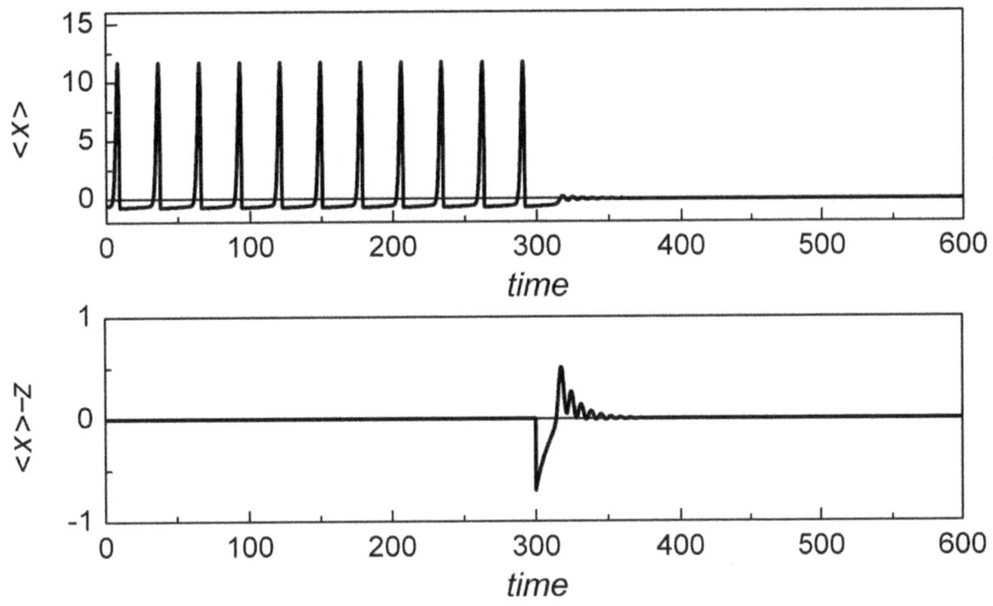

Figure 10. Waveforms from Eq. (14). $N = 24$, $\alpha = 1.5$, $\beta_i = 0.05 + 0.001i$, $\gamma = 1$, $\delta = 10^{-5}$, $\mu = 20$, $\omega_f = 0.1$, and $k = 1.6$. (Top) Mean-field variable $\langle x \rangle$, (bottom) control term $z - \langle x \rangle$. Control is switched on at $t = 300$ ($\langle x \rangle$ in the coupling term is replaced with z).

Once $\Delta_2 > 0$, the inequality for the third minor $\Delta_3 > 0$ can be replaced simply with $h_0 > 0$. This can be further simplified to $(1-\alpha\langle\beta\rangle) > 0$, since $\omega_f > 0$ by definition. Finally, we come to inequality $\alpha\langle\beta\rangle < 1$, which satisfied by itself, because it was already used as an assumption to derive the equilibrium coordinates; see formulas (19).

Numerical results from Eqs. (14), demonstrating dynamics of equilibrium stabilization, are presented in **Figure 10**.

5. Mean-field "nullifying" technique

A straightforward way to desynchronize the mean-field coupled oscillators is to "nullify" the mean field at the coupling node CN, i.e., to remove the reason of synchronization. The corresponding diagram is shown in **Figure 11**.

We repeat here Eq. (12) from Section 3 for clarity and for comparison with Eq. (30):

$$\begin{aligned}\dot{x}_i &= F(x_i) - y_i - \gamma + k(\langle x \rangle - x_i),\\ \dot{y}_i &= x_i - \beta_i y_i\end{aligned}$$

and emphasize that the mean-field value $\langle x \rangle$ by itself is not zero:

$$\langle x \rangle = \frac{1}{N}\sum_{i=1}^{N} x_i \neq 0. \tag{29}$$

The control technique implicates that the mean-field variable $\langle x \rangle$ is not fully nullified, but its value at the coupling node CN, $\langle x \rangle_{CN}$ is set zero:

$$\begin{aligned}\dot{x}_i &= F(x_i) - y_i - \gamma + k(0 - x_i),\\ \dot{y}_i &= x_i - \beta_i y_i.\end{aligned} \tag{30}$$

The coupling node is simply grounded, as sketched in **Figure 11**. Numerical results are shown in **Figure 12**. Note that when the control is switched on, the value of actual mean-field variable <x> becomes relatively small, but is not zero.

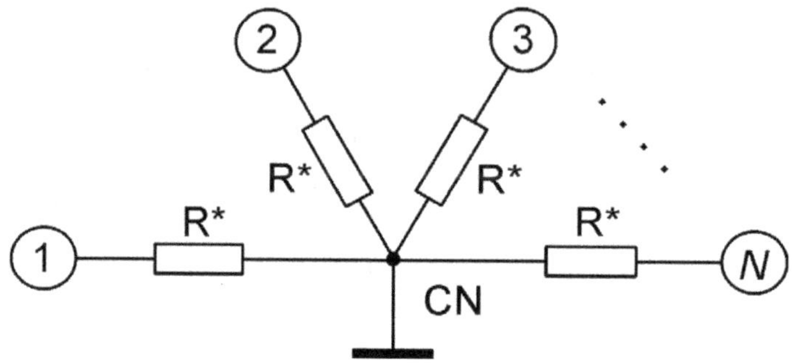

Figure 11. Diagram of mean-field coupled oscillators with the coupling node CN grounded.

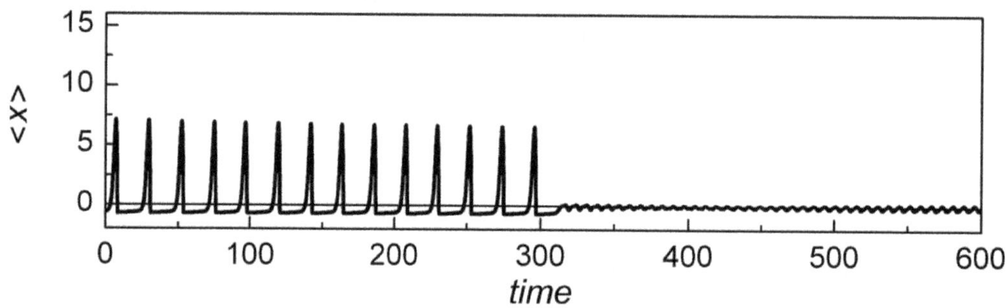

Figure 12. Waveform of the mean-field variable <x> from Eq. (30). $N = 24$, $\alpha = 1.5$, $\beta_i = 0.05 + 0.001i$, $\gamma = 1$, $\delta = 10^{-5}$, $\mu = 20$, and $k = 1$. Control is switched on at $t = 300$ (x_{CN} = <x> in the coupling term is replaced with $x_{CN} = 0$).

6. Repulsive coupling technique

An alternative method of desynchronization of coupled oscillators is the repulsive coupling, also called "repulsive synchronization" technique [26]. Diagram is sketched in **Figure 13**.

Voltage at the coupling node x_{CN} is found from the Kirchhoff's law for current:

$$\sum_{i=1}^{N} k(x_i - x_{CN}) - Gx_{CN} = 0, \tag{31}$$

$$x_{CN} = \frac{V_{CN}}{V^*}, \quad G = \frac{\rho}{R_n}. \tag{32}$$

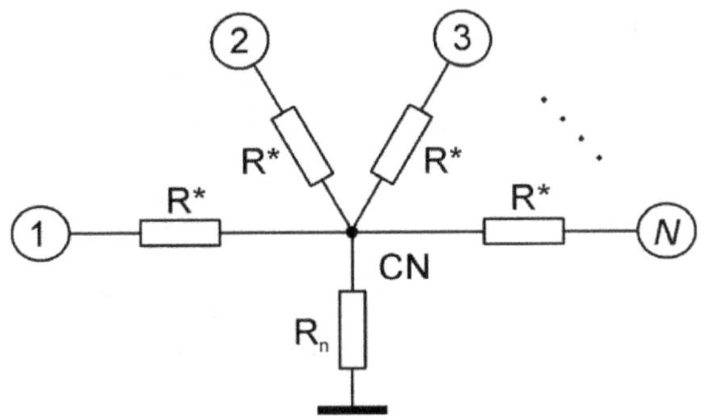

Figure 13. Diagram of mean-field coupled oscillators with coupling node CN, grounded via resistor R_n.

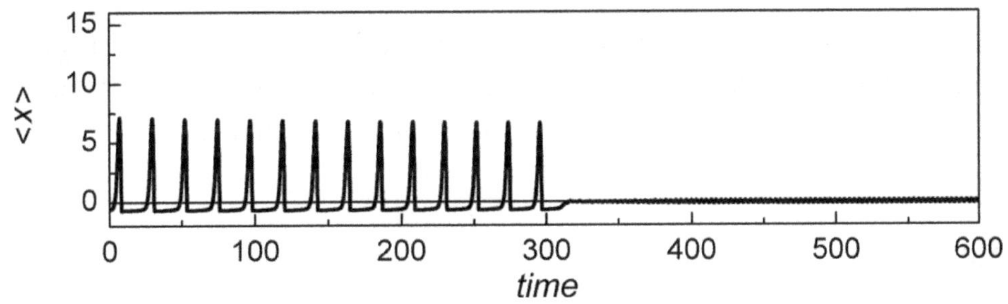

Figure 14. Waveform of the mean-field variable <x> from Eq. (34). $N = 24$, $\alpha = 1.5$, $\beta_i = 0.05 + 0.001i$, $\gamma = 1$, $\delta = 10^{-5}$, $\mu = 20$, and $k = 1$. Control is switched on at $t = 300$ ($x_{CN} = \langle x \rangle$ in the coupling term is replaced with $x_{CN} = -\langle x \rangle$).

Eq. (31) yields:

$$x_{CN} = kN\langle x\rangle/(kN + G). \quad (33)$$

Evidently, for $Rn = 0$, the $G \to \infty$ and $x_{CN} = 0$, as expected. It is the case considered in previous Section 5. If R_n is negative, say it provides value of $G = -2kN$, then $x_{CN} = -\langle x \rangle$. It is the case of so-called repulsive coupling:

$$\begin{aligned} \dot{x}_i &= F(x_i) - y_i - \gamma + k(-\langle x\rangle - x_i), \\ \dot{y}_i &= x_i - \beta_i y_i. \end{aligned} \quad (34)$$

Numerical results are presented in **Figure 14**. Similarly to the mean-field "nullifying" technique, the mean <x> becomes small, which is a typical feature of either non-synchronized or antiphase synchronized oscillators.

7. Conclusions

A modification of the FitzHugh–Nagumo (FHN) model of a spiking neuron has been proposed. In the original model, developed by FitzHugh [35], the cubic activation function $x - x^3/3$

has been replaced with a strongly asymmetric exponential one. This function provides more realistic shape of the membrane voltage spikes. Synchronization effect in an array of mean-field coupled non-identical FHN type oscillators has been demonstrated.

Three methods for controlling arrays of coupled FHN type oscillators have been described:

- Stable filter technique aimed to damp spikes in coupled oscillators. It is based on replacing the mean variable $<x>$ at the coupling node with its filtered version z.
- Mean field nullifying technique, $<x> = 0$ (grounding the coupling node).
- Repulsive coupling technique, following the idea described in [26] and shown for an array of Kuramoto 1D phase oscillators. It is based on replacing the mean-field variable $<x>$ at the coupling node with the inverse version "$- <x>$."

The above control techniques have different physical mechanisms behind, ranging from stabilization of the equilibrium states to desynchronization and antiphase synchronization. However, all of them ensure low value of mean-field variable in the array.

Author details

Elena Adomaitienė, Skaidra Bumelienė and Arūnas Tamaševičius*

*Address all correspondence to: arunas.tamasevicius@ftmc.lt

Department of Electronics, Center for Physical Sciences and Technology, Vilnius, Lithuania

References

[1] Kuo BC. Automatic Control Systems. Englewood Cliffs, New Jersey: Prentice-Hall; 1995

[2] Ogata K. Modern Control Engineering. Englewood Cliffs, New Jersey: Prentice-Hall; 2010

[3] Rulkov NF, Tsimring LS, Abarbanel HDI. Tracking unstable orbits in chaos using dissipative feedback control. Physical Review E. 1994;**50**(1):314-324. DOI: 10.1103/ PhysRevE. 50.314

[4] Namajūnas A, Pyragas K, Tamaševičius A. Stabilization of an unstable steady state in a Mackey–Glass system. Physics Letters A. 1995;**204**(3–4):255-262. DOI: 10.1016/0375-9601 (95)00480-Q

[5] Pyragas K, Pyragas V, Kiss IZ, Hudson JL. Stabilizing and tracking unknown steady states of dynamical systems. Physical Review Letters. 2002;**89**:244103. DOI: 10.1103/Phys RevLett.89.244103

[6] Pikovsky A, Rosenblum M, Kurths J. Synchronization: A Universal Concept in Nonlinear Sciences. Cambridge: Cambridge University Press; 2003

[7] Rosenblum MG, Controlling PAS. Synchronization in an ensemble of globally coupled oscillators. Physical Review Letters. 2004;**92**:114102. DOI: 10.1103/PhysRevLett. 92.114102

[8] Popovych OV, Hauptmann C, Effective TPA. Desynchronization by nonlinear delayed feedback. Physical Review Letters. 2005;**94**:164102. DOI: 10.1103/PhysRevLett. 94.164102

[9] Pyragas K, Popovych OV, Tass PA. Controlling synchrony in oscillatory networks with a separate stimulation-registration setup. Europhysics Letters. 2007;**80**:40002. DOI: 10.1209/0295-5075/80/40002

[10] Ratas I, Pyragas K. Controlling synchrony in oscillatory networks via an act-and-wait algorithm. Physical Review E. 2014;**90**:032914. DOI: 10.1103/PhysRevE.90.032914

[11] Bielawski S, Bouazaoui M, Derozier D, Glorieux P. Stabilization and characterization of unstable steady state in a laser. Physical Review A. 1993;**47**(4):3276-3279. DOI: 10.1103/PhysRevA.47.3276

[12] Johnston GA, Hunt ER. Derivative control of the steady state in Chua's circuit driven in the chaotic region. IEEE Transactions on Circuits and Systems. 1993;**40**(11):833-835. DOI: 10.1109/81.251822

[13] Parmananda P, Rhode MA, Johnson GA, Rollins RW, Dewald HD, Markworth AJ. Stabilization of unstable steady state in an electrochemical system using derivative control. Physical Review E. 1994;**49**(6):5007-5011. DOI: 10.1103/PhysRevE.49.5007

[14] Namajūnas A, Pyragas K, Tamaševičius A. Analog techniques for modeling and controlling the Mackey–Glass system. International Journal of Bifurcation and Chaos. 1997;**7**(4): 957-962. DOI: 10.1142/S0218127497000406

[15] Ciofini M, Labate A, Meucci R, Galanti M. Stabilization of unstable fixed points in the dynamics of a laser with feedback. Physical Review E. 1999;**60**(1):398-402. DOI: 10.1103/PhysRevE.60.398

[16] Schenck zu Schweinsberg A, Dressler U. Characterization and stabilization of the unstable fixed points of a frequency doubled Nd:YAG laser. Physical Review E. 2001;**63**(5): 056210. DOI: 10.1103/PhysRevE.63.056210

[17] Huijberts H. Linear controllers for the stabilization of unknown steady states of chaotic systems. IEEE Transactions on Circuits and Systems I. 2006;**53**(10):2246-2254. DOI: 10.1109/TCSI.2006.883157

[18] Pyragas K, Pyragas V, Kiss IZ, Hudson JL. Adaptive control of unknown unstable steady states of dynamical systems. Physical Review E. 2004;**70**(2):026215. DOI: 10.1103/PhysRevE.70.026215

[19] Braun DJ. Adaptive steady-state stabilization for nonlinear dynamical systems. Physical Review E. 2008;**78**(1):016213. DOI: 10.1103/PhysRevE.78.016213

[20] Tamaševičius A, Tamaševičiūtė E, Mykolaitis G, Bumelienė S. Switching from stable to unknown unstable steady states of dynamical systems. Physical Review E. 2008;**78**(2): 026205. DOI: 10.1103/PhysRevE.78.026205

[21] Tamaševičius A, Tamaševičiūtė E, Mykolaitis G, Bumelienė S, Kirvaitis R. Stabilization of saddle steady states of conservative and weakly damped dissipative dynamical systems. Physical Review E. 2010;**82**(2):026205. DOI: 10.1103/PhysRevE.82.026205

[22] Tamaševičius A, Tamaševičiūtė E, Mykolaitis G, Bumelienė S. Enhanced control of saddle steady states of dynamical systems. Physical Review E. 2013;**88**(3):032904. DOI: 10.1103/PhysRevE.88.032904

[23] Tamaševičiūtė E, Mykolaitis G, Bumelienė S, Tamaševičius A. Stabilizing saddles. Physical Review E. 2013;**88**(6):060901(R). DOI: 10.1103/PhysRevE.88.060901

[24] Adomaitienė E. Development of Methods for Controlling Equilibrium and Synchrony of Nonlinear Dynamical Systems [Thesis]. Vilnius: Vilnius University; 2017

[25] Adomaitienė E, Bumelienė S, Mykolaitis G, Tamaševičius A. Stabilization of a network of the FitzHugh–Nagumo oscillators by means of a single capacitor based RC filter feedback technique. Complexity. 2017;**2017**:4324879. DOI: 10.1155/2017/4324879

[26] Tsimring LS, Rulkov NF, Larsen ML, Repulsive GM. Synchronization in an array of phase oscillators. Physical Review Letters. 2005;**95**(1):014101. DOI: 10.1103/ PhysRevLett.95.014101

[27] Tukhlina N, Rosenblum M, Pikovsky A, Kurths J. Feedback suppression of neural synchrony by vanishing stimulation. Physical Review E. 2007;**75**(1):011918. DOI: 10.1103/PhysRevE.75.011918

[28] Tamaševičiūtė E, Mykolaitis G, Tamaševičius A. Feedback controller for destroying synchrony in an array of the FitzHugh–Nagumo oscillators. Applied Physics Letters. 2012; **101**(22):223703. DOI: 10.1063/1.4768938

[29] Tamaševičius A, Mykolaitis G, Tamaševičiūtė E, Bumelienė S. Two-terminal feedback circuit for suppressing synchrony of the FitzHugh–Nagumo oscillators. Nonlinear Dynamics. 2015;**81**(1–2):783-788. DOI: 10.1007/s11071-015-2028-y

[30] Benabid AL, Chabardes S, Mitrofanis J, Polak P. Deep brain stimulation of the subthalamic nucleus for the treatment of Parkinson's disease. The Lancet Neurology. 2009;**8**(1):67-81. DOI: 10.1016/S1474-4422(08)70291-6

[31] Pyragas K, Novičenko V, Tass PA. Mechanism of suppression of sustained neuronal spiking under high-frequency stimulation. Biological Cybernetics. 2013;**107**(6):669-684. DOI: 10.1007/s00422-013-0567-1

[32] Pyragas K, Tass PA. Suppression of spontaneous oscillations in high-frequency stimulated neuron models. Lithuanian Journal of Physics. 2016;**56**(4):223-238. DOI: 10.3952/physics.v56i4.3419

[33] Adomaitienė E, Mykolaitis G, Bumelienė S, Tamaševičius A. Inhibition of spikes in an array of coupled FitzHugh–Nagumo oscillators by external periodic forcing. Nonlinear Analysis: Modelling and Control. 2017;**22**(3):421-429. DOI: 10.15388/NA.2017.3.10

[34] Horowitz P, Hill W. Art of Electronics. 2nd ed. Cambridge, New York, Melbourne: Cambridge University Press; 1993

[35] FitzHugh R. Impulses and pstates in theoretical models of nerve. Biophysical Journal. 1961;**1**(6):445-466. DOI: 10.1016/S0006-3495(61)86902-6

[36] Tamaševičius A, Tamaševičiūtė E, Mykolaitis G, Bumelienė S, Kirvaitis R, Stoop R. Neural spike suppression by adaptive control of an unknown steady state. Lecture Notes in Computer Science. 2009;**5768**(Part I):618-627. DOI: 10.1007/978-3-642-04274-4_64

12

Fuzzy Fault Detection Filter Design for One Class of Takagi-Sugeno Systems

Dušan Krokavec, Anna Filasová, Jakub Kajan and Tibor Kočík

Abstract

The constrained unitary formalism to fuzzy fault detection filter synthesis for one class of nonlinear systems, representable by continuous-time Takagi-Sugeno fuzzy models, is presented in the chapter. In particular, a way to produce the special set of matrix parameters of the fuzzy filter is proposed to obtain the desired H_∞ norm properties of the filter transfer function matrix. The significance of the treatment in relation to the systems under influence of actuator faults is analyzed in this context, and relations to corresponding setting of singular values of filters are discussed.

Keywords: multiple models, continuous-time Takagi-Sugeno fuzzy models, fuzzy fault detection filters, fuzzy state observers

1. Introduction

Since the work of Hou and Patton [1], there has been much interest in the design of fault residuals for linear systems that use H_∞/H_- optimization principle in transfer function matrix of fault detection filter designed to scale up fault detection punctuality and high sensitivity to faults [2]. While retaining these features, a novel class of fault detection filters are proposed in [3, 4], preserving the unitary implementation of the fault detection filter transfer function matrix and receipting residual signal directional properties. However, the use of this methodology for Takagi-Sugeno (TS) fuzzy systems hits the boundaries of the working sectors and requires special adaptations.

Considering the properties of TS fuzzy models [5, 6], and some specifics in frequency characteristic evaluation of multiple model structures, the approach proposed in the chapter reformulates the H_∞ norm technique suitable in TS fuzzy fault detection filter design. The problem is solved via unitary modal technique when every linear TS fuzzy filter part is designed to have the same singular values of the transfer function matrix. Since working sector constraints may cause that the stable linear filter component cannot be obtained for a linear part in TS fuzzy model, to maintain H_∞ norm of the filter, the LQ modal control principle [7] is used for additional stabilization. Because additional stabilization aggravates directional properties of the applied linear part, in general, if additional stabilization is necessary, the residuals are only quasi-directional. It is immediately apparent that the formulated problem is related to forcing the singular values conditioned as state observer dynamics. The chosen model of the system is selected for this chapter to be sufficiently complex in illustration of all these specifics of synthesis.

Throughout the chapter, the following notations are used: $\boldsymbol{x}^T$ and $\boldsymbol{X}^T$ denote the transpose of the vector $\boldsymbol{x}$ and the matrix $\boldsymbol{X}$, respectively; for a square matrix $\boldsymbol{X} \geq 0$ means that $\boldsymbol{X}$ is a symmetric positive semi-definite matrix; the symbol $\boldsymbol{I}_n$ indicates the nth-order unit matrix; $\boldsymbol{R}$ denotes the set of real numbers; and $\boldsymbol{R}^n$ and $\boldsymbol{R}^{n \times r}$ refer to the set of all n-dimensional real vectors and $n \times r$ real matrices.

2. System description

The considered class of the Takagi-Sugeno dynamic systems with additive faults is described as the following:

$$\dot{\boldsymbol{q}}(t) = \sum_{i=1}^{s} \mathrm{h}_i(\boldsymbol{\theta}(t))\big(\boldsymbol{A}_i\boldsymbol{q}(t) + \boldsymbol{B}_i\boldsymbol{u}(t) + \boldsymbol{F}_i\boldsymbol{f}(t)\big) \tag{1}$$

$$\boldsymbol{y}(t) = \boldsymbol{C}\boldsymbol{q}(t) \tag{2}$$

where $\boldsymbol{q}(t) \in \boldsymbol{R}^n$, $\boldsymbol{u}(t) \in \boldsymbol{R}^r$, and $\boldsymbol{y}(t) \in \boldsymbol{R}^m$ stand for state, control input, and measurable output, respectively; $\boldsymbol{f}(t) \in \boldsymbol{R}^p$ is an additive fault vector; $\boldsymbol{A}_i \in \boldsymbol{R}^{n \times n}$, $\boldsymbol{B}_i \in \boldsymbol{R}^{n \times r}$, $\boldsymbol{F}_i \in \boldsymbol{R}^{n \times p}$, $\boldsymbol{C} \in \boldsymbol{R}^{m \times n}$, and $m = p$ and the matrix products $\boldsymbol{V}_i = \boldsymbol{C}\boldsymbol{F}_i$ and $\boldsymbol{V}_i \in \boldsymbol{R}^{m \times m}$ are regular matrices for all i.

The variables $\theta_j(t)$ and $j = 1, 2, \ldots, o$, bound with the sector TS model, span the o-dimensional vector of premise variables:

$$\boldsymbol{\theta}(t) = [\theta_1(t) \;\; \theta_2(t) \;\; \cdots \;\; \theta_o(t)] \tag{3}$$

and [8]

$$\sum_{i=1}^{s} \mathrm{h}_i(\boldsymbol{\theta}(t)) = 1 \tag{4}$$

where $\mathrm{h}_i(\boldsymbol{\theta}(t)), i = 1, 2, \ldots, s$ is the set of normalized membership function. It is supposed that the measurable premise variables, the nonlinear sectors, and the normalized membership functions are chosen in such a way that the pairs $(\boldsymbol{A}_i, \boldsymbol{B}_i)$ are controllable and the pairs $(\boldsymbol{A}_i, \boldsymbol{C})$ are observable for all i.

3. Basic preliminaries from linear systems

Let the state-space description of a linear continuous-time dynamic systems take the form with equivalent meanings and dimensions as they are described in Section 2. The nature of the characterization of expected solutions to the system [(5), (6)] is given by the following results.

$$\dot{\boldsymbol{q}}(t) = \boldsymbol{A}\boldsymbol{q}(t) + \boldsymbol{B}\boldsymbol{u}(t) + \boldsymbol{F}\boldsymbol{f}(t) \tag{5}$$

$$\boldsymbol{y}(t) = \boldsymbol{C}\boldsymbol{q}(t) \tag{6}$$

Definition 1 [9, 10] *If* $\boldsymbol{A}$ *has no imaginary eigenvalues, the* H_∞ *norm of the system transfer function matrix*

$$\boldsymbol{G}(s) = \boldsymbol{C}(s\boldsymbol{I}_n - \boldsymbol{A})^{-1}\boldsymbol{B} \tag{7}$$

is

$$\|\boldsymbol{G}(s)\|_\infty = \sup_{\omega \in R} \sigma_1(\boldsymbol{G}(j\omega)) = \sup_{\omega \in R} \sqrt{\varepsilon_1(\boldsymbol{G}^*(j\omega)\boldsymbol{G}(j\omega))} \tag{8}$$

while the kth singular value σ_h of the complex matrix $\boldsymbol{G}(j\omega)$ is the nonnegative square root of the kth largest eigenvalue ε_k of $\boldsymbol{G}^*(j\omega)\boldsymbol{G}(j\omega)$, $\boldsymbol{G}^*(j\omega)$ is the adjoint of $\boldsymbol{G}(j\omega)$, and σ_1 is the largest singular value. The singular values of the transfer function matrix $\boldsymbol{G}(s)$ are evaluated on the imaginary axis, and it is assumed that the singular values are ordered such that $\sigma_k \geq \sigma_{k+1}, \ \ k = 1, 2, \ldots, n-1$.

To apply in design methodology, the following result from [4] is quoted.

Lemma 1 *If* $m = p$ *and* $\boldsymbol{V} = \boldsymbol{C}\boldsymbol{F}$ *are regular matrices, then the system matrix factorization can be realized such that*

$$\boldsymbol{C} = [\boldsymbol{V} \quad \boldsymbol{0}]\boldsymbol{T}, \quad \boldsymbol{T}\boldsymbol{F} = \begin{bmatrix} \boldsymbol{I}_m \\ \boldsymbol{0} \end{bmatrix} \tag{9}$$

and the transform matrix $\boldsymbol{T} \in R^{n \times n}$ takes the form

$$\boldsymbol{T} = \begin{bmatrix} \boldsymbol{V}^{-1}\boldsymbol{C} \\ \boldsymbol{F}^{\perp} \end{bmatrix}, \tag{10}$$

where $\boldsymbol{V}^{-1}\boldsymbol{C} \in R^{m \times n}$, $\boldsymbol{F}^{\perp} \in R^{(n-m) \times n}$, and $\boldsymbol{F}^{\perp}$ are the left orthogonal complements to $\boldsymbol{F}$.

The idea of the following condition was derived originally as an approximation in the frequency domain for the fault transfer function matrix reflecting Eqs. (5) and (6) from [12]. Here, it is demonstrated that it can be simply adapted for fault residual filter design.

Theorem 1 *A linear fault detection filter to the system [(5), (6)] is stable and unitary if for regular* $\boldsymbol{V} = \boldsymbol{C}\boldsymbol{F}$ *and a given positive scalar* $s_o \in \mathbb{R}$ *the square transfer function matrix* $\boldsymbol{G}_r(s)$ *of the fault detection filter satisfies the conditions*

$$P(s) = \det(s\boldsymbol{I}_n - (\boldsymbol{A} - \boldsymbol{J}\boldsymbol{C})) = (s + s_o)^m P_o(s), \tag{11}$$

$$\sigma_1 = \sigma_2 = \cdots = \sigma_m, \qquad \lim_{\omega \to 0} \sigma_h = s_o, \tag{12}$$

$$\boldsymbol{G}_r(0) = \mathrm{diag}\left[s_o^{-1} \quad s_o^{-1} \quad \cdots \quad s_o^{-1} \right], \tag{13}$$

$$\boldsymbol{G}_r(s) = \boldsymbol{V}^{-1}\boldsymbol{C}(s\boldsymbol{I}_n - (\boldsymbol{A} - \boldsymbol{J}\boldsymbol{C}))^{-1}\boldsymbol{F} = (s + s_o)^{-1}\boldsymbol{I}_m, \tag{14}$$

$$\boldsymbol{A}_o = \boldsymbol{T}\boldsymbol{A}\boldsymbol{T}^{-1} = \begin{bmatrix} \boldsymbol{A}_{o11} & \boldsymbol{A}_{o12} \\ \boldsymbol{A}_{o21} & \boldsymbol{A}_{o22} \end{bmatrix}, \tag{15}$$

$$\boldsymbol{J} = \boldsymbol{T}^{-1}\boldsymbol{L}^o\boldsymbol{V}^{-1}, \qquad \boldsymbol{L}^o = \begin{bmatrix} s_o\boldsymbol{I}_m + \boldsymbol{A}_{o11} \\ \boldsymbol{A}_{o21} \end{bmatrix}, \tag{16}$$

where $\boldsymbol{J} \in \mathbb{R}^{n \times r}$ is the residual filter gain matrix, σ_1 is the maximal singular value of $\boldsymbol{G}_r(s)$, the polynomial $P_o(s)$ of order $(n - m)$ is stable, and $\boldsymbol{G}_r(0) \in \mathbb{R}^{m \times m}$.

Proof. Considering the fault transfer function matrix of dimension $m \times m$ as

$$\boldsymbol{G}_f(s) = \boldsymbol{C}(s\boldsymbol{I}_n - \boldsymbol{A})^{-1}\boldsymbol{F} \tag{17}$$

and then regrouping terms using Eqs. (9) and (10), it yields immediately the expressions

$$\boldsymbol{G}_f(s) = \boldsymbol{C}\boldsymbol{T}^{-1}\boldsymbol{T}(s\boldsymbol{I}_n - \boldsymbol{A})^{-1}\boldsymbol{T}^{-1}\boldsymbol{T}\boldsymbol{F} = \boldsymbol{C}\boldsymbol{T}^{-1}\left(\left(s\boldsymbol{I}_n - \boldsymbol{T}\boldsymbol{A}\boldsymbol{T}^{-1} \right)^{-1} \boldsymbol{T}\boldsymbol{F}, \tag{18}$$

$$\boldsymbol{G}_f(s) = [\boldsymbol{V} \quad \boldsymbol{0}](s\boldsymbol{I}_n - \boldsymbol{A}_o)^{-1} \begin{bmatrix} \boldsymbol{I}_p \\ \boldsymbol{0} \end{bmatrix}, \tag{19}$$

respectively, where $\boldsymbol{A}_o$ is given in Eq. (15).

Specifying the following matrix product $\boldsymbol{A}^o = \boldsymbol{T}\boldsymbol{M}\boldsymbol{V}^{-1}\boldsymbol{C}\boldsymbol{T}^{-1}$, where $\boldsymbol{M} \in \mathbb{R}^{n \times m}$ is a real matrix, it yields

$$\boldsymbol{A}^o = \boldsymbol{T}\boldsymbol{M}\boldsymbol{V}^{-1}\boldsymbol{C}\boldsymbol{T}^{-1} = \begin{bmatrix} \boldsymbol{V}^{-1}\boldsymbol{C} \\ \boldsymbol{F}^{\perp} \end{bmatrix} \boldsymbol{M}\boldsymbol{V}^{-1}[\boldsymbol{V} \quad \boldsymbol{0}] = \begin{bmatrix} \boldsymbol{V}^{-1}\boldsymbol{C}\boldsymbol{M} & \boldsymbol{0} \\ \boldsymbol{F}^{\perp}\boldsymbol{M} & \boldsymbol{0} \end{bmatrix} \tag{20}$$

and, with the block matrix structure of Eqs. (15) and (21), it can be defined as

$$\Delta\boldsymbol{A}_o = \boldsymbol{A}_o - \boldsymbol{A}^o = \begin{bmatrix} \boldsymbol{A}_{o11} - \boldsymbol{V}^{-1}\boldsymbol{C}\boldsymbol{M} & \boldsymbol{A}_{o12} \\ \boldsymbol{A}_{o21} - \boldsymbol{F}^{\perp}\boldsymbol{M} & \boldsymbol{A}_{o22} \end{bmatrix}. \tag{21}$$

Presetting

$$A_{o11} - V^{-1}CM = -s_o I_m, \qquad A_{o21} - F^{\perp}M = 0, \tag{22}$$

where $s_o \in \mathbb{R}$ is a prescribed positive real value. The plus sign is introduced for the purposes that come to light in the stability ensuing development of the observer system matrix.

Then,

$$\Delta A_o = \begin{bmatrix} -s_o I_m & A_{o12} \\ 0 & A_{o22} \end{bmatrix} \tag{23}$$

and it is evident that ΔA_o is stable if A_{o22} is Hurwitz, denoting here that

$$P_o(s) = \det(sI_{n-m} - A_{o22}). \tag{24}$$

Rewriting the set of Eq. (22) to admit a stable solution

$$\begin{bmatrix} s_o I_m + A_{o11} \\ A_{o21} \end{bmatrix} = \begin{bmatrix} V^{-1}C \\ F^{\perp} \end{bmatrix} M = TM = TT^{-1}L^o = L^o, \tag{25}$$

where

$$M = T^{-1}L^o, \tag{26}$$

then Eqs. (20) and (21) must satisfy the following conditions:

$$A^o = TMV^{-1}CT^{-1} = TJCT^{-1}, \tag{27}$$

$$\Delta \mathbf{A}_o = \mathbf{A}_o - \mathbf{A}^o = \mathbf{T}(\mathbf{A} - \mathbf{JC})\mathbf{T}^{-1} = \mathbf{T}\mathbf{A}_e\mathbf{T}^{-1}. \tag{28}$$

Therefore, the observer system matrix A_e takes the form

$$A_e = A - JC = A - MV^{-1}C \tag{29}$$

and

$$J = MV^{-1} = T^{-1}L^oV^{-1} \tag{30}$$

implies Eq. (16).

Regarding the transfer function matrix $G_e(s)$ of the state error estimate as follows

$$G_e(s) = C(sI_n - A_e)^{-1}F, \tag{31}$$

then with Eq. (29), it is

$$G_e(s) = CT^{-1}\left(sI_n - TA_eT^{-1}\right)^{-1}TF = [V \quad 0](sI_n - \Delta A_o)^{-1}\begin{bmatrix} I_p \\ 0 \end{bmatrix}. \tag{32}$$

Since

$$sI_n - \Delta A_o = \begin{bmatrix} (s+s_o)I_m & -A_{o12} \\ 0 & sI_{n-m} - A_{o22} \end{bmatrix}, \tag{33}$$

$$(sI_n - \Delta A_o)^{-1} = \begin{bmatrix} (s+s_o)^{-1}I_m & (s+s_o)^{-1}A_{o12}(sI_{n-m} - A_{o22})^{-1} \\ 0 & (sI_{n-m} - A_{o22})^{-1} \end{bmatrix}, \tag{34}$$

Substituting Eq. (34) into Eq. (32), it can obtain

$$G_e(s) = V(s+s_o)^{-1}I_m = \frac{V}{s+s_o}. \tag{35}$$

Thus, defining the fault detection filter transfer function matrix as $G_r(s) = V^{-1}G_e(s)$, then

$$G_r(s) = V^{-1}G_e(s) = (s+s_o)^{-1}I_m \tag{36}$$

and Eq. (36) implies Eq. (14). This concludes the proof.

Corollary 1 *Evidently, writing the fault residual vector as*

$$r(t) = V^{-1}Ce(t) = V^{-1}C\left(q(t) - q_e(t)\right), \tag{37}$$

where

$$e(t) = q(t) - q_e(t) \tag{38}$$

and $r(t) \in R^m$ is the vector of residual signals, then based on the following observer structure

$$\dot{q}_e(t) = Aq_e(t) + Bu(t) + JC\left(q(t) - q_e(t)\right), \tag{39}$$

$$y_e(t) = Cq_e(t), \tag{40}$$

the autonomous observer error equation is

$$\dot{e}(t) = (A - JC)e(t), \tag{41}$$

where $q_e(t) \in R^n$ is the observer state, $y_e(t) \in R^m$ is the estimated system output, and $J \in R^{n \times m}$ is the observer gain matrix; the fault detection filter (37), (39) is stable and unitary if for given positive scalar $s_o \in R$ and the Hurwitz matrix A_{o22} the conditions (15) and (16) are satisfied.

Practically, with understanding Eq. (30), the observer sensor subsystem for the fault detection filter can be designed as follows:

$$e_z(t) = z(t) - z_e(t) = V^{-1}C(q(t) - q_e(t)) \tag{42}$$

and, consequently, it yields

$$\dot{q}_e(t) = Aq_e(t) + Bu(t) + MC(q(t) - q_e(t)). \tag{43}$$

Another option is to design the observer sensor subsystem so that $V = I_m$.

With existence of the system parameter transformation, the above structures really mean that the subset of transformed state variables whose dynamics is explicitly affected by the additive fault $f(t)$ and the second one, whose dynamics is not affected explicitly by the fault $f(t)$, exists.

Remark 1 *It is important to note the fact that the eigenvalues of* A *and of* A_o *are the same whenever* A_o *is related to* A *as* $A_o = TAT^{-1}$ *for any invertible* T [11]. *But this does not mean that if eigenvalues of the matrix* A_o *are stable then eigenvalues of the matrix* A_{o22} *are also stable. Thus, as well as for a stable system, it can lead to an unstable matrix* A_{o22}, *and any additional stabilization is required.*

To apply the above results, it is necessary to be able to design fault residual filter if an unstable A_{o22} results such that A_e be stable without loss of unitarity.

Lemma 2 [7, 12] *To change signs of unstable eigenvalues of the system matrix* A, *the gain matrix* $K \in R^{n \times r}$ *of the state feedback additive stabilization*

$$u(t) = -Kq(t) \tag{44}$$

is a solution of the continuous-time algebraic Riccati equation (CARE)

$$PA + A^T P - PBR^{-1}B^T P + Q = 0, \tag{45}$$

where the matrix $Q \in R^{n \times n}$ is null matrix and $R \in R^{r \times r}$ and $R = R^T > 0$ are positive definite symmetric matrices.

Then, K *is given as*

$$K = R^{-1}B^T P. \tag{46}$$

It is in that form that is able to be exploit for specific properties of the problem in TS fuzzy fault detection filter design.

In view of the above, these results hold for continuous-time linear systems, and, in principle, Theorem 1 gives a practical method to design unitary fault residual filters for the given linear system. Similar results are obtained for unitary TS fuzzy fault detection filter design in the following section.

4. TS fuzzy fault detection filters

Using the same set of membership functions, the fuzzy fault detection filter is built on the TS fuzzy observer

$$\dot{\boldsymbol{q}}_e(t) = \sum_{i=1}^{s} \mathrm{h}_i(\boldsymbol{\theta}(t))\big(\boldsymbol{A}_i\boldsymbol{q}(t) + \boldsymbol{B}_i\boldsymbol{u}_i(t) + \boldsymbol{J}_i\boldsymbol{C}(\boldsymbol{q}(t) - \boldsymbol{q}_e(t))\big) \tag{47}$$

$$\boldsymbol{y}_e(t) = \boldsymbol{C}\boldsymbol{q}_e(t) \tag{48}$$

where $\boldsymbol{q}_e(t) \in \mathbb{R}^n$ is the observer state vector, $\boldsymbol{y}_e(t) \in \mathbb{R}^m$ is the estimated system output vector, and $\boldsymbol{J}_i \in \mathbb{R}^{n \times m}$ and $i = 1, 2, \dots, s$ are the sets of the observer gain matrices. Additionally, the output vector of the residual TS fuzzy filter is defined as

$$\boldsymbol{r}(t) = \sum_{i=1}^{s} \mathrm{h}_i(\boldsymbol{\theta}(t))\boldsymbol{r}_i(t) = \sum_{i=1}^{s} \mathrm{h}_i(\boldsymbol{\theta}(t))\boldsymbol{V}_i^{-1}\boldsymbol{C}\boldsymbol{e}(t) \tag{49}$$

$$\boldsymbol{r}_i(t) = \boldsymbol{V}_i^{-1}\boldsymbol{C}\boldsymbol{e}(t) \tag{50}$$

$$\boldsymbol{e}(t) = \boldsymbol{q}(t) - \boldsymbol{q}_e(t) \tag{51}$$

where $\boldsymbol{r}(t), \boldsymbol{r}_i(t) \in \mathbb{R}^m$, $\boldsymbol{V}_i \in \mathbb{R}^{m \times m}$. Evidently, $\boldsymbol{V}_i = \boldsymbol{C}\boldsymbol{F}_i$ has to be a regular matrix for all i.

Formally, the following result can be simply derived.

Theorem 2 *A TS fuzzy fault detection filter to the system [(1), (2)] is stable and unitary if for the set of regular matrices $\boldsymbol{V}_i = \boldsymbol{C}\boldsymbol{F}_i$ and $i = 1, 2, \dots, s$, and a given positive scalar $s_o \in \mathbb{R}$ every square transfer function matrix $\boldsymbol{G}_{ri}(s)$ of the fault detection filter satisfies for all i the conditions*

$$\sigma_1 = \sigma_2 = \cdots = \sigma_m, \qquad \lim_{\omega \to 0} \sigma_h = s_o, \tag{52}$$

$$\boldsymbol{G}_{ri}(0) = diag\left[\, s_o^{-1} \quad s_o^{-1} \quad \cdots \quad s_o^{-1} \,\right], \tag{53}$$

$$\boldsymbol{G}_{ri}(s) = \boldsymbol{V}_i^{-1}\boldsymbol{C}\big(s\boldsymbol{I}_n - (\boldsymbol{A}_i - \boldsymbol{J}_i\boldsymbol{C})\big)^{-1}\boldsymbol{F}_i = (s + s_o)^{-1}\boldsymbol{I}_m, \tag{54}$$

while

$$\boldsymbol{T}_i = \begin{bmatrix} \boldsymbol{V}_i^{-1}\boldsymbol{C} \\ \boldsymbol{F}_i^{\perp} \end{bmatrix}, \tag{55}$$

$$\boldsymbol{A}_{oi} = \boldsymbol{T}_i\boldsymbol{A}_i\boldsymbol{T}_i^{-1} = \begin{bmatrix} \boldsymbol{A}_{o11i} & \boldsymbol{A}_{o12i} \\ \boldsymbol{A}_{o21i} & \boldsymbol{A}_{o22i} \end{bmatrix}, \tag{56}$$

$$\boldsymbol{J}_i = \boldsymbol{T}_i^{-1}\boldsymbol{L}_i^o\boldsymbol{V}_i^{-1}, \qquad \boldsymbol{L}_i^o = \begin{bmatrix} s_o\boldsymbol{I}_m + \boldsymbol{A}_{o11i} \\ \boldsymbol{A}_{o21i} \end{bmatrix}, \tag{57}$$

$$P_i(s) = \det(s\boldsymbol{I}_n - (\boldsymbol{A}_i - \boldsymbol{J}_i\boldsymbol{C})) = (s + s_o)^m P_{oi}(s), \tag{58}$$

$$P_{oi}(s) = det(s\boldsymbol{I}_{n-m} - \boldsymbol{A}_{o22i}). \tag{59}$$

$\boldsymbol{J}_i \in \boldsymbol{R}^{n\times r}$ is the residual filter gain matrix, σ_1 is the maximal singular value of $\boldsymbol{G}_{ri}(s)$, the polynomial $P_{oi}(s)$ of order $(n-m)$ is stable, and $\boldsymbol{G}_{ri}(0) \in \boldsymbol{R}^{m\times m}$ and $\boldsymbol{F}_i^{\perp} \in \boldsymbol{R}^{(n-m)\times n}$ are left orthogonal complements to the fault input matrix $\boldsymbol{F}_i$.

Proof. Because every sub-model in Eq. (47) is described by linear equations, Eqs. (15) and (16) imply directly the conditions (56) and (57), and Eq. (58) is given by Eq. (11). This concludes the proof.

Corollary 2 *In practice, an additive fault typically enters through a matrix $\boldsymbol{F}$ that does not depend on the sectoral boundaries defining the TS model. In this case, the synthesis is substantially simplified because $\boldsymbol{V}$ is a constant matrix, and so it yields*

$$\boldsymbol{T} = \begin{bmatrix} \boldsymbol{V}^{-1}\boldsymbol{C} \\ \boldsymbol{F}^{\perp} \end{bmatrix}, \tag{60}$$

$$\boldsymbol{A}_{oi} = \boldsymbol{T}\boldsymbol{A}_i\boldsymbol{T}^{-1} = \begin{bmatrix} \boldsymbol{A}_{o11i} & \boldsymbol{A}_{o12i} \\ \boldsymbol{A}_{o21i} & \boldsymbol{A}_{o22i} \end{bmatrix}, \tag{61}$$

$$\boldsymbol{J}_i = \boldsymbol{T}^{-1}\boldsymbol{L}_i^o\boldsymbol{V}^{-1}, \qquad \boldsymbol{L}_i^o = \begin{bmatrix} s_o\boldsymbol{I}_m + \boldsymbol{A}_{o11i} \\ \boldsymbol{A}_{o21i} \end{bmatrix}. \tag{62}$$

Corollary 3 *Since, independently on i, the condition (52) is satisfied ($\sigma_1 = \sigma_2 = \cdots = \sigma_m$), all sub-filter transfer function matrices have the same H_∞ norm, i.e.,*

$$\|\boldsymbol{G}_{ri}(s)\|_\infty = \|\boldsymbol{G}_{ro}(s)\|_\infty \quad for\ all\ i. \tag{63}$$

Moreover, considering that $\sum_{i=1}^{s} h_i(\boldsymbol{\theta}(t)) = 1$, then

$$\|\boldsymbol{G}_r(s)\|_\infty = \sum_{i=1}^{s} h_i(\boldsymbol{\theta}(t))\|\boldsymbol{G}_{ri}(s)\|_\infty = \|\boldsymbol{G}_{ro}(s)\|_\infty \tag{64}$$

That is, the H_∞ norm of the transfer function matrix of such defined TS fuzzy fault detection filter is independent on the system working point. Of course, this cannot be said about the dynamics of the time response of the sub-filter components.

Moreover, $\boldsymbol{G}_{ri}(0)$ implies that all residual components of TS fuzzy fault detection filter have the same directional properties, which ensure unitary properties of the filter.

Remark 2 *Sectoral boundaries may cause a matrix $\mathbf{A}_i$ to be such, when transformed using $\mathbf{T}_i$ that $\mathbf{A}_{o22i}$ will not be Hurwitz matrix. Because the transfer function matrix of the corresponding filter linear component in this case is unstable, maintaining the unitary property requires changes in the signs of the unstable eigenvalues of the associated $\mathbf{A}_{ei}^{\circ} = \mathbf{A}_i - \mathbf{J}_i^{\circ}\mathbf{C}$.*

Applying the duality principle and inserting the additive observer gain component $\mathbf{K}_{si}^T$ obtained as a solution of the Riccati equation (45) for $\mathbf{A}_{ei}^{\circ T}$, according to the scheme given in Lemma 2, the observer gain matrix is changed as

$$\mathbf{J}_i^{\circ} = \mathbf{J}_i + \mathbf{K}_{si}^T, \quad \mathbf{A}_{ei}^{\circ} = \mathbf{A}_i - \mathbf{J}_i^{\circ}\mathbf{C}. \tag{65}$$

This additive stabilization results that the consequential characteristic polynomial, taking also the form

$$P_i(s) = \det(s\mathbf{I}_n - \mathbf{A}_{ei}) = (s + s_o)^m P_{oi}(s), \tag{66}$$

is stable since $P_{oi}(s)$ is now stable.

The price for such an additional stabilization is that if j signs are changing in eigenvalues of $\mathbf{A}_{o22i}$ to obtain the stable $\mathbf{A}_{o22i}$, also j eigenvalues s_o of $\mathbf{G}_{ri}(0)$ change their signs and the resulting matrix $\mathbf{G}_{ri}(0)$ will not be diagonal. According to Eq. (8), this does not result in a change in H_∞ norm, but such filter component will arrive at the unitary directional residual properties.

5. Illustrative example

The three-tank system is described by the set of Eqs. [13, 14] as

$$\frac{dq_1(t)}{dt} = \frac{u_1(t)}{F_1} - \frac{\alpha_1 \mathrm{sign}(q_1(t) - q_2(t))\sqrt{2g|q_1(t) - q_2(t)|}}{F_1 \sum_{i=1}^{3} \lambda_i q_i(t)} \sum_{i=1}^{3} \lambda_i q_i(t),$$

$$\begin{aligned}\frac{dq_2(t)}{dt} &= \frac{u_2(t)}{F_2} - \frac{\alpha_2\sqrt{2gq_2(t)}}{F_2 q_2(t)} q_2(t) + \frac{\alpha_1 \mathrm{sign}(q_1(t) - q_2(t))\sqrt{2g|q_1(t) - q_2(t)|}}{F_1 \sum_{i=1}^{3} \lambda_i q_i(t)} \sum_{i=1}^{3} \lambda_i q_i(t) \\ &+ \frac{\alpha_3 \mathrm{sign}(q_3(t) - q_2(t))\sqrt{2g|q_3(t) - q_2(t)|}}{F_3 \sum_{i=1}^{3} \eta_i q_i(t)} \sum_{i=1}^{3} \eta_i q_i(t),\end{aligned}$$

$$\frac{dq_3(t)}{dt} = \frac{u_3(t)}{F_3} - \frac{\alpha_3 \mathrm{sign}(q_3(t) - q_2(t))\sqrt{2g|q_3(t) - q_2(t)|}}{F_3 \sum_{i=1}^{3} \eta_i q_i(t)} \sum_{i=1}^{3} \eta_i q_i(t),$$

$$y_k(t) = F_k q_k(t), \quad k = 1, 2, 3,$$

where the measured output variables $y_k(t)$ are water levels in tanks $q_k(t)\ [m],\ k=1,2,3$ and the incoming flows are considered as the inputs variables $u_k(t)\ \left[m^3/s\right],\ k=1,2,3$; the bounds of the state and input variables are

$$q_1^{max}=q_3^{max}=1.00\ [m],\qquad q_2^{max}=0.90\ [m],\qquad u_{1,2,3}^{min}=0\ \left[m^3/s\right],$$
$$q_1^{min}=q_3^{min}=0.02\ [m],\qquad q_2^{min}=0.01\ [m],\qquad u_{1,2,3}^{max}=0.005\ \left[m^3/s\right].$$

$\lambda_k,\eta_k\in\mathbb{R}$ are positive scalars and $\text{sign}(\cdot)$ is the sign function.

The model parameters of the system are considered as:

g -the gravitational acceleration $9.80665\left[\text{m/s}^2\right]$,

F_k -the (same) section of tanks $0.25\left[\text{m}^2\right]$,

α_1 -the equivalent section of the pipe between the first and second tank $6.5\times10^{-4}\left[\text{m}^2\right]$,

α_3 -the equivalent section of the pipe between the third and second tank $6.5\times10^{-4}\left[\text{m}^2\right]$,

α_2 -the equivalent section of the outlet pipe from the second tank $6.5\times10^{-3}\left[\text{m}^2\right]$,

Minimizing the number of premise variables and excluding switching modes in controller work, the premise variables are chosen as follows

$$\theta_1(t)=\frac{\alpha_1\text{sign}(q_1(t)-q_2(t))\sqrt{2g|q_1(t)-q_2(t)|}}{F_1\sum_{i=1}^{3}\lambda_iq_i(t)},$$

$$\theta_2(t)=\frac{\alpha_2\sqrt{2gq_2(t)}}{F_2q_2(t)}=\frac{\alpha_2}{F_2}\sqrt{\frac{2g}{q_2(t)}},$$

$$\theta_3(t)=\frac{\alpha_3\text{sign}(q_3(t)-q_2(t))\sqrt{2g|q_3(t)-q_2(t)|}}{F_3\sum_{i=1}^{3}\eta_iq_i(t)}.$$

Computed from the input variable bounds, the sector bounds of the premise variables imply the numbering:

$$i=1\leftarrow\left(\theta_1^{max},\theta_2^{max},\theta_3^{max}\right),\qquad i=2\leftarrow\left(\theta_1^{max},\theta_2^{max},\theta_3^{min}\right),$$
$$i=3\leftarrow\left(\theta_1^{max},\theta_2^{min},\theta_3^{max}\right),\qquad i=4\leftarrow\left(\theta_1^{max},\theta_2^{min},\theta_3^{min}\right),$$
$$i=5\leftarrow\left(\theta_1^{min},\theta_2^{max},\theta_3^{max}\right),\qquad i=6\leftarrow\left(\theta_1^{min},\theta_2^{max},\theta_3^{min}\right),$$
$$i=7\leftarrow\left(\theta_1^{min},\theta_2^{min},\theta_3^{max}\right),\qquad i=8\leftarrow\left(\theta_1^{min},\theta_2^{min},\theta_3^{min}\right),$$

which is used in the system state matrix construction

$$A_i = \begin{bmatrix} -\lambda_1\theta_1^i & -\lambda_2\theta_1^i & -\lambda_3\theta_1^i \\ \lambda_1\theta_1^i + \eta_1\theta_3^i & \lambda_2\theta_1^i + \eta_2\theta_3^i - \theta_2^i & \lambda_3\theta_1^i + \eta_3\theta_3^i \\ -\eta_1\theta_3^i & -\eta_2\theta_3^i & -\eta_3\theta_3^i \end{bmatrix}, B = \begin{bmatrix} F_1^{-1} & 0 & 0 \\ 0 & F_2^{-1} & 0 \\ 0 & 0 & F_3^{-1} \end{bmatrix}, C = \begin{bmatrix} F_1 & 0 & 0 \\ 0 & F_2 & 0 \\ 0 & 0 & F_3 \end{bmatrix}$$

and prescribed, moreover, that the matrix C is given in such a way that the product CB is the identity matrix. This regularizes the residual design conditions if B and C are diagonal matrices.

The sector functions are trapezoidal, and the membership functions are constructed as product of three sector functions with the same ordering as A_i.

The set of real scalars, λ_k, η_k, and $k = 1, 2, 3$, is interactively optimized under limitations that all couples (A_i, B) and (A_i, C) are controllable and observable for the given set of indices i, where

$$\begin{aligned} \lambda_1 &= 0.1992, & \lambda_2 &= 0.6894, & \lambda_3 &= 0.1618, \\ \eta_1 &= 0.6891, & \eta_2 &= 0.3646, & \eta_3 &= 0.0569. \end{aligned}$$

Consequently, the TS model matrix parameters are

$$A_1 = \begin{bmatrix} -0.0163 & -0.0563 & -0.0132 \\ 0.1225 & -1.0392 & 0.0220 \\ -0.1062 & -0.0562 & -0.0088 \end{bmatrix}, \quad A_2 = \begin{bmatrix} -0.0163 & -0.0563 & -0.0132 \\ -0.0054 & -1.1069 & 0.0114 \\ 0.0217 & 0.0115 & 0.0018 \end{bmatrix},$$

$$A_3 = \begin{bmatrix} -0.0163 & -0.0563 & -0.0132 \\ 0.1225 & -0.0089 & 0.0220 \\ -0.1062 & -0.0562 & -0.0088 \end{bmatrix}, \quad A_4 = \begin{bmatrix} -0.0163 & -0.0563 & -0.0132 \\ -0.0054 & -0.0766 & 0.0114 \\ 0.0217 & 0.0115 & 0.0018 \end{bmatrix},$$

$$A_5 = \begin{bmatrix} 0.0034 & 0.0119 & 0.0028 \\ 0.1028 & -1.1073 & 0.0060 \\ -0.1062 & -0.0562 & -0.0088 \end{bmatrix}, \quad A_6 = \begin{bmatrix} 0.0034 & 0.0119 & 0.0028 \\ -0.0251 & -1.1750 & -0.0046 \\ 0.0217 & 0.0115 & 0.0018 \end{bmatrix},$$

$$A_7 = \begin{bmatrix} 0.0034 & 0.0119 & 0.0028 \\ 0.1028 & -0.0771 & 0.0060 \\ -0.1062 & -0.0562 & -0.0088 \end{bmatrix}, \quad A_8 = \begin{bmatrix} 0.0034 & 0.0119 & 0.0028 \\ -0.0251 & -0.1447 & -0.0046 \\ 0.0217 & 0.0115 & 0.0018 \end{bmatrix}.$$

$$B = \begin{bmatrix} 4 & 0 & 0 \\ 0 & 4 & 0 \\ 0 & 0 & 4 \end{bmatrix}, \quad C = \begin{bmatrix} 0.25 & 0 & 0 \\ 0 & 0.25 & 0 \\ 0 & 0 & 0.25 \end{bmatrix}.$$

Since the orthogonal complement to a square matrix does not exist, three fault detection filters can be considered for single actuator fault detection. To illustrate the design procedure, the TS fuzzy fault detection filter for the pair (C_{23}, B_{23}) is considered, i.e.,

$$C \Leftarrow C_{23} = \begin{bmatrix} 0 & 0.25 & 0 \\ 0 & 0 & 0.25 \end{bmatrix}, F \Leftarrow B_{23} = \begin{bmatrix} 0 & 0 \\ 4 & 0 \\ 0 & 4 \end{bmatrix},$$

with the derived parameters

$$V = CF = \begin{bmatrix} 1 & 0 \\ 0 & 1 \end{bmatrix}, \quad V^{-1}C = \begin{bmatrix} 0 & 0.25 & 0 \\ 0 & 0 & 0.25 \end{bmatrix}, \quad F^{\perp} = [1 \quad 0 \quad 0], \quad T = \begin{bmatrix} 0 & 0.25 & 0 \\ 0 & 0 & 0.25 \\ 1 & 0 & 0 \end{bmatrix}.$$

Note that in this case all A_i with index higher than 4 lead to an unstable structure of $\overset{\circ}{A}_{o22i}$ and the resulting observer matrices A_{ei} need to be additionally stabilized, applying the principle given in Lemma 2.

Applying Eq. (56), the following structure of A_{o1} for the initial matrix A_1 is computed:

$$A_{o1} = \begin{bmatrix} -1.0392 & 0.0220 & 0.0306 \\ -0.0562 & -0.0088 & -0.0266 \\ -0.2250 & -0.0528 & -0.0163 \end{bmatrix}, \quad A_{o111} = \begin{bmatrix} -1.0392 & 0.0220 \\ -0.0562 & -0.0088 \end{bmatrix}, \quad A_{o121} = \begin{bmatrix} 0.0306 \\ -0.0266 \end{bmatrix},$$
$$A_{o211} = [-0.2250 \quad -0.0528], \quad A_{o221} = [-0.0163],$$

and $A_{o221} = -0.0163$ implies that the associated TS fuzzy fault detection filter linear component can be designed directly.

Choosing $s_o = 5$, it is resulting from Eqs. (57) and (58) that

$$\overset{\circ}{L}_1 = \begin{bmatrix} 3.9608 & 0.0220 \\ -0.0562 & 4.9912 \\ -0.2250 & -0.0528 \end{bmatrix}, J_1 = \begin{bmatrix} -0.2250 & -0.0528 \\ 15.8432 & 0.0879 \\ -0.2248 & 19.9649 \end{bmatrix}, A_{e1} = \begin{bmatrix} -0.0163 & 0 & 0 \\ 0.1225 & -5.0 & 0 \\ -0.1062 & 0 & -5.0 \end{bmatrix},$$

where the eigenvalue spectrum of A_{e1} and the steady-state value of the TS fuzzy fault detection filter transfer function matrix $G_{r1}(0)$ are

$$\rho(A_{e1}) = \{-0.0163 \quad -5.0 \quad -5.0\}, \quad G_{r1}(0) = -V^{-1}CA_{e1}^{-1}F = \begin{bmatrix} 0.2 & \\ & 0.2 \end{bmatrix},$$

respectively. It is evident that all diagonal elements of $G_{r1}(0)$ take the value $s_o^{-1} = 0.2$. The same structure of $G_{r*}(0)$ is obtained solving with A_l for $l = 1, 2, 3, 4$.

Analogously, designing for the matrix A_5, it can be seen that

$$A_{o5} = \begin{bmatrix} -1.1073 & 0.0060 & 0.0257 \\ -0.0562 & -0.0088 & -0.0266 \\ 0.0475 & 0.0111 & 0.0034 \end{bmatrix}, \quad A_{o511} = \begin{bmatrix} -1.1073 & 0.0060 \\ -0.0562 & -0.0088 \end{bmatrix}, \quad A_{o512} = \begin{bmatrix} 0.0257 \\ -0.0266 \end{bmatrix},$$
$$A_{o521} = [0.0475 \quad 0.0111], \quad A_{o522} = [0.0034].$$

Since $A_{o222} = 0.0034$, evidently, the associated TS fuzzy fault detection filter linear component with the unitary transfer function matrix has to be stabilized additively.

Solving also for $s_o = 5$, then

$$\overset{\circ}{L}_5 = \begin{bmatrix} 3.8927 & 0.0060 \\ -0.0562 & 4.9912 \\ 0.0475 & 0.0111 \end{bmatrix}, J_5 = \begin{bmatrix} 0.0475 & 0.0111 \\ 15.5707 & 0.0239 \\ -0.2248 & 19.9649 \end{bmatrix}, A_{e5} = \begin{bmatrix} 0.0034 & 0 & 0 \\ 0.1028 & -5.0 & 0 \\ -0.1062 & 0 & -5.0 \end{bmatrix}.$$

It is evident that matrix F_{e5} is not Hurwitz and has to be additively stabilized.

Thus, defining the weighting matrices of appropriate dimensions as

$$Q = 0, \;\; S = 0, \;\; R = VV^T = I_2$$

and solving the dual LQ control problem to change the sign of unstable eigenvalue of F_{e5} using the MATLAB function $K_{s5} = care(F_{e2}^T, Q, R, S, I_3)$, then

$$K_{s5}^T = \begin{bmatrix} 0.6456 & -0.6671 \\ 0.0133 & -0.0137 \\ -0.0137 & 0.0142 \end{bmatrix}, \quad A_{es5} = A_{e5} - K_{s5}^T C = \begin{bmatrix} 0.0034 & -0.1614 & 0.1668 \\ 0.1028 & -5.0033 & 0.0034 \\ -0.1062 & 0.0034 & -5.0035 \end{bmatrix}.$$

It can be easily verified that

$$\rho(A_{es5}) = \{ -0.0034 \quad -5.0 \quad -5.0 \},$$

$$G_{r5}(0) = -V^{-1} C A_{es5}^{-1} F = \begin{bmatrix} -0.0066 & -0.1999 \\ -0.1999 & 0.0066 \end{bmatrix}, \quad \rho(G_{r5}(0)) = \{ -0.2000 \quad 0.2000 \}.$$

while, evidently, $G_{r5}(0)$ is not diagonal and the eigenvalues of $G_{r5}(0)$ are $\pm 0.2 = \pm s_0^{-1}$.

Note that the same structure of $G_{rl}(0)$ is obtained solving with the system matrices A_l and $l = 5, 6, 7, 8$ when additional stabilization is required. Evidently, elements of this set of TS fuzzy residual filter linear components are stable, non-unitary, and without directional residual properties. Nevertheless, these properties guarantee the same singular values of the linear transfer function matrix components; as follows the result of Definition 1, the TS fuzzy residual filter will have all the singular values the same. To document this, the singular value plot of the TS fuzzy fault detection filter, as well as of all its linear parts, is equal to that presented in **Figure 1**. With respect to the structure of the matrices B and C, the comparable results are obtainable for the matrix pairs (C_{12}, B_{12}) and (C_{13}, B_{13}).

The rest of gain matrices of the stable TS fault detection filter is as follows:

$$J_2 = \begin{bmatrix} -0.2250 & -0.0528 \\ 15.5725 & 0.0456 \\ 0.0459 & 20.0072 \end{bmatrix}, J_3 = \begin{bmatrix} -0.2250 & -0.0528 \\ 19.9643 & 0.0879 \\ -0.2248 & 19.9649 \end{bmatrix}, J_4 = \begin{bmatrix} -0.2250 & -0.0528 \\ 19.6935 & 0.0456 \\ 0.0459 & 20.0072 \end{bmatrix},$$

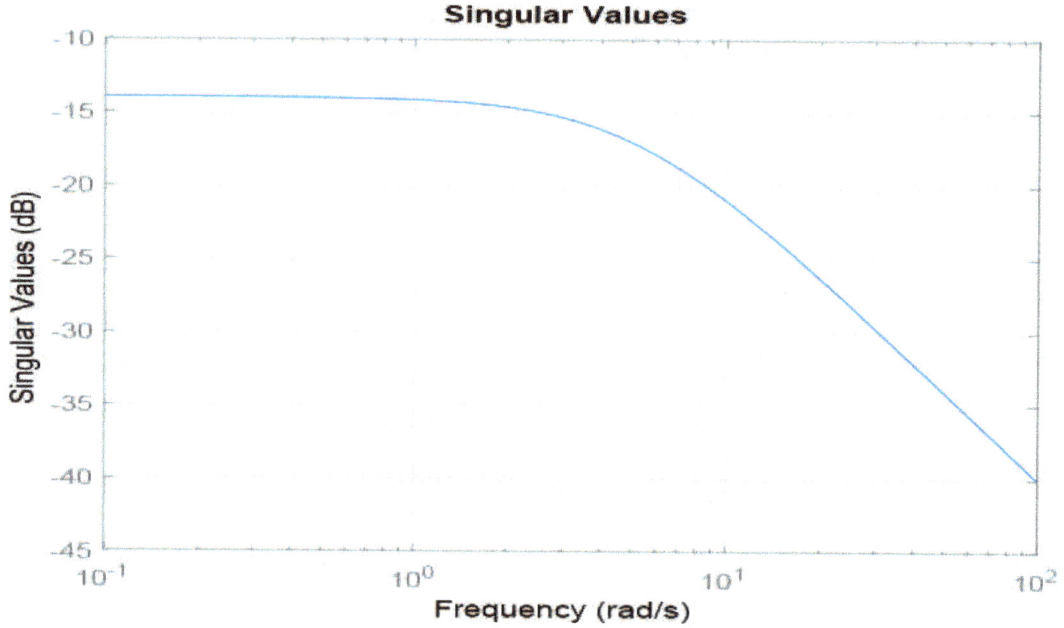

Figure 1. TS fuzzy fault detection filter singular value plot.

$$J_5 = \begin{bmatrix} 0.6930 & -0.6560 \\ 15.5840 & 0.0102 \\ -0.2385 & 19.9791 \end{bmatrix}, J_6 = \begin{bmatrix} -3.0809 & 2.7126 \\ 15.3157 & -0.0319 \\ 0.0324 & 20.0189 \end{bmatrix},$$

$$J_7 = \begin{bmatrix} 0.6930 & -0.6560 \\ 19.7051 & 0.0102 \\ -0.2385 & 19.9791 \end{bmatrix}, J_8 = \begin{bmatrix} -3.0809 & 2.7126 \\ 19.4367 & -0.0319 \\ 0.0324 & 20.0189 \end{bmatrix}.$$

Since the matrices $\boldsymbol{A}_i$ of the TS fuzzy system are not Hurwitz, the system in simulations is stabilized using the local-state feedback control laws, acting in the forced modes. Adapting the method presented in [14] to design the control law parameters, the local controller parameters are computed as

$$\boldsymbol{K}_1 = \begin{bmatrix} 0.1780 & 0.0083 & -0.0150 \\ 0.0083 & -0.0701 & -0.0041 \\ -0.0150 & -0.0043 & 0.1798 \end{bmatrix}, \boldsymbol{K}_2 = \begin{bmatrix} 0.1780 & -0.0079 & 0.0008 \\ -0.0075 & -0.0869 & 0.0028 \\ 0.0008 & 0.0027 & 0.1824 \end{bmatrix},$$

$$\boldsymbol{K}_3 = \begin{bmatrix} 0.1780 & 0.0084 & -0.0150 \\ 0.0082 & 0.1842 & -0.0041 \\ -0.0150 & -0.0042 & 0.1798 \end{bmatrix}, \boldsymbol{K}_4 = \begin{bmatrix} 0.1780 & -0.0078 & 0.0008 \\ -0.0076 & 0.1675 & 0.0027 \\ 0.0008 & 0.0028 & 0.1824 \end{bmatrix},$$

$$\boldsymbol{K}_5 = \begin{bmatrix} 0.1829 & 0.0142 & -0.0131 \\ 0.0141 & -0.0870 & -0.0061 \\ -0.0130 & -0.0063 & 0.1798 \end{bmatrix}, \boldsymbol{K}_6 = \begin{bmatrix} 0.1829 & -0.0020 & 0.0027 \\ -0.0017 & -0.1037 & 0.0008 \\ 0.0027 & 0.0007 & 0.1824 \end{bmatrix},$$

$$K_7 = \begin{bmatrix} 0.1829 & 0.0143 & -0.0131 \\ 0.0140 & 0.1674 & -0.0061 \\ -0.0130 & -0.0063 & 0.1798 \end{bmatrix}, K_8 = \begin{bmatrix} 0.1829 & -0.0019 & 0.0027 \\ -0.0018 & 0.1507 & 0.0007 \\ 0.0027 & 0.0007 & 0.1824 \end{bmatrix},$$

$$W_1 = \begin{bmatrix} 0.1821 & 0.0224 & -0.0117 \\ -0.0223 & 0.1897 & -0.0096 \\ 0.0115 & 0.0098 & 0.1820 \end{bmatrix}, W_2 = \begin{bmatrix} 0.1821 & 0.0062 & 0.0041 \\ -0.0061 & 0.1899 & -0.0001 \\ -0.0047 & -0.0002 & 0.1819 \end{bmatrix},$$

$$W_3 = \begin{bmatrix} 0.1821 & 0.0225 & -0.0117 \\ -0.0224 & 0.1865 & -0.0096 \\ 0.0115 & 0.0098 & 0.1820 \end{bmatrix}, W_4 = \begin{bmatrix} 0.1821 & 0.0063 & 0.0041 \\ -0.0062 & 0.1867 & -0.0001 \\ -0.0047 & -0.0001 & 0.1819 \end{bmatrix},$$

$$W_5 = \begin{bmatrix} 0.1820 & 0.0112 & -0.0138 \\ -0.0116 & 0.1899 & -0.0076 \\ 0.0135 & 0.0077 & 0.1820 \end{bmatrix}, W_6 = \begin{bmatrix} 0.1820 & -0.0049 & 0.0021 \\ 0.0046 & 0.1901 & 0.0019 \\ -0.0027 & -0.0022 & 0.1819 \end{bmatrix},$$

$$W_7 = \begin{bmatrix} 0.1820 & 0.0114 & -0.0138 \\ -0.0117 & 0.1867 & -0.0076 \\ 0.0135 & 0.0078 & 0.1820 \end{bmatrix}, W_8 = \begin{bmatrix} 0.1820 & -0.0048 & 0.0021 \\ 0.0045 & 0.1869 & 0.0019 \\ -0.0027 & -0.0021 & 0.1819 \end{bmatrix},$$

where

$$W_i = -\left(C(A_i - BK_i)^{-1}B\right)^{-1},$$

$$u_i(t) = -K_i q(t) + W_i w_o,$$

while $w_o \in R^n$ is the vector of the desired steady-state system outputs.

If necessary for any more complex system, PDS controller principle can be applied to stabilize the plant (see, e.g., authors' publications [15, 16] or other references [17, 18]).

To display simulations in the MATLAB and Simulink environment, the forced mode control is established with local controller parameter given as above for the system initial conditions $q^T(0) = [0.2 \;\; 0.3 \;\; 0.2]$ and $w_o^T = [0.6 \;\; 0.5 \;\; 0.4]$. Fault detection filter is constructed on the couple (C_{23}, B_{23}) and the set of matrices A_i and $i = 1, 2..., 8$.

As the results, **Figure 2** shows the TS fuzzy system output responses, illustrating their asymptotic convergence to the steady states, and **Figure 3** presents the TS fuzzy fault detection filter response, reflecting a steplike 90% gain loss of the second actuator at the time instant $t = 60s$. These examples illustrate the power that can be invoked through the prescribed H_∞ norm properties.

It can verify that TS fuzzy fault detection filters created for the couple pairs (C_{12}, B_{12}) and (C_{13}, B_{13}) have similar properties as that defined for the couple (C_{23}, B_{23}). The difference is, for

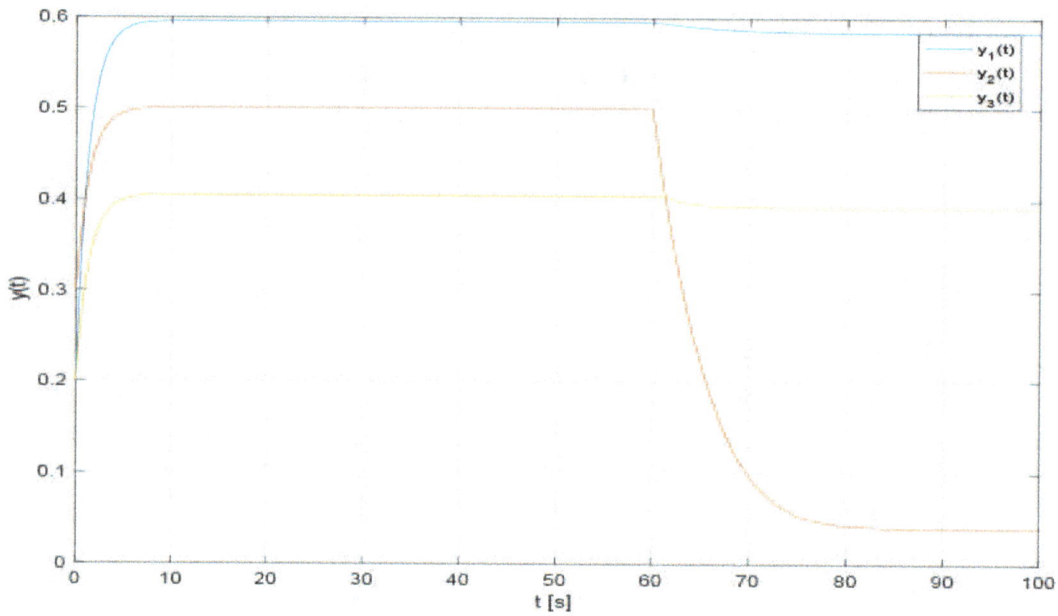

Figure 2. System output responses.

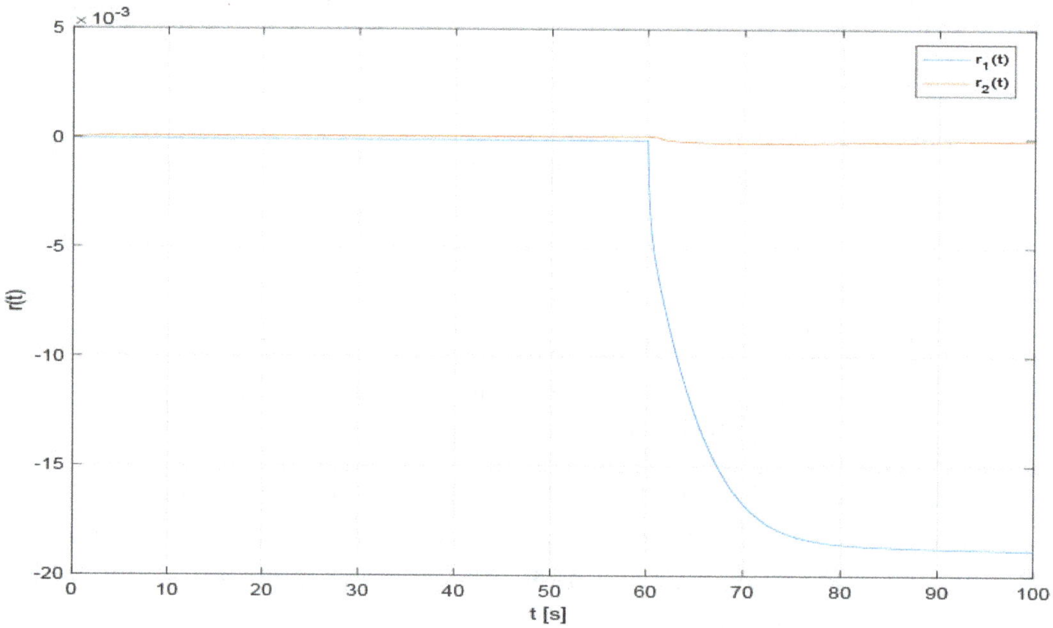

Figure 3. Residual signal responses.

example, that in the occurrence of a single fault of the second actuator the responses of TS fuzzy fault detection filter defined for the couple (C_{13}, B_{13}) naturally do not have directional properties, since the second column of $\boldsymbol{K}$ is not included in its construction.

As can be seen from the solution, the sector functions defined in this way cannot create a unitary TS fuzzy fault detection filter, but the obtained orthogonal properties of the residual signals are sufficient to detect and isolate actuator faults.

6. Concluding remarks

The problem of designing the TS fuzzy fault detection filters for highly nonlinear mechanical systems representable by the TS fuzzy model is considered, to achieve the desired filter H_∞ norm property in all working point belonging to the assigned work sectors. The proposed method exploits features offered in TS fuzzy system models to design TS fuzzy fault detection filters. The rules and formulation are developed to generate residual signals with quasi-directional properties and to make the TS filter transfer function matrix with prescribed H_∞ norm properties. By a convenient choose of the sector functions, this purpose is reached using a relative small number of membership functions. If unitary definition for TS fuzzy fault detection filters is satisfied, the design methodology provides new opportunities for fault detection and isolation rules in fault tolerant nonlinear control systems, their analysis, and optimization.

Acknowledgements

The work presented in this chapter was supported by VEGA, the Grant Agency of the Ministry of Education, and the Academy of Sciences of Slovak Republic, under Grant No. 1/0608/17. This support is very gratefully acknowledged.

Author details

Dušan Krokavec*, Anna Filasová, Jakub Kajan and Tibor Kočík

*Address all correspondence to: dusan.krokavec@tuke.sk

Department of Cybernetics and Artificial Intelligence, Faculty of Electrical Engineering and Informatics, Technical University of Košice, Košice, Slovakia

References

[1] Hou M, Patton RJ. An LMI approach to H_-/H_∞ fault detection observers. In: Proceedings of the UKACC International Conference on Control; 2–5 September 1996; Exeter, England, pp. 305-310. DOI: 10.1049/cp:19960570

[2] Noura H, Theilliol D, Ponsart JC, Chamseddine A. Fault-Tolerant Control Systems. Design and Practical Applications. London: Springer-Verlag; 2009. 233 p

[3] Zhao Z, Xie WF, Hong H, Zhang Y. Unitary system I. Constructing a unitary fault detection observer. IFAC Proceedings. 2011;**44**(1):7725-7730. http://doi.org/10.3182/20110828-6-IT-1002.01555

[4] Krokavec D, Filasová A, Liščinský P. On fault detection filters design with unitary transfer function matrices. Journal of Physics. Conference Series. 2015;**659**:1-12. DOI: 10.1088/1742-6596/659/1/012036

[5] Takagi T, Sugeno M. Fuzzy identification of systems and its applications to modeling and control. IEEE Transactions on Systems, Man and Cybernetics. 1985;**15**(1):116-132. DOI: 10.1109/TSMC.1985.6313399

[6] Chadli M, Borne P. Multiple Models Approach in Automation. Takagi-Sugeno Fuzzy Systems. Hoboken: John Wiley & Sons; 2013. p. 256

[7] Solheim OA. Design of optimal control systems with prescribed eigenvalues. International Journal of Control. 1972;**15**(1):143-160. http://doi.org/10.1080/00207177 208932136

[8] Tanaka T, Wang HO. Fuzzy Control Systems Design and Analysis. A Linear Matrix Inequality Approach. New York: John Wiley & Sons; 2001. 320 p

[9] Boyd D, Balakrishnan V. A regularity result for the singular values of a transfer matrix and a quadratically convergent algorithm for computing its L_∞-norm. Systems & Control Letters. 1990;**15**(1):1-7. DOI: 10.1109/CDC.1989.70267

[10] Green M, Limebeer DJN. Linear Robust Control. Englewood Cliffs: Prentice Hall; 1995. 558 p

[11] Fairman FW. Linear Control Theory. The State Space Approach. Chichester: John Wiley & Sons; 1998. 318 p

[12] Krokavec D, Filasová A, Liščinský P. Unitary approximations in fault detection filter design. Mathematical Problems in Engineering. 2016;**2016**:1-14. http://dx.doi.org/10.1155/2016/7249803

[13] Nagy AM, Marx B, Mourot G, Schutz G, Ragot J. State estimation of the three-tank system using a multiple model. In: Proceedings of the Joint 48th IEEE Conference on Decision and Control and 28th Chinese Control Conference; 15–18 December 2009; Shanghai, China, pp. 7795-7800. DOI: 10.1109/CDC.2009.5400889

[14] Krokavec D, Filasová A, Serbák V. Virtual actuator based fault tolerant control design for Takagi-Sugeno fuzzy systems. In: Proceedings of the 14th IEEE International Symposium on Applied Machine Intelligence and Informatics SAMI 2016; 21–23 January 2016; Herl'any, Slovakia, pp. 63-68. DOI: 10.1109/SAMI.2016.7422983

[15] Krokavec D, Filasová A. Stabilizing fuzzy output control for a class of nonlinear systems. Advances in Fuzzy Systems. 2013;**2013**:1-9. http://dx.doi.org/10.1155/2013/294971

[16] Krokavec D, Filasová A. Stabilizing fuzzy control via output feedback. In: Ramakrishnan S, editor. Modern Fuzzy Control Systems and Its Applications. Rijeka: InTech; 2017. pp. 3-25. DOI: 10.5772/68129

[17] Zhang K, Jiang B, Shi P. Observer-Based Fault Estimation and Accommodation for Dynamic Systems. Berlin: Springer-Verlag; 2013. 181 p

[18] Lam HK. Polynomial Fuzzy Model-Based Control Systems. Stability Analysis and Control Synthesis Using Membership Function-Dependent Techniques. Cham: Springer-Verlag; 2016. 307 p

Permissions

All chapters in this book were first published in NSMES, by InTech Open; hereby published with permission under the Creative Commons Attribution License or equivalent. Every chapter published in this book has been scrutinized by our experts. Their significance has been extensively debated. The topics covered herein carry significant findings which will fuel the growth of the discipline. They may even be implemented as practical applications or may be referred to as a beginning point for another development.

The contributors of this book come from diverse backgrounds, making this book a truly international effort. This book will bring forth new frontiers with its revolutionizing research information and detailed analysis of the nascent developments around the world.

We would like to thank all the contributing authors for lending their expertise to make the book truly unique. They have played a crucial role in the development of this book. Without their invaluable contributions this book wouldn't have been possible. They have made vital efforts to compile up to date information on the varied aspects of this subject to make this book a valuable addition to the collection of many professionals and students.

This book was conceptualized with the vision of imparting up-to-date information and advanced data in this field. To ensure the same, a matchless editorial board was set up. Every individual on the board went through rigorous rounds of assessment to prove their worth. After which they invested a large part of their time researching and compiling the most relevant data for our readers.

The editorial board has been involved in producing this book since its inception. They have spent rigorous hours researching and exploring the diverse topics which have resulted in the successful publishing of this book. They have passed on their knowledge of decades through this book. To expedite this challenging task, the publisher supported the team at every step. A small team of assistant editors was also appointed to further simplify the editing procedure and attain best results for the readers.

Apart from the editorial board, the designing team has also invested a significant amount of their time in understanding the subject and creating the most relevant covers. They scrutinized every image to scout for the most suitable representation of the subject and create an appropriate cover for the book.

The publishing team has been an ardent support to the editorial, designing and production team. Their endless efforts to recruit the best for this project, has resulted in the accomplishment of this book. They are a veteran in the field of academics and their pool of knowledge is as vast as their experience in printing. Their expertise and guidance has proved useful at every step. Their uncompromising quality standards have made this book an exceptional effort. Their encouragement from time to time has been an inspiration for everyone.

The publisher and the editorial board hope that this book will prove to be a valuable piece of knowledge for researchers, students, practitioners and scholars across the globe.

List of Contributors

Horacio Florez and Miguel Argáez
Computer Science Department, The University of Texas at El Paso, USA

Sachin Kumar
Department of Mathematics, Central University of Punjab, Bathinda, Punjab, India

Zhiguo Wang, Xiaojing Shen and Yunmin Zhu
Department of Mathematics, Sichuan University, Chengdu, Sichuan, China

Abdulrahman H. Bajodah
Aeronautical Engineering Department, King Abdulaziz University, Jeddah, Saudi Arabia

Ye-Hwa Chen
School of Mechanical Engineering, Georgia Institute of Technology, Atlanta, Georgia, United States of America

Sami Elmadssia
Higher Institute of Applied Sciences and Technology, Electrical Engineering, Gafsa, Tunisia

Mohamed Benrejeb
National Engineering School of Tunis, Tunis, Tunisia

Jiří Náprstek and Cyril Fischer
Institute of Theoretical and Applied Mechanics CAS, Prague, Czech Republic

Emil Anton
Department Obstetrics Gynecology, Gr. T. Popa University of Medicine and Pharmacy, Iasi, Romania

Anna Gavrilut
Faculty of Mathematics, Alexandru Ioan Cuza University from Iasi, Romania

Maricel Agop
Department of Physics, Gheorghe Asachi Technical University of Iasi, Romania

Daniel Timofte
Department of Surgery, Gr. T. Popa University of Medicine and Pharmacy, Iasi, Romania

Yury A. Rossikhin and Marina V. Shitikova
Voronezh State Technical University, Voronezh, Russian Federation

Ehsan Izadpanahi and Pezhman Mardanpour
Department of Mechanical and Materials Engineering, Florida International University, Miami, FL, USA

Valeriy E. Arkhincheev
Theoretical Physics Research Group, Advanced Institute of Materials Science, Ton Duc Thang University, Ho Chi Minh City, Vietnam
Faculty of Applied Sciences, Ton Duc Thang University, Ho Chi Minh City, Vietnam

Lubsan V. Budazapov
Buryat Research Institute of Agriculture, Ulan-Ude, Russian Federation

Elena Adomaitienė, Skaidra Bumelienė and Arūnas Tamaševičius
Department of Electronics, Center for Physical Sciences and Technology, Vilnius, Lithuania

Dušan Krokavec, Anna Filasová, Jakub Kajan and Tibor Kočík
Department of Cybernetics and Artificial Intelligence, Faculty of Electrical Engineering and Informatics, Technical University of Košice, Košice, Slovakia

Index

www.ingramcontent.com/pod-product-compliance
Lightning Source LLC
LaVergne TN
LVHW060554210926
848485LV00005B/1312
* 9 7 8 1 6 3 9 8 7 3 9 9 9 *